Cover photo, clockwise from top, replica of oakwood pounding mallet used in Mino, Japan; ironwood tapa beater from Samoa; milkweed fibers from Ellensburg, Washington; coil of inner mulberry bark from Tonga; and gampi fibers from Japan.

Karen C. Hall
Spring '00

PLANT FIBERS FOR PAPERMAKING

by

Lilian A. Bell

With a foreword by Marcia Morse and botanical drawings by Virginia Kerwin

Liliaceae Press
McMinnville, Oregon

Frontispiece: Nepalese sheet formation, drawing by the author

Page 11: Japanese postcard, stencil print on mulberry paper

Liliaceae Press
P.O. Box 1235
McMinnville, Oregon 97128

Printed in the United States of America by
BookCrafters, Chelsea, Michigan 48118

Eighth printing, 1995

ISBN 0-9625076-5-2

ACKNOWLEDGEMENTS

I am grateful for technical assistance and guidance from Lucille Houseley, Linfield College Biology Department, McMinnville, Oregon; Jack Thompson, Thompson Restoration Laboratory, Portland, Oregon; and Rob Stephenson, Cascadia Landscaping, McMinnville, Oregon.

Photography was done by Bill Bachhuber, photographer, Portland, Oregon; poetry was written by William I. Elliott, Yokohama, Japan; photomicrographs were done by Tony Bell; and the manuscript was typed by Vicki Yonally.

Thanks go to the many artists and papermakers across the country who have shared information and sent fiber and paper samples. A special note of thanks to Elaine Koretsky of Carriage House Paper for taking time to review the entire manual and suggest revisions for updating.

Lilian A. Bell
McMinnville, Oregon, 1981 and 1989

LIST OF CONTENTS

FOREWORD

The art of making paper by hand, whether with utilitarian or expressive intent, has enjoyed a popular resurgence in the last decade. There have been times when attitudes of faddish experimentation and commercial exploitation have obscured the strong and quiet continuity which has sustained the tradition of handmade paper.

Signs are strong that this renewal of interest has now entered a new phase--a phase of consolidation, of thoughtful evaluation and sharing of information. In this context, Lilian A. Bell's book is a timely and valuable addition to an ongoing tradition. By virtue of its format and content, the book is both serious and useful--an all-too-rare combination. It is a paradigm, an invitation for further exploration by individual artist-craftsmen who work in a sustained way with paper.

The methods and materials emphasized by Bell have several implications, both practical and aesthetic. First, they possess the advantages of simplicity in the equipment, needing perhaps only slight modification of or additions to a basic, pre-existing workshop. Second, they restore to the artist-craftsman a wholeness of control and involvement in an extended process of papermaking by hand, beginning with selection of raw plant materials (many of which may never have been considered for this use), through appropriate methods of cooking and beating, to casting and sheet formation. Third, as Bell's research reveals, the visual and tactile qualities of papers made from plant fibers are rich, diverse, and highly distinctive. Variations of surface quality, sheet density, subtle coloration (all of which come from the fiber itself rather than from subsequent embellishment) add substantially to the palette of the papermaker. Fourth, the use of plant fibers as described here speaks in a dual way to both the regional and the cross-cultural impulses in contemporary handpapermaking: the regional, in that it encourages the artist-craftsman to use plant materials readily available in a specific area of residence and work, bound only by possible seasonal variations; the cross-cultural, in that the pulps which are prepared in ways that retain maximum fiber length are ideally suited to modifications of the oriental methods of sheet formation, producing thin, translucent papers of exceptional beauty. Finally, and most importantly, this manual reminds us of an ideal balance of concerns in which we retain and honor the integrity of our materials even as we extend the ways in which they may be used.

Marcia Morse
Honolulu, Hawaii
December, 1980

PREFACE

In 1977 I began my research on plant fibers with the only intention of using these fibers for my own art work. Since I was not especially interested in investing in machinery, I was looking for simple ways of processing the fibers into quality papers.

The subject became so absorbing that I found myself traveling to Japan and Polynesia to study oriental papermaking techniques and tropical fibers. Information on fiber plants accumulated so rapidly that I decided to incorporate it into the papermaking workshops that I was teaching; and as a result, I found that many others were interested in learning more about the use of these fibers.

While the book is designed primarily for those who already have some basic papermaking background, it will also be useful to those who are just beginning to work with handmade paper.

IN MINO

Sky is descending over

bamboo and hinoki

ascending the hills around Mino.

By a rain-rapid river,

beneath a tile roof,

Mieko dips into milky pulp;

dips and tilts,

dips and tilts;

lets the screen drip;

and waits:

a sheet of paper holds 400 years together.

William I. Elliott

INTRODUCTION

A renewed interest, beginning in the 1970's, in the use of handmade paper as a medium has not only brought about a renaissance of western style rag papermaking but has also generated interest in using plant materials for oriental and other papermaking methods, which require fibers that have not been overly shortened by beating action. There is nothing new about the use of plant fibers for papermaking: the use of papyrus by the Egyptians in approximately 3,000 B.C. and the use of mulberry by the Chinese in the second century A.D. is, of course, well known.

Why are contemporary artists interested in using plant fibers for their work? For one thing many plant fibers have unique properties like lustre, transparency and great strength in very thin paper. Papers made from these fibers are quite different from those made from rags and call for a different aesthetic approach. As artists have gone beyond the paper surface, a desire to manipulate the pulp into aesthetic statements has resulted in the need for a broader spectrum of pulp qualities than has been offered by the use of rag pulps alone. Another reason for an interest in the use of plant fibers is a practical one. Artists living in states where climate allows for luxuriant growth of a wide variety of fiber plants have realized that plant fibers are often more easily available than rags. Depending on one's point of view, plant fibers are often easier to process than rag fibers, especially at the beating stage. The focus of the medium on rags as a material for paper has resulted in the need for expensive heavy machinery requiring large studio spaces. Consequently, plant fibers have been overlooked, especially by those artists focussed on a thick white paper. For those who are already well equipped with machinery but desire to expand their paper vocabulary by making use of a greater variety of pulps, plant fibers offer the perfect solution. There are many long plant fibers that can be successfully beaten in the Hollander.

THE FIBER DATA SHEETS

Both tropical and temperate fibers have been included in this work. While most of the states in this country are in the temperate zone, there are several states that lie in tropical and semi-tropical areas. Data on sheets with line drawings is based on actual tests, while data on those sheets without line drawings is untested. I have included untested fibers because readers living in other parts of the country may have access to them for testing.

The fiber catalog of the book has been divided into three sections: bast fibers, leaf fibers, and grass fibers. The fiber length range for each fiber category has been given at the beginning of each section which has been alphabetized by plant family name and botanical name. 6 mm = 1/4" approximately

Botanical names have been used to identify the exact plant in use

Common names have been given as an aid to those unfamiliar with Latin names, as indeed I was just a few years ago!

Plant family name is included because if one member of the family yields a good fiber, it is quite possible that another member may do so too.

Fiber length, where known, is included because it can be an indicator of paper strength. In general although there are always exceptions, longer fibers make a stronger sheet of paper. In cases where a fiber is found to be desirable but weak, it can be "reinforced" by mixing it with a stronger fiber. Fiber lengths vary from one fiber type to another, as well as within a particular plant. Fiber width, although not noted, also varies. A thicker fiber gives a coarser paper, while a thinner fiber results in a more flexible smoother paper that offers a better printing surface. Variations in fiber width can be seen in the micrographs on page 31.

Fiber type indicates whether the fiber is a bast, leaf, or grass fiber.

Plant location in the U.S. indicates roughly where a plant may be found. Many plants are cultivated in areas which are not necessarily best for their optimum growth. Many plants are listed as growing in a particular location but grow there in such small quantities that they are not really useable for the paper artist.

Plant part used refers to the portion of the plant that contains the fiber that is going to be used.

Season harvested indicates a suggested time to harvest the plant parts.

Fiber purchase indicates whether the fiber can be purchased by mail order or by direct purchase from a store.

Pre-cooking preparation

Cooking

Beating method

Sheet formation

These sections give indications of what I have found to work under certain conditions. Since the reader may be harvesting plant parts and preparing them under different conditions, results may vary; and, consequently, these are to be thought of only as guidelines.

Paper color indicates what I obtained from a given batch of fiber and may vary.

Historical references indicate how the plant has been used historically from the point of view of its fiber.

Further notes contains additional information about the fiber.

SEARCHING FOR PLANT FIBERS TO USE

Taking into consideration the fact that one can make some kind of paper from almost any plant and that there are thousands of fiber yielding plants in the world, how does the artist make the decision about which fibers to use? One way to start is to examine historical uses of fibers growing in one's own neighborhood or state. Other criteria include the fiber length, whether the fiber can be easily harvested and properly prepared, and whether it is grown in large enough quantities for one's artistic needs. For those who desire to avoid the use of bleach, the final paper color is also of importance. Others may prefer going to a store or mail ordering partially prepared raw fibers.

Artists who enjoy doing their own research should be prepared to do a lot of reading. Information on plant fibers is scattered, sketchy, and only rarely specific with respect to fiber preparation for papermaking. In addition to garden encyclopedias and textile and horticultural dictionaries, one can consult with botanists, landscape architects, and horticulturists. Visits can also be made to nearby herbariums and arboretums.

Sources for raw plant material are quite plentiful: home garden plants, weeds, food crops, wild shrubs and trees, gardeners' prunings, roadsides, vacant lots, building sites, farms, industrial sites, and along railroad tracks. If you can't obtain plant material, either dead or alive, it can be purchased at florists, nurseries or garden stores. If you don't wish to grow your own, make sure that your friends grow plants which are suitable for papermaking. Want something that doesn't grow in your area? Write a papermaking friend and ask for a dry shipment--dry because after a week in transit you might be the recipient of a lot of mold!

After the plant material has been obtained and the fiber category of bast, leaf, or grass has been established, experiments can be done with the various separation, cooking, beating, and sheet formation methods to see which are best suited to the fiber and whether indeed the fiber is suited for the artist's papermaking purpose. For those who are unsure about which procedures to follow, it is suggested that they follow directions for a plant that is in the same or a similar plant family. A blank data sheet has been included so readers can copy it for their own fiber research files.

HARVESTING OF PLANT FIBERS

All harvesting is done from fresh plants unless otherwise stated. Bast fibers, like milkweed for example, can be harvested after they have retted naturally in the field and the stems are dry. Fibers can be extracted from either fresh or dry plant parts. In cases where crop plants are being used for fiber, the fiber is usually harvested after the crop has been harvested. With regards to flowering garden plants, the fiber is usually harvested after the flowers have been enjoyed. Harvesting of fibers from shrubs or trees can be done while pruning. The branches to be harvested should be between 1/2 in. and 1 in. in diameter and about 6 ft. to 9 ft. in length. In the case of leaf fibers, the length of the fiber varies with the length of the leaf, so it is best to harvest from plants at least 2 years old with leaves 20 in. long or longer. It is difficult to give hard and fast rules because of climate differences: a Japanese mulberry stem might be harvested after 2 years of growth, while a Polynesian mulberry could be harvested after 9 months growth. Harvesting for an annual plant is probably best done in the late summer when one can assume that most of its growth has taken place. Obviously thicker stems will give a better fiber yield. They also contain a higher percentage of secondary fibers which are the softer finer fibers that lie next to the woody core. There are also primary fibers which lie between the secondary fibers and the outer bark; but these are more brittle.

When dealing with plant material, the most important thing to remember is that everything is variable. The season harvested affects the fiber length and paper color. The soil in which the plant grows affects the plant's growth, which in turn affects the fiber quality. Other influences are geographic location, differences in soil types, site quality, availability of moisture, and age of plant, shrub, or tree from which the fiber is being harvested. In addition to effects on fiber length, yield, and quality, the following items have an effect on the final paper color: using fiber from fresh or dry plant material, pre-cooking preparations like retting and fermentation, and the type of alkali used in the cooking process. It is quite possible to get several different paper colors from the same fiber, using different harvesting times and different preparation procedures.

At this point questions will arise about the identification of plants. Books like horticultural dictionaries and manuals of the flora of particular states with line drawings and color photos are very helpful, but they are no good to those of us who don't know what it is we are supposed to be looking for! A visit to a local herbarium, botanical garden, or arboretum might be helpful. If you have a plant that you think might be the one you are looking for, it can be taken to the home extension office at your county courthouse for identification. If the horticulturist there cannot identify it, he or she can send it to the botany department of your state university.

PURCHASING PLANT FIBERS FOR PAPERMAKING

U.S.SOURCES - Mail Order

1. Carriage House Paper, Donna Koretsky, 79 Guernsey St., Brooklyn, New York 11222. Phone & Fax: 718-599-7857, Orders: 1-800-669-8781. Carriage House Research Center, Elaine Koretsky, 8 Evans Rd., Brookline, Massachusetts 02146, Phone 617-232-1636, Fax 617-277-7719

2. Dieu Donne Paper, 433 Broome St., New York, New York 10013-2622, Phone 212-226-0573,Fax 212-226-6088

2. Lee S. McDonald Inc., P.O.Box 264, Charlestown, Massachusetts 02129, Phone 617-242-2505, Fax 617-242-8825

3. Magnolia Editions, David Kimball & Don Farnsworth, 2527 Magnolia St., Oakland,California 94607, Phone 510-839-5268, Fax 510-893-8334

4. Twinrocker, Kathryn & Howard Clark, P.O.Box 413, Brookston, Indiana 47923, Phone 317-563-3119

JOURNALS

1. Hand Papermaking Magazine, P.O.Box 77027, Washington DC 20013-7027

MEMBERSHIPS - For current contact information, please call one of the above suppliers

1. International Association of Hand Papermakers & Paper Artists
2. Friends of the Dard Hunter Paper Museum

BY DIRECT PURCHASE

1. Hardware stores for Manila and Sisal twine or rope
2. Weaving supply shops carry a wide variety of fibers including raffia, possibly true hemp, flax, jute, Manila, ramie, sea grass and sisal.

Please note: even though these purchased fibers are partially prepared, they still need cooking before beating even if they are going to be used in the Hollander beater. These raw fibers still contain small amounts of fat and pectin which prevent proper hydration of the fiber in the beating process. The only exceptions to the cooking requirement are the dry pulp sheets of fibers which have already been cooked, and in some cases partly beaten. These are also called half-stuff. The availability of the above mentioned fibers fluctuates. Fibers other than those listed may also be available. Please write to the individual supplier for complete catalogs.

FIBER STORAGE

Fibers can be stored dry at various stages of their preparation. However, different paper colors can result, according to whether the fiber is used immediately after harvesting or is dried for future use when the rest of the preparation is completed.

1. Plant parts can be dried after harvesting and stored (leaves, stems/branches, grasses).

2. Bast fibers can be separated from their woody cores, then dried and stored; leaf fibers can be scraped, then dried and stored.

3. Bast fibers, after separation from their woody cores, can further have their outer barks removed by scraping or peeling, then can be dried and stored.

4. Once the fiber has been dried in any form, it should be soaked for 24 hours prior to continuing the preparation.

5. Drying a fiber that has been cooked with alkali is not recommended because the once dried fiber will never again absorb the same amount of water upon re-wetting.

6. Left-over cooked fibers can be stored in plastic bags or containers in the refrigerator for a week or in the freezer until needed.

YIELDS

Since many of the methods described in this work are not intended for production papermaking, priority has not been given to the final yield from a given amount of dry fiber. The amount of raw fiber to harvest is, of course, dependent on the needs of the final project and varies with the fiber type.

One-half pound of inner bast fibers will yield approximately 20 sheets of thin paper 8" x 11" in size, while one-half pound of grass fibers will yield approximately 3 or 4 sheets of thin paper 8" x 11" in size.

FIBER SEPARATION

General Information

The portion of a plant that papermakers are interested in is the cellulose fiber. All plants contain cellulose, some containing a higher percentage than others. In order to be made suitable for papermaking, the cellulose portion of the plant must first be separated from substances that are not cellulose; for example, lignin, pith, woody shives and cores, fats, etc. This is done by the procedures which I have outlined in the following pages. A fiber having a high cellulose content is generally considered more advantageous than one with a lower cellulose content because the fiber can be purified and made into pulp more easily. The presence of non-cellulose materials like lignin is considered undesirable because the lignin rejects water and resists bonding. In papermaking we are aiming for just the opposite: the absorption of water and the promotion of bonding, both of which make for a stronger sheet of paper. Fibers contained in the various plants fall into the following categories: (1) fruit fibers like cotton seed hairs, (2) wood fibers like those fibers from deciduous and coniferous trees, (3) bast fibers, (4) leaf fibers, and (5) grass fibers. I have restricted my research to the last three mentioned categories.

After the fibers have been separated, they can either be used immediately, stored in the refrigerator or freezer, or dried. Drying can be done by laying the fiber strips on an old window screen or by hanging over a clothes line or a pole either indoors or outdoors. There are various conflicting reports as to whether, when drying outdoors, the fiber should be dried in the sun to promote whitening of the fiber or in the shade to prevent darkening of the fiber due to the action of the ultraviolet rays on the lignin still contained in the cell walls.

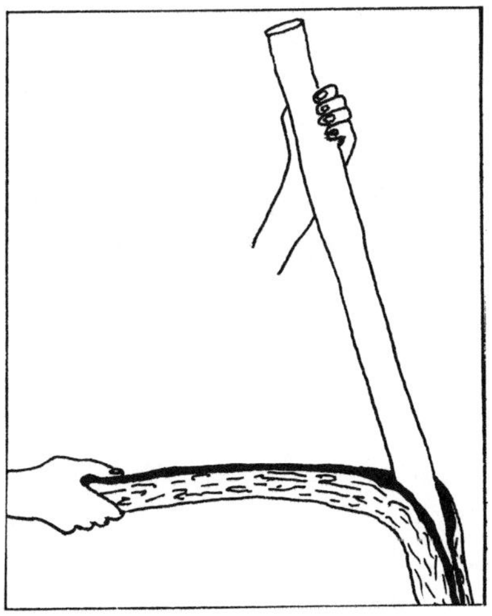

Bast Fiber Separation

Retting

The harvested plant part is soaked in water or spread on the grass and exposed to the dew. Retting allows enzyme action to loosen the bark containing the fiber so that it can easily be stripped off the woody core. This is said to work best with a pH of 6.8 and above. The retting process can take several months; and if one is not careful, can result in moldy fibers. Some fibers are naturally retted by being left standing in the field; for example, milkweed and nettles. For more complete details on retting, see Mauersberger and Kirby listed in the bibliography.

Stripping

Using freshly harvested plants, a vertical cut is made immediately after harvesting down the center of the stem with a knife; and the bark is peeled from the woody core in one piece. The outer bark is then removed from the inner bark by peeling or soaking and scraping with a paring knife. The stripping method is only successful when the bark contains a high percentage of moisture, as in the spring, or in very humid climates. However, there are always exceptions: fig and willow barks always seem easy to strip from their woody cores.

Steaming

The harvested fresh or dry stems are cut into pieces and packed into a stainless steel pan. Water is added to cover, brought to a boil, and the stems are cooked over a medium heat for one or two hours. The stems are then rinsed and cooled to handle; and the fibers are stripped off the cores which are discarded. If the fibers cannot be stripped easily, further cooking is needed. This strip of fiber now needs to have the outer bark removed from the inner bark, and this is done by scraping or peeling it off with a paring knife. In some cases the outer bark is so thin that it is too difficult to remove: it is left on and usually is dissolved in the alkaline cooking process.

An alternative method of removing the outer bark is to peel or scrape it off with a potato peeler or paring knife <u>before</u> the bark is stripped or steamed from the woody core.

Leaf Fiber Separation

Many leaf fibers need no special separation other than cooking with a strong alkali in the regular cooking process. However, there are many leaf fibers with thin strong outer skins which need removing before the inner fibers can be released. This process is known as decortication and is done commercially by a machine called a decorticator. The leaves are fed into the machine no later than 24 hours after harvesting; otherwise the leaves become too dry and fiber separation becomes too difficult. The machine scrapes off the upper and lower skin, thus releasing the fibers. The machine decorticated leaves can be further treated by retting in water for several days. More ancient methods of leaf fiber separation involve scraping each leaf with a knife or shell till all chlorophyll and skin is removed, after which the fibers are washed, rinsed, and hung up to dry. The leaves can also be individually pounded with a rubber mallet or crushed between two stones and retted for a few weeks in fresh or salt water. A more contemporary version of this ancient process is to lay out the leaves across your driveway, and the action of parking in and leaving your driveway will decorticate your leaf fibers. Retting, which is optional, can also be done in alkaline solutions for a week or so.

Grass Fiber Separation

Generally no special separation is required for grass fibers. Some grass fibers which are high bulk and rather buoyant can be soaked and precooked in plain water for 1/2 hour. Bamboo, which is a woody grass, needs to have its stems pounded with a hammer or crushed between rollers in order to aid chemical penetration and free the large amount of air that the stems hold. Chinese papermakers used to ret timber bamboo stems in a pit for several months by placing lime between layers of stems and covering the whole stack with water. Today they use young bamboo with solid stems, harvested before six months of growth.

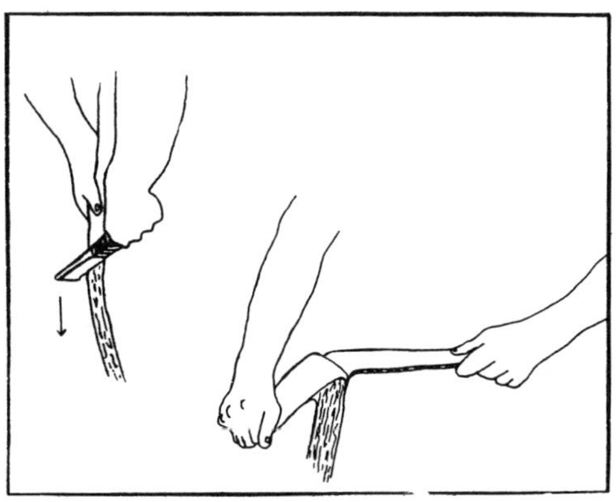

Scraping or peeling the outer bark off the inner bark of a bast fiber

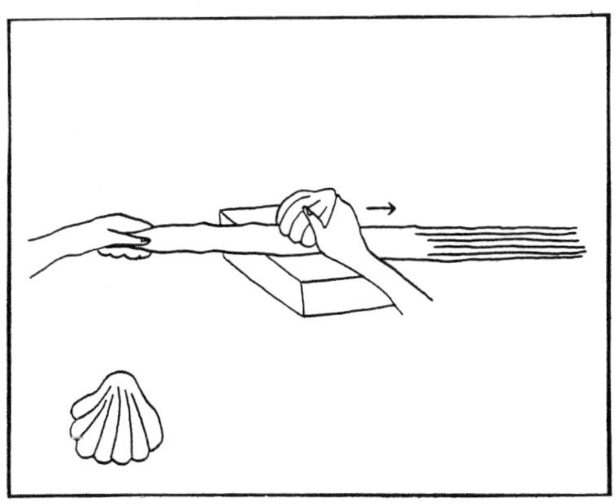

Leaf fiber separation by scraping off the outer skin with a shell

FERMENTATION

This is a process that takes the separated fiber and allows it to soak in a milk or water solution for several weeks. Air, warmth, and moisture are needed for bacterial action to eliminate any non-fibrous substances and can be used before cooking or in lieu of cooking. The soaking solution starts out by having an acid pH and ends up by being neutral or slightly on the alkaline side. Fermentation was used in lieu of cooking by early Chinese papermakers for the making of hemp paper and by the Hawaiians for the making of mulberry bark paper. The process is described in greater detail by Winifred Lutz in her chapter about fiber types in Timothy Barrett's book on the theory and practice of Japanese papermaking, listed in the bibliography. It is based on her own extensive tests with a number of fibers.

PRE-COOKING PREPARATION

After the fiber has been separated by retting, steaming, stripping, or decortication, soak it in water 24 hours prior to cooking. If a fiber has been dried for storage, it too needs to be soaked in clear water for 24 hours or more prior to cooking.

COOKING

All fibers need to undergo a cooking process prior to being beaten into pulp. Cooking frees and purifies the fibers by eliminating appreciable amounts of lignin, and other non-cellulose substances. As with the soaking process, the cooking process also helps swell the walls of the cellulose fibers so that they can take up even more water during the beating process, resulting in a stronger sheet of paper.

In order to avoid adding any trace minerals to the pulp, it is essential that the cooking be done in a non-reacting pan, which is usually stainless steel. Pack the damp fibers into the pan and add cold water by the quart to cover the fibers. For each quart of water add 1 tablespoon of lye or soda ash according to the data sheet. Bring the water, with lye and fiber, slowly to the boil, and turn heat to low, cover the pan, and simmer for approximately 2 hours or as indicated on the data sheet. An important reminder here about the use of lye: protect eyes and hands when using and do all work in a well-ventilated area. A more accurate way of determining the amount of alkali to use is based on the dry weight of the fiber. For every 100 grams of dry fiber use 18 grams of lye. The fiber would need to be weighed prior to the soaking process. In cases when soda ash is being used instead of lye, 20 grams are used for every 100 grams of dry fiber. The pH of the alkaline solution needs to be between 12 and 14 in order to eliminate the non-cellulose substances.

For some fibers which require a longer cook, it may be advantageous to use a pressure cooker for cooking the fibers. Again, use a stainless steel pan. Since small fibers can clog the gasket and air vent, it is advisable to continually supervise the pressure cooking process and to use soda ash instead of lye, or dangerous explosions may occur.

After cooking is completed, the fiber is rinsed thoroughly to remove all traces of alkali. Wear rubber gloves and either rinse under a continuous slow stream of water, leaving the fiber in the pan under the faucet, or tip the fiber into a colander lined with organdy or sheer polyester to prevent the loss of the fine fibers, returning the fiber to a bucket of clean water and repeating the process four times or more until the water runs clear. Fine nylon paint strainer bags come in various sizes and are very convenient for rinsing fibers.

The rinsed, damp fibers can be dyed at this stage with either natural or synthetic dyes. If the fibers, either dyed or undyed, are not going to be used immediately, they can be stored in plastic bags in the refrigerator for a few days or in the freezer.

Further notes about alkalis

1. The most powerful alkali is caustic soda = sodium hydroxide = supermarket lye. Potassium hydroxide is similar in chemical properties to sodium hydroxide and can also be used.
2. A moderately strong alkali, but less powerful than caustic soda, is trisodium phosphate which is available in hardware stores as a cleaner for walls and woodwork. It is known as TSP.
3. A less powerful alkali is soda ash = sodium carbonate.
4. Another less powerful alkali is lye of wood ashes = a solution made by treating wood ashes with water = potassium carbonate which is similar in chemical properties to sodium carbonate and can also be used.

The following information on beating methods, sheet formations, and casting techniques is an introductory outline to those not familiar with these papermaking processes. It is not intended to be an in-depth discussion. For more detailed information on these subjects, please consult books listed in the bibliography.

BEATING

In order for the cooked fiber to be made into a sheet of paper, it needs to undergo some kind of beating process to form a slurry or pulp. The damp fibers are taken from the refrigerator, and any remaining black specks and other impurities are removed. The fibers are then ready for beating. The beating process causes a roughening of the outside of the fiber surface, and the fiber swells as it takes up water. Fibers that have been properly beaten to achieve the right amount of roughening and swelling bond together better, thus forming a stronger sheet of paper. The length of time a fiber is beaten determines the kind of paper that is made. If the fiber is beaten a short time, a blotter type of paper results. If the fiber is beaten for a moderate amount of time, a crisp paper results. If the fiber is beaten for a long time, a tracing-type paper is produced.

Pounding Mallets

This non-mechanical method is a very ancient method of beating fibers into pulp and is best suited to the long bast fibers like milkweed and mulberry.

The damp fibers are placed on a hardwood or stone slab and pounded with wooden mallets or sticks for 1/2 to 2 hours depending on the amount of fiber. Water is sprinkled on the fiber from time to time to prevent the fibers from drying out. The mallets draw out, roughen, and help the long bast fibers soak up water as they are pounded. Very little cutting of the fiber occurs. The mallets should be made from a hardwood to avoid getting splinters in the pulp and may have grooved or smooth surfaces. A baseball bat makes an ideal pounding stick.

Mechanical Beating Devices

The Blender

The blender doesn't do much hydrating and very little roughening and cutting of the fibers but rather acts to disperse them in water. Shorter fibers, like most grass and some leaf fibers, can be "beaten" in the blender. Blend 1 cup of water and 2 tablespoons of fiber for one minute. Bast fibers like gampi and daphne can also be "beaten" in the blender, but the longer bast fibers, like milkweed, need care and supervision to make sure that they don't get tangled under the blades and burn out the blender motor. For these fibers blend 1 cup water and 1 tablespoon fiber on the lowest blender speed. In all cases the damp fiber should be cut into 1/2 in. lengths before blending. In general a softer sheet of paper results from fibers prepared in the blender. The blender has a bad reputation in the papermaking world, but scientific tests have shown that it does an adequate job of roughening and hydrating some plant fibers for oriental methods of papermaking. A variable speed electric drill with a paint mixer attachment or a "whiz mixer" consisting of a 1/3 horsepower motor, stainless steel shaft, and propeller-type blade can be used for doing a better job of hydrating larger batches of dry pulp sheets like abaca, but these mixing devices are not suitable for blending damp fiber strands which tangle and spin around the blades.

The Hollander Beater

The Hollander beater cuts and hydrates large quantities of very long fibers as they pass between the beater roll and bedplate, while circulating in water around the beater tub. After the tub is filled with water, the fibers, which have been cut into 1/2 in. lengths, are fed slowly into the tub to prevent clogging as the motor runs. The beating time varies with the fiber type. Old cotton and linen rags do not need cooking or soaking before cutting and beating but should be laundered. All raw plant fibers, however, should be cooked prior to cutting and beating in the Hollander. Because this machine is so expensive, many artists have experimented with various substitutes like washing machines, garbage disposals, and hand-operated lawnmowers!

The Ball Mill

This beating method is generally more suitable for longer leaf fibers like sisal, Manila, and iris, and for some bast fibers like hops and hollyhock. The ball mill acts in a similar way to the old stampers by roughening, hydrating, and cutting the fibers as the pebbles pound them in water between the pebbles and the side of the rotating jar. The ball mill consists of a jar, grinding pebbles, a turntable, and a motor and can be purchased from ceramic and lapidary supply houses. Several jars can be rotated at the same time on one large stand and motor. The ball mill is similar to the rock tumbler. Porcelain jars are being used instead of rubber barrels; and water, porcelain pebbles, and fiber are used instead of grit, rocks, and water.

Ball Mill Data

One oz. fiber measured while dry, now cooked and damp, and 5 lb. porcelain pebbles are placed in a 1 1/4 gallon porcelain jar. Water is added to fill the jar three quarters full. The jar is sealed and rotated for 1 to 6 hours or longer, depending on the type and amount of fiber. Manila takes approximately 4 to 5 hours while beach grass takes 1 1/2 hours. After the beating has been completed, the contents of the jar are tipped into a colander resting over a plastic pail. The water with the fine fibers drains into the pail, and the pulp and pebbles remain in the colander. The pebbles are removed and the pulp is tipped into the pail and is ready for use. Further dilution may be required for Japanese papermaking. The disadvantage of the ball mill is that it produces only a small amount of pulp. It is also very difficult to do a proper job of beating very long fibers like linen and ramie.

Ball Mill Parts

1. Two 1 1/4 gallon porcelain jars, each jar 8 1/2" diameter, 11" height

2. 10 lb. 1 1/4" diameter porcelain grinding pebbles, 5 lb. for each jar

3. Standard lapidary stand for rotating jars, 30 1/2" x 16"

4. 1/3 H.P. motor, R.P.M. 1725

The jars and pebbles are easy to keep clean. The machine, which needs very little maintenance, is designed to run for days. However, all moving parts should be oiled each time it is used.

Ball Mill References

1. Chapter and further references on pebble mills are included in SUTERMEISTER "Chemistry of Pulp and Papermaking." Please see bibliography.

2. Information on ball mills and rod mills are included in BRITT "Handbook of Pulp and Paper Technology." Please see bibliography.

3. Paper Trade Journal, October 15, 1981, pp. 36-38, "A Comparison between Flint Pebbles and Porcelain Balls as a Means of Hydrating in the Pebble Mill Test" by F. C. Peterson, S. C. Jeffrey, M. Maughan, TAPPI section,p. 178.

4. Paper Trade Journal, April 7, 1932, pp. 35-38, "Pulp Consistency Variables in Pebble Mill Test" by F. C. Peterson and Frank Makara, TAPPI section, p. 182.

5. Paper Trade Journal, May 17, 1934, pp. 33-42, "Processing Variables in Evaluating Pulps by the Pebble and Rubber Covered Ball Methods" by F. A. Simmonds and P. K. Baird, TAPPI section, p. 249.

SHEET FORMATION

In order for the pulp to be made into a sheet of paper, a screen of some sort is needed. Apart from the Western style of sheet forming, all sheet formation techniques described below are suitable only for the artist papermaker who does not desire to make many sheets of paper or to go into production papermaking.

Nepalese/Indian Sheet Formation

This is a combination of the laboratory leaf casting procedure sometimes used in paper company test labs and the more ancient methods of oriental papermaking formerly used in China, Thailand, and Vietnam, and still being used in India and Nepal. It works best with leaf and grass fibers and shorter bast fibers. The pulp is poured onto a screen with a deep deckle floating in water. The pulp and water mixture is stirred by hand. As the screen is lifted out of the water, the excess water in the pulp drains, leaving a sheet of matted fibers on the screen. For a full account of how to build the screens and use this method, please see "Handpapermaking Manual" by Peter Sarjeant, listed in the bibliography. Couching is done onto a cotton cloth.

Western Sheet Formation

The papermaking mold is dipped into a vat of pulp, shaken to distribute and cross the fibers; and the sheet forms as the water drains. The wet paper is then transferred to a felt, and drying is done by pressing in a press.

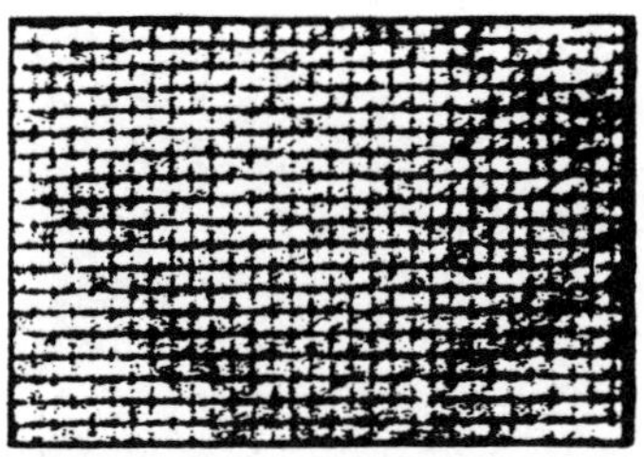

1

3

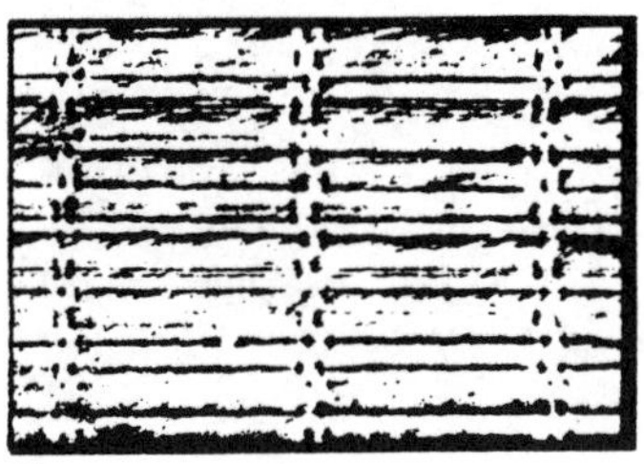

2

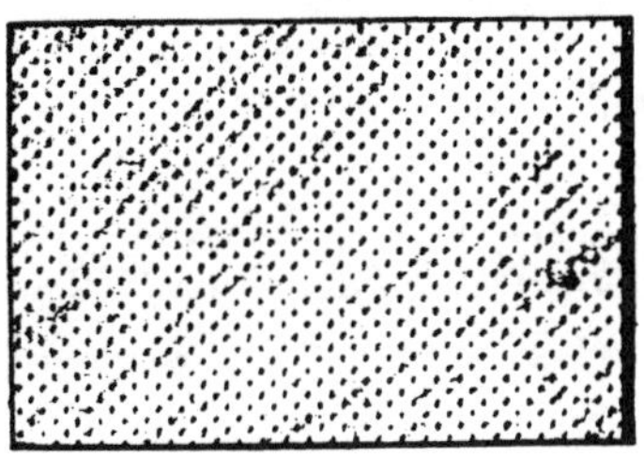

4

SCREENING, shown actual size

1. Fiberglass
2. Bamboo placemat
3. Traditional Japanese bamboo and silk laid
4. 40 mesh bronze woven wire

Modified Japanese Sheet Formation

Prior to any Japanese sheet formation techniques, the screens should be soaked in cold water for at least five minutes. This is done to allow the pulp to flow evenly over the screen, resulting in an even fiber distribution in the finished sheet of paper.

If you already have a Western style mold with a fixed metal woven screen, it can be adapted for Japanese papermaking by using a cotton cloth for couching. The mold is either dipped into the vat of pulp, or the pulp is poured onto the screen slowly and gently using a cup. Sheets are formed by rocking the pulp across the surface of the screen in a wave-like action, casting off the excess. The dipping or pouring is repeated several times until the desired thickness of the sheet is reached. The deckle edge of the mold is removed. The cotton couching cloth, now soaking wet and cut to the size of the screen, is slowly laid on top of the wet paper, taking care that no air bubbles form between the cloth and the wet paper. Blotting is done once, quickly, with a piece of terry towelling; and the cotton cloth, with the wet paper stuck to it, is peeled off the screen. It is placed directly onto a terry cloth towel paper side up.

For those who wish to build a separate mold for this technique, an ancient style Japanese mold can be made by building two identical frames of 3/4 in. by 3/4 in. lumber, mitered at the corners, and secured with bronze boat nails. Pre-cut narrow stretcher bars are available in fabric stores and also make ideal frames. A natural undyed bamboo placemat and a piece of fiberglass window screening are cut to the size of the frames. To assemble the mold, set the fiberglass screening on top of the bamboo placemat and lay these onto the first frame. The second frame is placed on top of the fiberglass/bamboo placemat combination and acts as the deckle.* A frame that is larger than 9 in. by 12 in. will need center ribs to support the placemat and prevent it from sagging. For the dipping method, hold both frames firmly together and dip the mold into the pulp. For the pouring method, set the frames onto wood supports resting over a tub of some kind so that the water can drain into it. Use one hand to press down onto the frames to keep them in place while the pulp is poured onto the screen. Build up the sheet by adding pulp, layer after layer, until the desired sheet thickness is reached. After the sheet has been formed, remove the deckle or top frame. Peel the placemat with the fiberglass screening and the wet sheet of paper from the bottom frame and lay it wet paper side down onto a piece of dry cotton cloth cut to the size of the screen and resting on a terry cloth towel. Remove the placemat; press the back of the fiberglass screen with the palms of your hands and quickly peel the fiberglass screen off, leaving the wet sheet of paper which is now stuck to the cotton cloth. This method is not as odd as it first appears! In some parts of Japan a piece of silk gauze is placed over the bamboo screen for the making of gampi paper. Occasionally finer bamboo placemats or scrolls become available at import shops, and these have been used successfully without a fiberglass overlay.

* To prevent pulp from seeping under the deckle, apply 1/4 in. thick and 1/2 in. wide Macklanburg Duncan weatherstripping. Cut 4 pieces to fit the dimensions of the deckle and remove the paper strips. Coat both the sticky side of the weatherstripping and the underside of the deckle with rubber cement and let dry. Then stick the weatherstripping onto the deckle, making sure there are no gaps. The rubber cement makes for a more permanent backing. Apply it in a well ventilated area.

DRYING TECHNIQUES FOR NEPALESE/INDIAN AND MODIFIED JAPANESE SHEET FORMATION

After the wet papers have been on the terry cloth towels for about 2 minutes, take them one by one and lay them paper side down onto wood drying boards. Lay a piece of terry cloth over each one and roll them one at a time firmly with a wood rolling pin, starting from the center and taking care not to deform the wet sheets by using too much pressure. Rolling must be done firmly so that the paper sticks to the board and dries flat. After the paper and the cloth are completely dry, they are peeled from the boards; and the cloth is removed from the paper. Drying is done without artificial heat and takes about 24 hours. Drying boards should be wood or formica covered wood and not less than 3/4 in. thick to prevent warping. Other surfaces that can be used for drying include pebble grain translucent plastic ceiling panels available in 2 ft. by 4 ft. sizes at the lumberyard and used silk screens. The latter would be advantageous in a humid climate since air circulation would facilitate drying.

PULP PREPARATION FOR JAPANESE SHEET FORMATION

In order to make paper in the Japanese style, mucilage of some kind must be added to the pulp. The mucilage is added to the pulp to slow drainage and to disperse the very long fibers being used. In Japan the mucilage is obtained from the root of the Hibiscus manihot or from the bark of a Hydrangea shrub. Other natural substitutes for American artists include Carageen gum, Hollyhock roots, and fresh or frozen Okra pods. A synthetic mucilage powder called formation aid is also available through supply catalogs listed on page 17. One teaspoon of the synthetic powder is sprinkled onto 1/2 gallon of cold water, stirring vigorously. The mixture is allowed to stand at least 24 hours before using. If there are lumps in the mixture, let it stand longer as the lumps will eventually dissolve. When mixed, the solution will keep well at cool room temperatures in a covered container. There are 2 types of synthetic formation aid available: one where additives like dyes are added to the pulp (PMP) and one when no additions are made to the pulp (PNS).

For pulp preparation add several tablespoons of the mucilage solution to the beaten diluted pulp. The amount of mucilage added depends on the fiber used; mulberry needs more mucilage than does gampi. A rough guide is 1/4 to 1/2 cup of mucilage to 1 to 2 gallons of diluted pulp. After the mucilage has been added to the pulp, mix it vigorously with a stick. High temperatures and too vigorous a mixing diminish the potency of the mucilage solution.

Those readers who are interested in knowing more about the theory and practice of Japanese papermaking should consult Timothy Barrett's book on the subject. Please see bibliography.

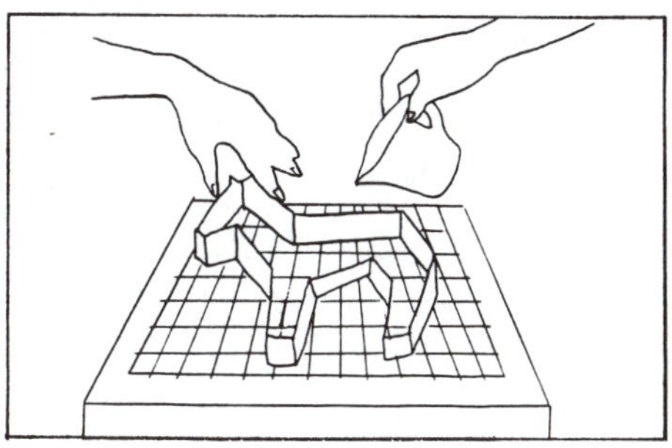

Shallow casting by pouring pulp with mucilage through shaped metal molds 1" deep

CASTING WITH PLANT FIBERS

Basically, all sheet formation methods are a casting process of sorts with the papermaking screen and deckle acting as a mold. I have already discussed how plant fibers can be used for shallow casting in a variety of rectangular molds. The same plant pulps that are used for these methods can be used for shallow casting through shaped metal molds. An example of a shaped metal mold is the standard metal cookie cutter. Molds shaped to the artist's taste can be designed out of sheet aluminum by cutting the sheet into one inch strips and shaping it into the desired form. For casting the metal mold is set on a loose piece of fiberglass screen resting on a metal papermaking screen. Pulp containing mucilage is poured into the metal shape with one hand, while the other holds the form to prevent it from slipping. After the casting is completed, the metal form is removed, and the shaped thin casting is couched from the fiberglass onto a cotton cloth. The cotton cloth is laid, paper side down, onto a wood board and rolled in the same way as for other papers made Japanese style.

For the deep casting process plaster, wood, or styrofoam molds are used. It is much more difficult to use plant fiber pulps for this process because the pulp shrinks and takes a long time to dry. In general shorter fibered pulps are easier to work with for the deep casting process. Plant fibers that I have worked with successfully for deep casting are Manila and cedar. Dry pulp sheets that work especially well for deep casting include European hemp, abaca, and Spanish flax. After the pulp has been beaten, it is strained through a piece of fiberglass screening to remove excess water. Then it is placed in the mold by the handful, and the water is sponged out until the pulp clings, damp dry, around all surfaces of the mold. The casting is dried in the mold and separates quite easily when completely dry. The drying process can be speeded up by directing an air flow from a fan over the castings. Heat is not recommended because it causes high shrinkage and warping.

Casting can also be done by using paper sheets cast by the Japanese method of sheet formation using either 100 percent mulberry fibers or a combination of 50 percent mulberry and 50 percent dry pulp sheets. Large papers up to 26 in. by 36 in. can easily be made by an artist working alone using this sheet forming method on a fixed 40 mesh brass wove screen. These papers can be used wet or dry by molding over armatures or into and over molds and forms.

Other casting methods involve the pulp sprayer and the vacuum table. For pulp spraying a pattern pistol is used to spray thin layers of pulp over and into forms or into sheets. Pulp with formation aid is sprayed from a 1 1/2-gallon plastic hopper using a 2 horsepower (H.P.) air compressor with an air pressure set at 40 pounds per square inch (p.s.i.) for abaca. Both the air pressure and pulp consistency vary with the type of fiber used. An approximate pulp mixture is 6 oz. dry weight beaten abaca, 1 cup premixed formation aid and water added to fill a 5-gallon bucket. Vacuum casting is another method that provides an easy way to form large heavy weight paper sheets, to remove water from shallow and high relief castings as well as facilitating a variety of techniques like collage/assemblage, deep embossments, and pulp painting and laminations.

For specific references about casting techniques, please see page 124 in the bibliography.

FIBER ELECTRON MICROGRAPHS

Magnification 130X, JEOL JSM 35 Scanning Electron Microscope

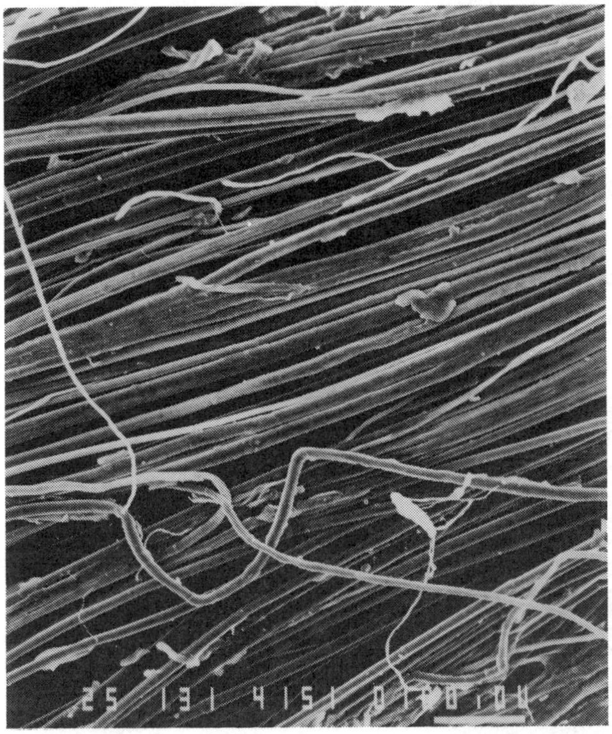

1. Cooked unbeaten milkweed bast fibers

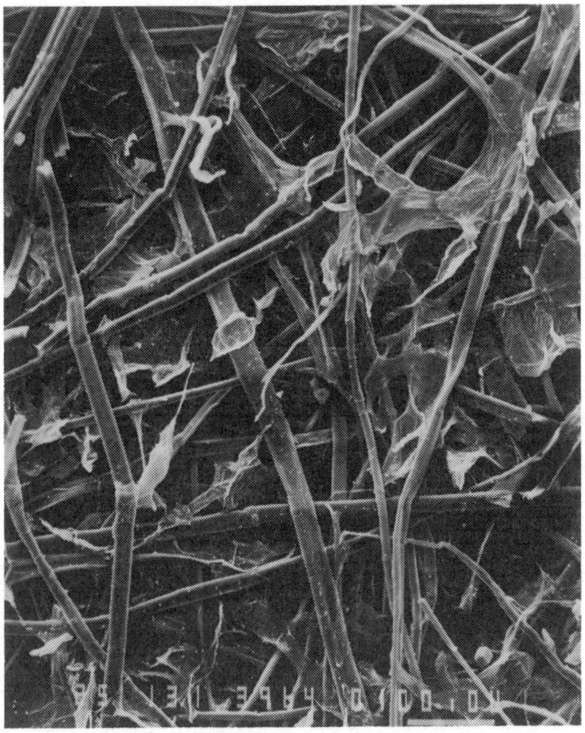

2. Hand beaten milkweed bast fibers in a paper sample

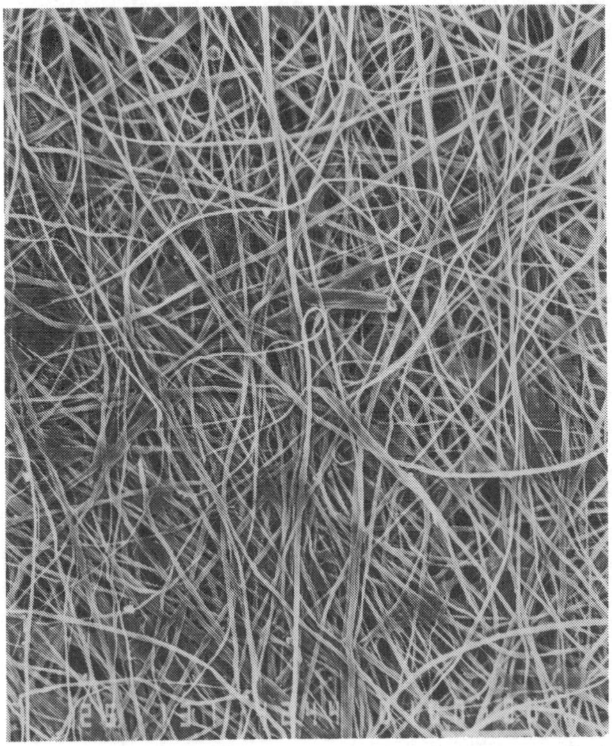

3. Blender beaten pineapple leaf fibers in a paper sample

4. Ball mill beaten cornhusk grass fibers in a paper sample

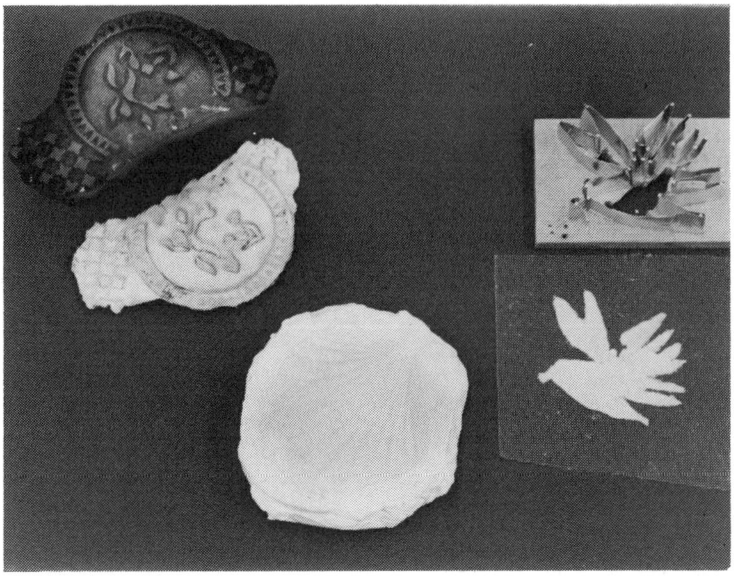

Wood, plaster and metal molds for deep and shallow casting

The author couching a sheet of paper, using a cotton cloth to remove the wet paper from a fixed Nepalese/ Indian screen

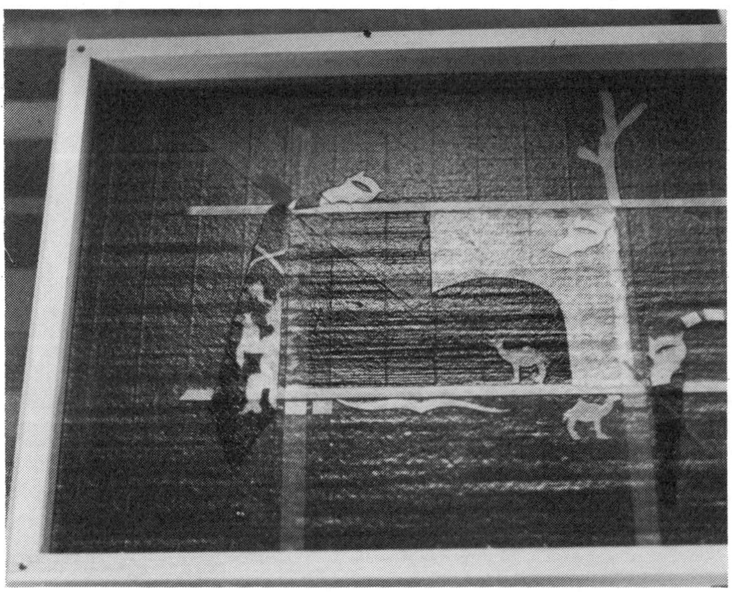

Taking advantage of the transparent properties of the gampi fiber: handmade paper cutouts are embedded between layers of transparent gampi pulp in the final stages of Japanese sheet formation. LILIAN A. BELL "Ambiguity," 1982, 12"H x 17"W, Fibers: mitnan, gampi and redwood

PAPERMAKING EQUIPMENT

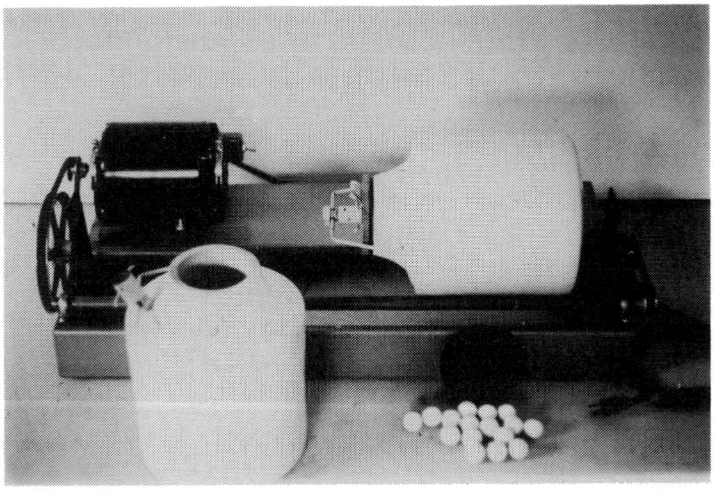

1. The ball mill

2. Indian/Nepalese style mold

3. Adapted Japanese mold

SUMMARY

The use of plant fibers for contemporary art with paper is in keeping with long standing traditions and is appropriate for current hand papermaking methods.

While plant fibers may not be of interest to all papermakers because of time-consuming preparations and small yields, they do offer interesting additions and alternatives to those who desire an expanded pulp palette.

Because of the tremendous range of plants available and the variety of papers that can be made from their fibers, the artist, by choice of fiber and its pulping treatment, has great control and flexibility with the resulting papers that comprise the final aesthetic work.

LILIAN A. BELL
"Still Life 1," 1989
Cast paper, Mixed media, 21"H x 19"W x 7¼"D
Photo: Glenn Hashitani

A CATALOG OF PLANT FIBERS

BAST FIBERS

Length Range: 3 mm to 120 mm

Common name	Indian hemp, Common dogbane
Botanical name	*Apocynum cannabinum*
Plant family	Apocynaceae
Fiber length	12-18 mm
Fiber type	Bast
Plant location in the U.S.	Throughout mainland U.S.
Plant part used	Stems from perennial herb
Season harvested	Summer or fall
Fiber purchase	Not available
Pre-cooking preparation	Can be field retted by leaving stalks to stand till dry in winter; if using fresh plants, remove leaves, steam stems and strip fiber.
Cooking	2 hours with lye
Beating method	Hand pound with mallets, Hollander beater
Sheet formation	Japanese
Paper color	Red/brown
Historical references	Used by native Americans for nets and ropes. In 1780 the Swedes in the Delaware River Colonies purchased ropes from local natives made from Indian hemp. Also used by prehistoric American Indians for fiber.

Further notes

Difficult to remove outer bark from inner – possibility we did not boil long enough. After lye bath, hand beat for ~2 hrs – We ended up blender beating ~1/2 of pulp. Paper beautiful – Pulp went a very long way. One trash bag full of plant material = 2 large vats of pulp!

Common name	Milkweed
Botanical name	*Asclepias speciosa*
Plant family	Asclepiadaceae
Fiber length	10-45 mm
Fiber type	Bast
Plant location in the U.S.	Throughout mainland U.S.
Plant part used	Stems from perennial herb
Season harvested	Fall after pods have formed
Fiber purchase	Not available
Pre-cooking	Can be field retted by leaving the stalks to stand till dry in winter; if using fresh plants, remove leaves, steam stems, strip fiber.
Cooking	2 hours with lye
Beating method	Hand pound with mallets, Hollander beater
Sheet formation	Japanese
Paper color	Whites to creamy green, varies with season fiber is harvested, whether retted or fermented
Historical references	Used by American pioneers for papermaking and by native Americans for weaving into muslin cloth.
Further notes	When the fiber is harvested in the summer, the latex from the milky sap often finds its way into the pulp and is hard to remove.

Common name	Crown flower
Botanical name	*Calotropis gigantea*
Plant family	Asclepiadaceae
Fiber length	7-30 mm
Fiber type	Bast
Plant location in the U.S.	Hawaii
Plant part used	Branches from shrub or tree
Season harvested	Summer
Fiber purchase	Not available
Pre-cooking preparation	Remove leaves, steam branches, strip fiber.
Cooking	
Beating method	
Sheet formation	
Paper color	
Historical references	Fiber used in India for ropes, nets. Flowers which resemble miniature crowns were enjoyed by Hawaii's Queen Liliuokalani and were made into leis.
Further notes	

Common name	Paper birch
Botanical name	*Betula papyrifera*
Plant family	Betulaceae
Fiber length	0.8 to 2.7 mm*
Fiber type	Bast
Plant location in the U.S.	Northern states
Plant part used	Branches from tree
Season harvested	Spring or summer
Fiber purchase	Not available
Pre-cooking preparation	Remove leaves, steam branches, strip fiber. Remove outer bark from inner bark by scraping or peeling. Discard woody cores.
Cooking	
Beating method	
Sheet formation	
Paper color	
Historical references	Bark used without processing before paper in Nepal in the Himalayan regions for wood block prints, charms and amulets. Also used as a paper substitute by U.S. Chippewa Indians for scrolls and cut-out figures.
Further notes	Heartwood used commercially for machine-made paper.

*Fiber length for heartwood fibers, bast fibers probably longer.

Common name	Hazelnut, Filbert
Botanical name	*Corylus californica*
Plant family	Betulaceae
Fiber length	Not known
Fiber type	Bast
Plant location in the U.S.	Various species throughout the U.S.
Plant part used	Branches from shrub or tree
Season harvested	Fall
Fiber purchase	Not available
Pre-cooking preparation	Remove leaves, steam branches and strip fiber.
Cooking	2 hours with lye
Beating method	Ball mill 1 1/2 hours, or blender
Sheet formation	Japanese, Nepalese/Indian or Western
Paper color	Brown
Historical references	Used by native Americans for cordage. Sample of hazelnut paper made by Leorier De Lisle for a book printed in 1786.
Further notes	

Common name	Burdock	
Botanical name	*Arctium minus*	
Plant family	Compositae	
Fiber length	0.91 mm approximately	
Fiber type	Bast	
Plant location in the U.S.	Throughout most of the U.S.	
Plant part used	Stems from perennial herb	
Season harvested	Late summer	
Fiber purchase	Not available	
Pre-cooking preparation	Remove leaves, steam stems, strip fiber.	
Cooking	2 hours with soda ash	
Beating method	Ball mill 2 hours	
Sheet formation	Nepalese/Indian	
Paper color	Light tan/brown	
Historical references	Sample of Burdock paper included in 1765 book by Jacob Schaffer on his research on plant fibers for paper-making.	
Further notes		

Common name	Sagebrush
Botanical name	*Artemisia tridentata*
Plant family	Compositae
Fiber length	1.36 mm approximately
Fiber type	Bast
Plant location in the U.S.	Western states and parts of Northeastern U.S.
Plant part used	Stems from perennial herb or shrub
Season harvested	Late summer
Fiber purchase	Not available
Pre-cooking preparation	Remove leaves, steam stems and strip fiber.
Cooking	2 hours with lye
Beating method	Ball mill 4 hours
Sheet formation	Japanese, Nepalese/Indian
Paper color	Light gold tan
Historical references	Used for fiber by native Americans in 9,000 year old sandal discovered in Fort Rock, Oregon.
Further notes	Other Artemisia species yield a similar paper.

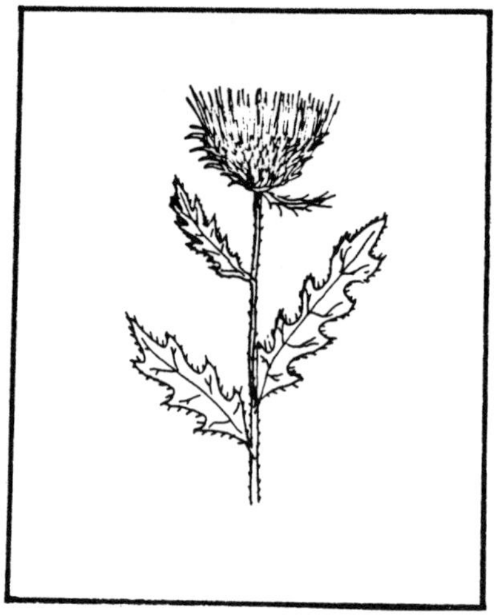

Common name	Thistle
Botanical name	*Cirsium vulgare*
Plant family	Compositae
Fiber length	0.90 mm approximately
Fiber type	Bast
Plant location in the U.S.	Northern half of the U.S.
Plant part used	Stems from perennial herb
Season harvested	Late summer
Fiber purchase	Not available
Pre-cooking preparation	Remove leaves, steam stems, strip fiber.
Cooking	2 hours with lye
Beating method	Ball mill 3 hours
Sheet formation	Nepalese/Indian
Paper color	Light brown tan
Historical references	Sample of thistle paper included in 1765 book by Jacob Schaffer on his research on plant fibers for papermaking.
Further notes	

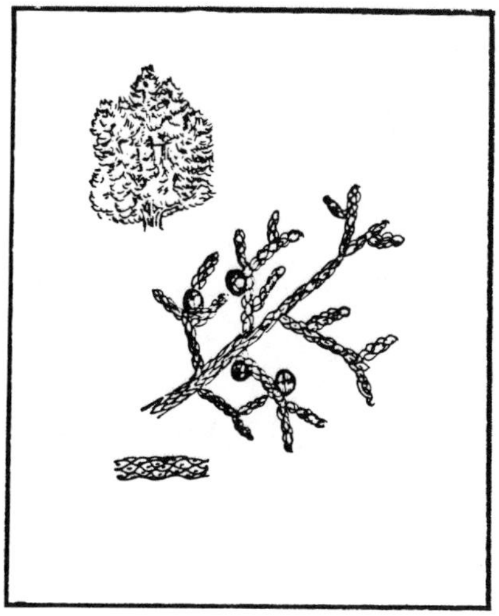

Common name	Juniper
Botanical name	*Juniperus spp.*
Plant family	Cupressaceae
Fiber length	Not known
Fiber type	Bast
Plant location in the U.S.	Wild species mainly in southwestern parts of U.S., cultivated species throughout U.S.
Plant part used	Branches of shrub or tree
Season harvested	Summer or fall
Fiber purchase	Not available
Pre-cooking preparation	Remove leaves, steam branches, and strip fiber.
Cooking	2 hours or more with lye
Beating method	Ball mill 2 hours or blender
Sheet formation	Nepalese/Indian, Japanese
Paper color	Deep brown
Historical references	Used for mats in 1100-1425 A.D. by Pueblo Indians of the southwestern U.S.
Further notes	

Common name	Western red cedar
Botanical name	*Thuja plicata*
Plant family	Cupressaceae
Fiber length	3.8 mm approximately*
Fiber type	Bast
Plant location in the U.S.	Northwest coastal regions of the U.S.
Plant part used	Bark from branches or trunk of tree
Season harvested	Anytime from logged trees
Fiber purchase	Possibly cedar bark chips from gardening store or cedar shingle factories
Pre-cooking preparation	Cut into useable pieces and pre-soak in clear water prior to cooking.
Cooking	6 hours or more with lye, difficult to rinse to clear water because seems to be a dye material
Beating method	Hand pounding with mallets, blender, or 6 hours in ball mill, difficult to hydrate properly
Sheet formation	Nepalese/Indian, Japanese or Western
Paper color	Rich deep brown/red
Historical references	Used by Northwest Coast Indians for mats, clothing, blankets, sails, ropes. Bark was soaked, pounded, and the fiber was rolled into yarn, then woven in wefts of nettle string.
Further notes	Makes an excellent brown pulp for deep casting.

*Fiber length for heartwood fibers, bast fibers probably longer.

Common name	Acacia
Botanical name	*Acacia spp.*
Plant family	Leguminosae
Fiber length	Not known
Fiber type	Bast
Plant location in the U.S.	Warmer parts of the U.S. including Hawaii and California
Plant part used	Branches of tree or shrub
Season harvested	
Fiber purchase	Not available
Pre-cooking preparation	Remove leaves, steam branches, strip fiber. Remove outer bark from inner bark by scraping or peeling.
Cooking	
Beating method	
Sheet formation	
Paper color	
Historical references	A. jacquemontii used for papermaking in India, A. spirocarpa used for fiber purposes in Africa. Acacia cornigera used by Aztecs for bark paper.
Further notes	A. leucophloea fiber used for ropes and nets in Sri Lanka.

Common name	
Botanical name	Bauhinia racemosa syn. vahlii
Plant family	Leguminosae
Fiber length	1.5 to 4.0 mm
Fiber type	Bast
Plant location in the U.S.	Other species in Hawaii, Arizona, California
Plant part used	Branches from shrub or tree
Season harvested	
Fiber purchase	Not available
Pre-cooking preparation	Remove leaves, steam branches, strip fiber. Remove outer bark from inner bark by scraping or peeling.
Cooking	
Beating method	
Sheet formation	
Paper color	
Historical references	Used in Africa for ropes and clothing, and in India for sails.
Further notes	

Common name	Sunn hemp, Bengal hemp, Indian hemp, Madras hemp, Sun pat
Botanical name	*Crotolaria juncea*
Plant family	Leguminosae
Fiber length	5-7 mm
Fiber type	Bast
Plant location in the U.S.	Other Crotolaria species grow in Hawaii and southern U.S. states
Plant part used	Stem from annual herb
Season harvested	Late summer
Fiber purchase	Not available
Pre-cooking preparation	If using fresh plants, remove leaves, steam stems, strip fiber. If using dry fiber sheets, soak the sheets 24 hours prior to beating.
Cooking	2 hours with lye for fresh fiber. No cooking is necessary for dry fiber sheets.
Beating method	Blender for dry fiber sheets. Ball mill for fresh fiber.
Sheet formation	Japanese, Nepalese/Indian or Western
Paper color	Buff. White for dry fiber.
Historical references	Grown in India and Pakistan since prehistoric times, used for fiber needs, for fiber export, and formerly for papermaking.
Further notes	Still used in Asia for papermaking and for filter papers in the U.S. and Europe.

Common name	Scotch broom, Broom
Botanical name	*Cytisus scoparius* (formerly Spartium scoparium)
Plant family	Leguminosae
Fiber length	2-9 mm
Fiber type	Bast
Plant location in the U.S.	California and northwestern states, as a weed; elsewhere as a cultivated plant
Plant part used	Branches from shrub
Season harvested	Late summer, fall
Fiber purchase	Not available
Pre-cooking preparation	Remove leaves, steam stems, strip fiber.
Cooking	3 hours with lye
Beating method	Ball mill 3 hours
Sheet formation	Nepalese/Indian
Paper color	Pale tan
Historical references	Another species related to Cytisus, Spartium junceum, was used in the south of France for cordage and cloth. Other uses include baskets, mats and papermaking.
Further notes	Genista spp. is also related to Spartium spp. and Cytisus spp. and was used for a paper sample made by Jacob Schaffer for his book on plant fibers in 1767.

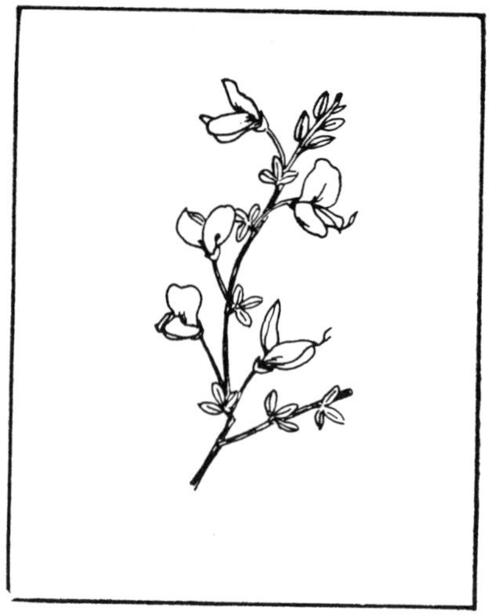

Common name	Kudzu, Ko hemp
Botanical name	*Pueraria lobata* (formerly P. thunbergiana)
Plant family	Leguminosae
Fiber length	2-3 mm
Fiber type	Bast
Plant location in the U.S.	Southeastern states
Plant part used	Straight first year 6 to 9 ft. stems from perennial woody vine
Season harvested	Early June to late July
Fiber purchase	Not available
Pre-cooking preparation	Remove leaves, steam stems, strip fiber.
Cooking	Cook 2 hours with lye for tender vines; 4 hours for tougher vines.
Beating method	Ball mill 3 hours, or possibly Hollander
Sheet formation	Nepalese/Indian or Japanese
Paper color	Greenish cream
Historical references	Used for weaving by Mongols prior to 3rd century A.D. Used by Japanese for cloth 200 A.D.
Further notes	Formerly cultivated, now a persistent weed.

Common name	Colorado river hemp
Botanical name	*Sesbania macrocarpa*
Plant family	Leguminosae
Fiber length	2.35 mm approximately
Fiber type	Bast
Plant location in the U.S.	New York to Florida and west to southern California
Plant part used	Stems from annual herb
Season harvested	Summer, fall
Fiber purchase	Not available
Pre-cooking preparation	Remove leaves, steam stems, strip fiber.
Cooking	
Beating method	
Sheet formation	
Paper color	
Historical references	Important fiber plant of the native Americans
Further notes	Other species include S. aculeata, a tree yielding an Indian fiber called Dundee. Other species are shrubs.

Common name	Wisteria
Botanical name	*Wisteria* spp.
Plant family	Leguminosae
Fiber length	1.3 to 3.7 mm
Fiber type	Bast
Plant location in the U.S.	Widely cultivated throughout the U.S. as a garden plant
Plant part used	Stems from woody vine
Season harvested	Summer
Fiber purchase	Not available
Pre-cooking preparation	Remove leaves, steam stems, strip fiber.
Cooking	2 hours with lye
Beating method	Ball mill 3 hours
Sheet formation	Japanese or Nepalese/Indian
Paper color	Buff
Historical references	Wisteria bark was used in China for papermaking in the 3rd to 6th centuries A.D. W. brachybotrus used in Japan for cloth.
Further notes	Wisteria chinensis yields a white, fine fiber, and was used in Italy for cloth.

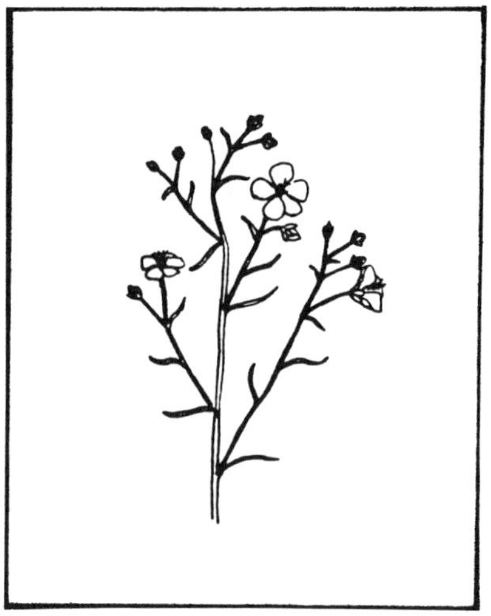

Common name	Flax, Linen
Botanical name	*Linum usitatissimum*
Plant family	Linaceae
Fiber length	33 mm approximately
Fiber type	Bast
Plant location in the U.S.	Various species, now mainly wild throughout the U.S.
Plant part used	Stems from annual herb or perennial weed
Season harvested	Summer/fall when stems are two thirds yellow
Fiber purchase	Available from weaving supply stores, either unbleached or bleached or by mail order.
Pre-cooking preparation	Commercial flax/linen is usually retted; if using fresh plants, remove leaves, steam stems and strip fiber; if using dry fiber, pre-soak in clear water for 24 hours before cooking.
Cooking	2 hour cook with lye, or longer
Beating method	Hollander beater
Sheet formation	Western
Paper color	Grey, buff, tan or white depending on variety and whether bleached
Historical references	Linen textile fragments date from 2500 B.C. First used for paper by the Chinese in 2nd to 4th centuries.
Further notes	

Common name	Indian mallow
Botanical name	*Abutilon avicennae*
Plant family	Malvaceae
Fiber length	1.30 mm approximately
Fiber type	Bast
Plant location in the U.S.	Southern part of the U.S., other species as weeds throughout the U.S. (A. theophrasti, called Velvet leaf)
Plant part used	Stems from annual herb or branches from shrub, depending on species used
Season harvested	Summer
Fiber purchase	Not available
Pre-cooking preparation	Remove leaves, steam stems or branches and strip fiber.
Cooking	
Beating method	
Sheet formation	
Paper color	
Historical references	Formerly cultivated in China for fiber, used for cordage, paper, cloth, sails.
Further notes	A. bedfordianum, a shrub, used in Brazil for fiber; A. asiaticum used for paper and cordage. Fiber also called China Jute because of similarities to Jute.

Common name	Hollyhock
Botanical name	Althea rosea
Plant family	Malvaceae
Fiber length	1.93 mm approximately
Fiber type	Bast
Plant location in the U.S.	Throughout the U.S.
Plant part used	Stems from perennial herb
Season harvested	Late summer
Fiber purchase	Not available
Pre-cooking preparation	Remove leaves, steam stems and strip fiber.
Cooking	2 hours with lye
Beating method	Ball mill 3 hours, or hand pounding with mallets
Sheet formation	Japanese, or Nepalese/Indian
Paper color	Light tan
Historical references	The hollyhock is one of the oldest cultivated garden flowers.
Further notes	Other species may be suitable for paper, i.e. Althea officinalis (marsh mallow). The hollyhock roots may be a substitute for tororo-aoi mucilage; both stems and roots exude a sticky substance. The root from A. officinalis contains 25% to 35% mucilage.

8 Dec. 2001 Easy to strip. hardest part is collection of stems - very itchy. After stripping (no outer bark removal necessary) + lye boil, we hand beat. The fibers were still way too long though they had spread apart. We ended up cutting the fiber w/ scissors and blender beating. The fiber was translucent - light gold! After all this manipulation, a thin sheet required 2-3 pours. It's possible that these stalks had retted the woody layer off of the outside (outer bark) - they had been drying for several mos.

Common name	Cotton
Botanical name	*Gossypium hirsutum* L.
Plant family	Malvaceae
Fiber length	0.7 to 4.2 mm
Fiber type	Bast
Plant location in the U.S.	Southeastern and southwestern states
Plant part used	Stem from annual or perennial herb
Season harvested	Summer, or after the cotton seed hair crop has been harvested
Fiber purchase	Not available
Pre-cooking preparation	Remove leaves, steam stems and strip fiber.
Cooking	2 hours with lye
Beating method	Blender, or ball mill
Sheet formation	Japanese
Paper color	Brown
Historical references	Cotton seed hairs have been used for fiber purposes since 3rd millenium B.C. Paper was made from cotton rags in Xativa, Spain in 1151.
Further notes	The stems are a crop residue and have been used for commercial papermaking.

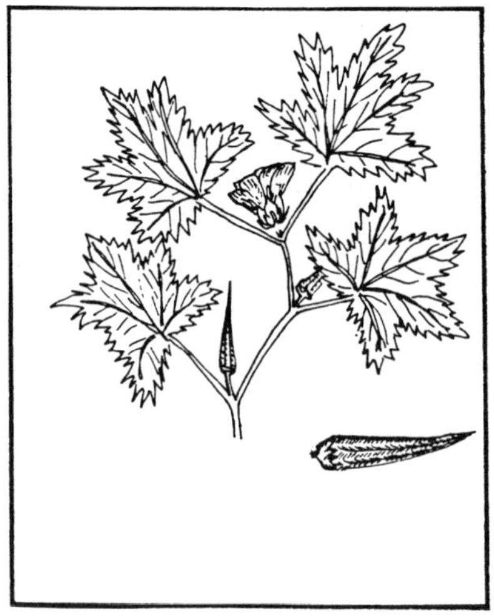

Common name	Okra
Botanical name	*Hibiscus esculentus*
Plant family	Malvaceae
Fiber length	2.43 mm approximately
Fiber type	Bast
Plant location in the U.S.	Southern states
Plant part used	Stem from annual or perennial herb
Season harvested	Summer/fall after the pod crop has been harvested
Fiber purchase	Not available, although it might be possible to glean stems from harvested fields
Pre-cooking preparation	Remove leaves, steam stems and strip fiber.
Cooking	2 hours with lye
Beating method	Ball mill 3 hours, or Hollander beater
Sheet formation	Japanese or Nepalese/Indian
Paper color	Cream
Historical references	Used for commercial papermaking in 1869 in the South. The *Mobile Register* newspaper was printed on Okra paper in 1870. Other Hibiscus species H. sabdariffa and H. cannabinus (kenaf) possibly domesticated in western Sudan before 4000 B.C. Kenaf produces a paper similar in color to Okra.
Further notes	All parts of the Okra, including the pods, may be a possible substitute for tororo-aoi inasmuch as they exude a similar sticky substance.

Common name	Hau
Botanical name	*Hibiscus tiliaceus*
Plant family	Malvaceae
Fiber length	2-6 mm
Fiber type	Bast
Plant location in the U.S.	Hawaii, Florida
Plant part used	Branches from woody shrub or small tree
Season harvested	Anytime
Fiber purchase	Not available
Pre-cooking preparation	Remove leaves, steam branches, and strip fiber. Purchased fiber needs soaking in clear water for 24 hours prior to cooking.
Cooking	2 hours with lye
Beating method	Ball mill 4 hours
Sheet formation	Japanese, Nepalese or Western
Paper color	Light tan
Historical references	Formerly used by Hawaiians for sandals and for making tapa; used in Polynesia for twine and grass skirts.
Further notes	Other species of Hibiscus grow in California, while others can be grown from seed as garden annuals. A hardy species H. syriacus might produce fiber that is similar to the Hau fiber. It is also possible that Hau branches and bark could be good substitutes for tororo-aoi mucilage, since the bark yields a similar sticky substance when fresh.

Common name	Queensland hemp, Sida, Tea plant
Botanical name	*Sida rhombifolia*
Plant family	Malvaceae
Fiber length	1.8 mm approximately
Fiber type	Bast
Plant location in the U.S.	Hawaii, various other Sida spp. in eastern portion of the U.S.
Plant part used	Stems from perennial herb
Season harvested	Summer or fall
Fiber purchase	Not available
Pre-cooking preparation	Remove leaves, steam stems and strip fiber.
Cooking	
Beating method	
Sheet formation	
Paper color	
Historical references	Cultivated for fiber in the Philippines, India, and S.E. Asia. Other Sida spp. formerly cultivated in southern parts of U.S. as a forage plant.
Further notes	Sida acute cultivated in Mexico, Sida augustifolia in Hawaii and S. micrantha in India, Brazil and Australia.

Common name	Milo
Botanical name	*Thespesia populnea*
Plant family	Malvaceae
Fiber length	5.2 to 8.0 mm
Fiber type	Bast
Plant location in the U.S.	Hawaii, formerly cultivated in southern California
Plant part used	Branches from small tree
Season harvested	Summer
Fiber purchase	Not available
Pre-cooking preparation	Remove leaves, steam branches and strip fiber.
Cooking	
Beating method	
Sheet formation	
Paper color	
Historical references	Used by the Hawaiians and the Marquesans for Tapa.
Further notes	

Common name	Cesar weed
Botanical name	*Urena lobata*
Plant family	Malvaceae
Fiber length	1.5 to 5.1 mm
Fiber type	Bast
Plant location in the U.S.	Florida
Plant part used	Stems from shrub
Season harvested	Late summer
Fiber purchase	Not available
Pre-cooking preparation	Remove leaves, steam stems and strip fiber.
Cooking	
Beating method	
Sheet formation	
Paper color	
Historical references	Fiber has been used from prehistoric times in Brazil, cultivated as a fiber plant in Africa where it is used for rope. Used for papermaking in the East Indies.
Further notes	

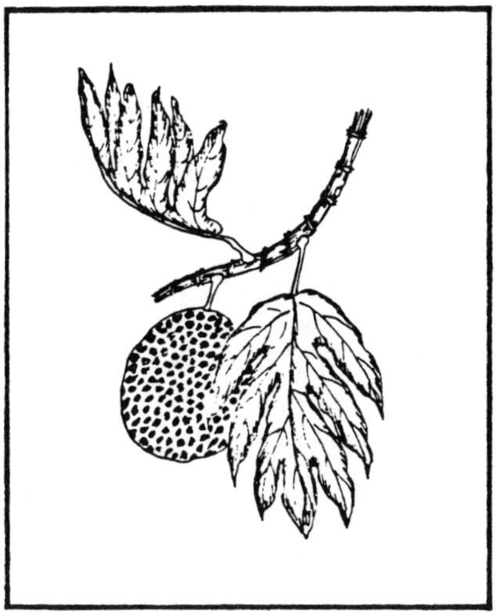

Common name	Breadfruit
Botanical name	*Artocarpus spp.*
Plant family	Moraceae
Fiber length	9-27 mm
Fiber type	Bast
Plant location in the U.S.	Hawaii and possibly Florida
Plant part used	Branches from tropical tree
Season harvested	Summer
Fiber purchase	Sometimes available
Pre-cooking preparation	Remove leaves, steam branches, strip fiber. Separate the outer bark from the inner bark by scraping or peeling.
Cooking	2 hours with lye
Beating method	Hand pounding with mallets
Sheet formation	Japanese
Paper color	1. White (inner bark only) 2. Brown (inner and outer bark)
Historical references	Artocarpus atilis and A. incisa were among the species used in the Polynesian islands for the making of bark paper. The breadfruit is a multi-purpose plant used for food, fiber, latex and timber, and probably was domesticated in Oceania approximately 1000-2000 B.C.
Further notes	Indirect cause of mutiny on Capt. Bligh's ship while on a mission for the King of England to collect and transplant the breadfruit tree to the West Indies.

Common name	Mulberry
Botanical name	*Broussonetia papyrifera*
Plant family	Moraceae
Fiber length	10 mm approximately
Fiber type	Bast
Plant location in the U.S.	Hawaii, other species grow throughout U.S. (Morus spp.)
Plant part used	Branches from tree
Season harvested	Late fall after leaves have dropped
Fiber purchase	Dry fiber bundles can be mail ordered
Pre-cooking preparation	Remove leaves, steam stems, strip fiber. In humid areas fiber can sometimes be stripped without steaming. Separate the outer bark from the inner bark by scraping. In humid areas this can be done by peeling. Dry fiber bundles need to be soaked for 24 hours prior to cooking.
Cooking	2 hours with lye
Beating method	Hand pounding with mallets
Sheet formation	Japanese
Paper color	1. White (inner bark only) 2. Tan with flecks (inner and outer bark) 3. Greenish white (inner bark that does not have the green layer between the inner and outer bark scraped off)
Historical references	Fiber first used for paper by Chinese in 2nd to 10th centuries, also used for clothing by the Chinese in the same period and by Polynesians where it was prepared differently and called "Tapa" or bark paper.
Further notes	Broussonetia kajinoki is species used in Japan for papermaking. The Polynesian Broussonetia tends to be tougher than the Japanese perhaps because of different growing conditions which include a great deal more moisture and much more rapid growth. In 1298 Marco Polo noted that Morus alba was the preferred fiber source for making Chinese paper money.

We used Morus rubra - easy to strip & remove outer bark. We removed outer on 1/2, Left on for the other. Paper is beautiful and whitish. We have Broussonetia sp. in Anderson County.

Common name	True hemp
Botanical name	Cannabis sativa
Plant family	Moraceae
Fiber length	25 mm approximately
Fiber type	Bast
Plant location in the U.S.	Throughout mainland U.S. as a weed and illegal cultivation as a drug plant
Plant part used	Stems from annual herb or weedy perennial
Season harvested	Fall
Fiber purchase	Available as woven cloth, fiber, or dry pulp sheets
Pre-cooking preparation	Remove leaves, steam stems and strip fiber. Commercially produced hemp is usually retted.
Cooking	2 hours or more with lye. If using cloth, cut into 1 in. squares and ret one month in lye solution described on page 23 in lieu of cooking.
Beating method	Hollander or ball mill
Sheet formation	Japanese
Paper color	Off-white
Historical references	Hemp paper was made in China 200 B.C. to 10th cent. A.D. In Japan hemp was the most important papermaking fiber during the Nara period (710-793 A.D.). Hemp is an ancient crop plant cultivated in China since 2800 B.C.
Further notes	The Chinese sometimes fermented the fiber in lieu of cooking for their ancient paper. Used today for currency paper in Europe and for fine paper in China.

Common name	Fig, Banyan
Botanical name	*Ficus bengalensis*
Plant family	Moraceae
Fiber length	1-3 mm
Fiber type	Bast
Plant location in the U.S.	Hawaii, Gulf States, California, Arizona, New Mexico, Texas, S.E. states
Plant part used	Branches or aerial roots from tree
Season harvested	Summer or fall
Fiber purchase	Not available
Pre-cooking preparation	Remove leaves, strip fiber from the woody core immediately after cutting branches. If branches have been allowed to dry, steam them, then strip off fiber. Scrape or peel outer bark off inner bark.
Cooking	2 hours with lye
Beating method	Hand pound with mallets, or blender
Sheet formation	Japanese or Nepalese/Indian
Paper color	Brown
Historical references	In the 5th century A.D. the Mayans in Mexico developed bark paper from fig trees and called the paper Huun. Fig bark paper is currently being made in San Pablito, Mexico, where it is used for tourist trade and witchcraft. Various species of the fig were used in Polynesia and Africa for bark paper, and in Burma for papermaking.
Further notes	The following were used in Polynesia for Tapa: Ficus obliqua, F. prolixa, F. tinctoria. The Aztecs and Mayans used approximately 25 species of fig for their bark paper, called Amate, including F. nitida for a brown paper, F. padifolia for a white paper, F. petiolaris, F. tecolutensis, F. involuta, F. cotinifolia. The edible fig, Ficus carica, yields a khaki-colored paper.

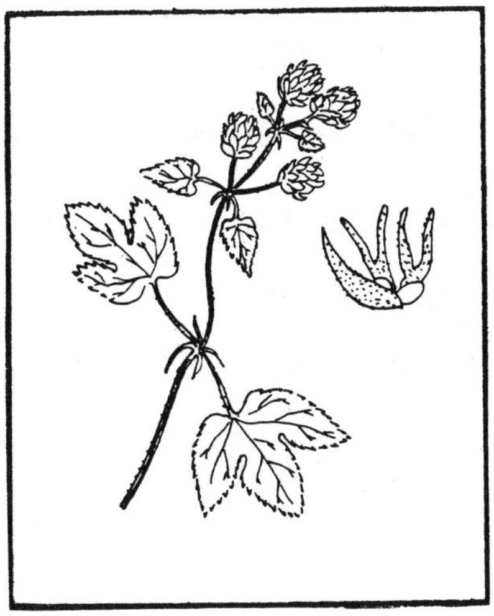

Common name	Hop
Botanical name	*Humulus lupulus*
Plant family	Moraceae
Fiber length	4-19 mm
Fiber type	Bast
Plant location in the U.S.	Northwestern and northern states, Florida and Arizona
Plant part used	Stems from perennial vine
Season harvested	Fall after the crop has been harvested
Fiber purchase	Not available
Pre-cooking preparation	Remove leaves, steam stems and strip fiber.
Cooking	2 hours with lye
Beating method	Hand pounding with mallets, or ball mill 2 1/2 hours
Sheet formation	Japanese or Nepalese/Indian
Paper color	Brown
Historical references	Sample of hop paper included in 1786 book printed on paper by Leorier De Lisle. Fiber also used for textiles in Sweden.
Further notes	Cultivation of hops as a crop was well established in central Europe by the 13th century.

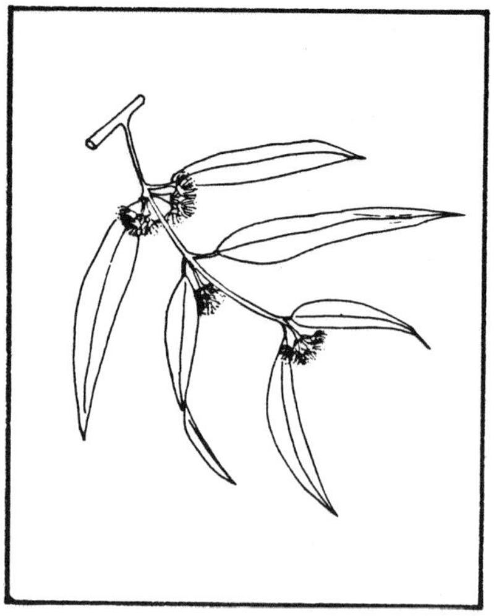

Common name	Eucalyptus
Botanical name	*Eucalyptus spp.*
Plant family	Myrtaceae
Fiber length	0.8 to 1.0 mm
Fiber type	Bast
Plant location in the U.S.	California, southwestern states, Hawaii
Plant part used	Branches from tree
Season harvested	Summer, fall
Fiber purchase	Not available
Pre-cooking preparation	Remove leaves, steam branches and strip fiber.
Cooking	2 hours with soda ash
Beating method	Ball mill 1 1/2 hours, or blender
Sheet formation	Nepalese/Indian or Japanese
Paper color	Reddish brown
Historical references	Used for papermaking in Australia, Brazil, Africa and other countries.
Further notes	Species used for commercial papermaking are: E. globulus (Bluegum), E. rostrata (Redgum), and E. obliqua (stringybark) used for ropes and bark paintings by Australian aborigines.

Common name	Raspberry
Botanical name	*Rubus idaeus*
Plant family	Rosaceae
Fiber length	Not known
Fiber type	Bast
Plant location in the U.S.	Throughout mainland U.S. and R. Macraei, R. Hawaiiensis in Hawaii
Plant part used	Stems from shrub
Season harvested	Summer after berries have been harvested
Fiber purchase	Not available
Pre-cooking preparation	Remove leaves, steam stems, strip fiber.
Cooking	2 hours with lye
Beating method	Hand pounding with mallets, or ball mill 3 hours
Sheet formation	Japanese, Nepalese/Indian or Western
Paper color	Light brown
Historical references	Hawaiian species, Akala, formerly used by Hawaiians for making bark paper/Tapa.
Further notes	

Common name	Willow
Botanical name	*Salix spp.*
Plant family	Salicaceae
Fiber length	1-3 mm
Fiber type	Bast
Plant location in the U.S.	Different species throughout the U.S.
Plant part used	Branches from tree
Season harvested	Spring or summer
Fiber purchase	Not available
Pre-cooking preparation	Remove leaves, steam branches and strip fiber.
Cooking	2 hours with lye
Beating method	Blender or hand pound with mallets
Sheet formation	Japanese
Paper color	Red/brown
Historical references	Several species used for string by native Americans. Willow paper made in 1781 by Thomas Greaves of England.
Further notes	Species used for commercial papermaking are Salix nigra and S. alba. In central Europe willow bark is steeped in strong lye solution and used for twine and bags.

Common name	Redwood
Botanical name	*Sequoia sempervirens*
Plant family	Taxodiaceae
Fiber length	7 mm approximately*
Fiber type	Bast
Plant location in the U.S.	Pacific Northwest
Plant part used	Bark from branches or trunk of tree
Season harvested	Anytime from available logged trees
Fiber purchase	Not generally available, but sometimes available as bleached dry pulp sheets from a paper company in over 100 lb. lots
Pre-cooking preparation	If using fresh bark, cut into useable pieces and pre-soak in clear water prior to cooking. Dry pulp sheets need pre-soaking only.
Cooking	6 hours or more with lye for fresh bark, no cooking for dry pulp sheets
Beating method	Hand pounding, blender, or ball mill 6 hours; difficult to hydrate properly for fresh bark; dry pulp sheets can be beaten in the blender, or hydrated with a paint mixer.
Sheet formation	Japanese, Nepalese/Indian, Western
Paper color	Brown if fresh bark; white if bleached dry pulp sheets
Historical references	Formerly used for insulation, and in feltmaking where it was mixed with wool.
Further notes	Curled fiber that felts very readily. Paper could possibly be made from redwood bark chips sold in garden supply stores. Paper company dry pulp sheets are prepared from the wood.

*Fiber length for heartwood fibers, bast fibers probably longer.

Common name	
Botanical name	*Dais glaucescens*
Plant family	Thymelaeaceae
Fiber length	Not known
Fiber type	Bast
Plant location in the U.S.	Florida, southern California
Plant part used	Branches of shrub
Season harvested	
Fiber purchase	Not available
Pre-cooking preparation	Remove leaves, steam branches, strip fiber. Remove outer bark from inner bark by scraping or peeling.
Cooking	
Beating method	
Sheet formation	
Paper color	
Historical references	Used in Madagasgar for twine.
Further notes	Dais cotinifolia, African button flower, cultivated in Australia for fiber.

Common name	Daphne
Botanical name	*Daphne spp.*
Plant family	Thymelaeaceae
Fiber length	2-7 mm
Fiber type	Bast
Plant location in the U.S.	Western portion of the U.S.
Plant part used	Branches of shrub
Season harvested	Spring, early summer
Fiber purchase	Sometimes available
Pre-cooking preparation	Remove leaves, steam branches and strip the fibers. Separate the outer bark from the inner bark by peeling or scraping.
Cooking	1 1/2 to 2 hours with soda ash
Beating method	Hand pounding with mallets or in blender
Sheet formation	Japanese or Nepalese/Indian
Paper color	1. Creamy buff 2. Flecked tan if outer bark is not removed
Historical references	Used in Nepal by Himalayan papermakers in the 11th to 13th centuries. Also used in Italy for fiber.
Further notes	Daphne produces a translucent, silky paper similar to Gampi. Daphne paper, from Daphne bholua is still being made today throughout the Himalayan region from Nepal to Bhutan.

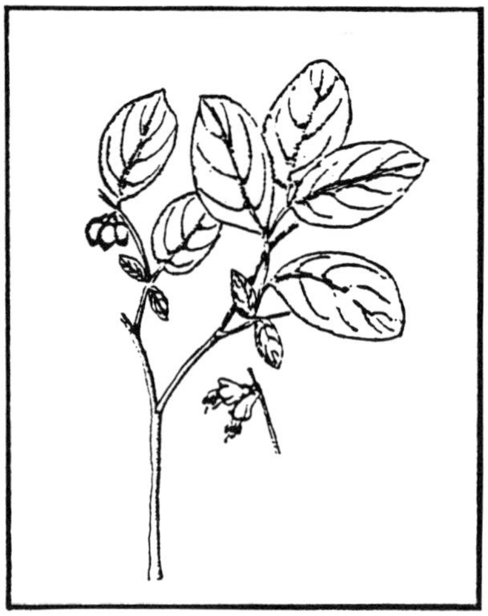

Common name	Leatherwood, Swampwood
Botanical name	*Dirca palustris*
Plant family	Thymelaeaceae
Fiber length	Not known
Fiber type	Bast
Plant location in the U.S.	Eastern U.S. New Brunswick to Minnesota and south to Florida. Dirca occidentalis in California
Plant part used	Branches from shrub
Season harvested	Summer
Fiber purchase	Not available
Pre-cooking preparation	Remove leaves, steam branches and strip fiber. Remove outer bark from inner bark by scraping or peeling.
Cooking	2 hours or less with soda ash
Beating method	Hand pounding with mallets, or blender
Sheet formation	Japanese
Paper color	Greenish cream
Historical references	Fiber was used by native Americans for ropes.
Further notes	Thymelaea hirsuta, another member of this plant family, growing in the Negev desert of Israel, was used by the Bedouins for rope. The plant is called Mitnan in Arabic and Hebrew.

Possibly in Rabun Gap, GA

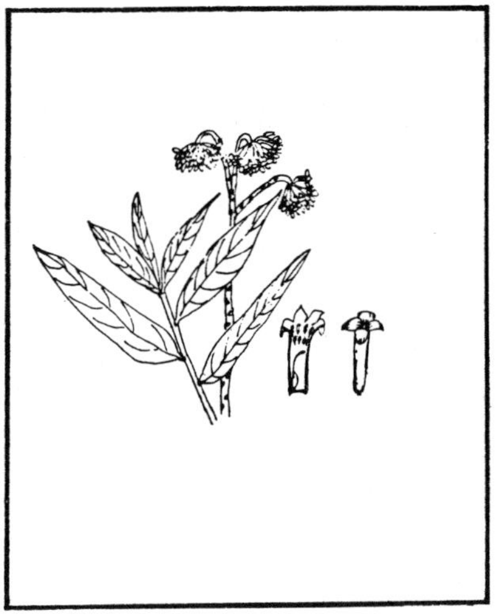

Common name: Mitsumata

Botanical name: *Edgeworthia papyrifera*

Plant family: Thymelaeaceae

Fiber length: 3 mm approximately

Fiber type: Bast

Plant location in the U.S.: None to my knowledge

Plant part used: Branches of shrub

Season harvested: Spring, early summer

Fiber purchase: Dry fiber bundles can sometimes be purchased by mail order

Pre-cooking preparation: Remove leaves, steam branches, strip fiber. Outer bark is removed from inner bark by scraping or peeling. If using dry fiber, soak fiber in clear water 24 hours prior to cooking.

Cooking: 2 hours with soda ash

Beating method: Hand pounding with mallets, or blender

Sheet formation: Japanese

Paper color: Off-white

Historical references: 1597 is the first recorded use of Mitsumata paper in Japan. The fiber was also used by the Nepalese papermakers.

Further notes:

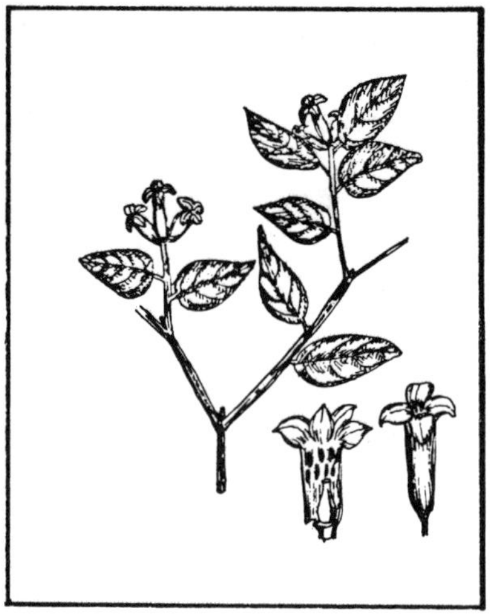

Common name	Gampi
Botanical name	*Wikstroemia retusa* or *Diplomorpha sikokiana*
Plant family	Thymelaeaceae
Fiber length	2.5 to 5.3 mm
Fiber type	Bast
Plant location in the U.S.	Hawaii
Plant part used	Branches from shrub
Season harvested	Spring
Fiber purchase	Dry fiber bundles can be mail ordered
Pre-cooking preparation	Remove leaves, and strip fiber off the woody core immediately after harvesting branches. Remove outer bark from inner bark by scraping or peeling. If using dry fiber, soak 24 hours in clear water before cooking.
Cooking	2 hours or less with soda ash
Beating method	Hand pounding with mallets, or blender
Sheet formation	Japanese or Nepalese/Indian
Paper color	1. Greenish cream (inner bark only) 2. Flecked tan (outer and inner bark)
Historical references	Fiber first used by Japanese papermakers in the Nara period (710-793 A.D.). Hawaiians used another species, Wikstroemia oahuensis, for making tapa.
Further notes	Gampi is a translucent and lustrous fiber that yields a silky paper. It is also supposed to be insect resistant, and is used in restoration work. Even after cooking it retains its own mucilage, and consequently needs less tororo-aoi mucilage added to the pulp for Japanese sheet formation.

Common name	Jute
Botanical name	*Corchorus capsularis*
Plant family	Tiliaceae
Fiber length	1-5 mm
Fiber type	Bast
Plant location in the U.S.	Florida
Plant part used	Stem from annual herb
Season harvested	Summer, fall
Fiber purchase	Available from hardware or weaving supply stores
Pre-cooking preparation	Commercial jute is usually retted; if using fresh plants, remove leaves, steam stems and strip fiber; if using dry fiber, pre-soak in clear water 24 hours prior to cooking.
Cooking	2 hours with lye
Beating method	Ball mill 4 1/2 hours
Sheet formation	Japanese, Nepalese/Indian or Western
Paper color	Grey/buff
Historical references	Fiber has been used since prehistoric times, for sackcloth of biblical times in Mediterranean areas and in India for fiber needs, export and papermaking.
Further notes	The problem with remaining lignin in uncooked fiber for textiles is perhaps mostly eliminated by cooking for papermaking. The USDA attempted production of the plant in 1870 setting out plants from Texas to South Carolina.

Common name	
Botanical name	*Grewia populifolia*
Plant family	Tiliaceae
Fiber length	1.0 to 1.6 mm
Fiber type	Bast
Plant location in the U.S.	Other species in California and southern Florida
Season harvested	
Fiber purchase	Not available
Pre-cooking preparation	Remove leaves, steam branches, strip fiber. Remove outer bark from inner bark by scraping or peeling.
Cooking	
Beating method	
Sheet formation	
Paper color	
Historical references	G. occidentalis, Kaffir hemp, used in S. Africa, G. oppositifolia used in the Himalayas, said to be a strong white fiber.
Further notes	Grewia occidentalis, usually sold as G. caffra, Lavendar star flower, is species in California.

Common name	Little leaf linden
Botanical name	*Tilia cordata*
Plant family	Tiliaceae
Fiber length	1.3 to 5.0 mm
Fiber type	Bast
Plant location in the U.S.	Throughout North America
Plant part used	Branches of tree
Season harvested	Spring or summer
Fiber purchase	Not available
Pre-cooking preparation	Remove leaves, steam branches, strip fiber. Remove outer bark from inner bark by scraping or peeling.
Cooking	2 hours with lye
Beating method	Ball mill
Sheet formation	Japanese
Paper color	Beige
Historical references	Bark used for cordage. T. japonica used by Ainu in Japan for ropes, T. tuan used for sandals by mountaineers in China.
Further notes	Tilia europa (Lime tree) used in Russia for cordage, mats and shoes. T. americana was used by the Chippewa Indians for twine.

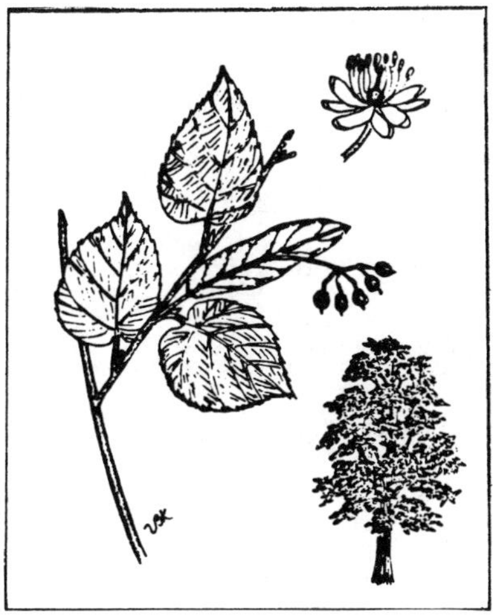

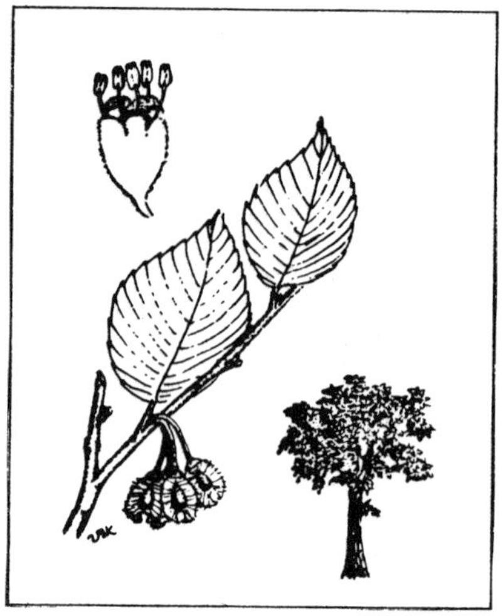

Common name	Elm
Botanical name	*Ulmus americana*
Plant family	Ulmaceae
Fiber length	1.5 to 7.5 mm
Fiber type	Bast
Plant location in the U.S.	N. America
Plant part used	Branches of tree
Season harvested	Spring
Fiber purchase	Not available
Pre-cooking preparation	Remove leaves, steam branches, strip fiber. Remove outer bark from inner bark by scraping or peeling.
Cooking	2 hours with lye
Beating method	Hand pounding with mallets
Sheet formation	Japanese
Paper color	Beige
Historical references	U. japonica used in Japan by Ainu women for a hard rough fabric for clothing called Attush. After soaking the bark for 7-10 days in water the inner and outer barks are separated, and the inner bark is stripped into strands and made into thread by chewing it.
Further notes	U. fulva was used by the Chippewa Indians for twine.

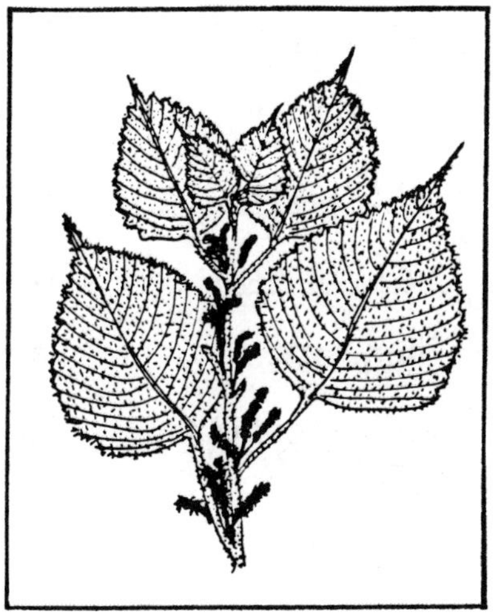

Common name	China grass, Ramie
Botanical name	*Boehmeria nivea*
Plant family	Urticaceae
Fiber length	120 mm approximately
Fiber type	Bast
Plant location in the U.S.	Formerly grown for fiber in California and Louisiana
Plant part used	Stems from perennial herb
Season harvested	Summer, fall
Fiber purchase	Purchase dry bleached fiber from weaving supply store.
Pre-cooking preparation	Remove leaves, steam stems and strip fiber, if fresh plants are available. When using dry fiber, pre-soak for 24 hours in clear water prior to cooking.
Cooking	2 hours with lye, fresh plant material may require a longer or two-stage cook.
Beating method	Hollander beater
Sheet formation	Japanese or Western
Paper color	White if fiber is bleached.
Historical references	Used for paper in China 200 B.C. to 10th cent. A.D. , one of the most common materials of paper manufactured in northern China. Used in Egypt 5000 to 3300 B.C. for cloth to wrap mummies.
Further notes	B. lobata used for fiber in India.

Common name	Mamaki
Botanical name	*Pipturis albidus*
Plant family	Urticaceae
Fiber length	3.8 to 6.0 mm
Fiber type	Bast
Plant location in the U.S.	Hawaii
Plant part used	Branches from shrub
Season harvested	Spring or summer
Fiber purchase	Not available
Pre-cooking preparation	Remove leaves, steam stems and strip fiber.
Cooking	
Beating method	
Sheet formation	
Paper color	
Historical references	Formerly used by Hawaiians for the making of brown bark paper, also used for fiber in Samoa, Tahiti and New Caledonia. Pipturis argenteus known as Queensland grass cloth is an Australian fiber used for making cloth.
Further notes	

Common name	Olona
Botanical name	*Touchardia latifolia*
Plant family	Urticaceae
Fiber length	Not known
Fiber type	Bast
Plant location in the U.S.	Hawaii
Plant part used	Branches from shrub
Season harvested	Spring or summer
Fiber purchase	Not available
Pre-cooking preparation	Remove leaves, steam branches and strip fiber.
Cooking	
Beating method	
Sheet formation	
Paper color	
Historical references	Used by the Hawaiians for cordage and helmets.
Further notes	

Common name	Opuhe, Nettle Bush
Botanical name	*Urera sandwicensis*
Plant family	Urticaceae
Fiber length	165-175 mm
Fiber type	Bast
Plant location in the U.S.	Hawaii
Plant part used	Branches from shrub or tree
Season harvested	
Fiber purchase	Not available
Pre-cooking preparation	Remove leaves, steam branches and strip fiber. Separate outer bark from inner bark by peeling or scraping.
Cooking	
Beating method	
Sheet formation	
Paper color	
Historical references	Urera baccifera used by Aztecs for making bark paper. U. subpelata used by Brazilian Indians for making nets and clothing. U. sandwicensis used by Hawaiians for ropes.
Further notes	

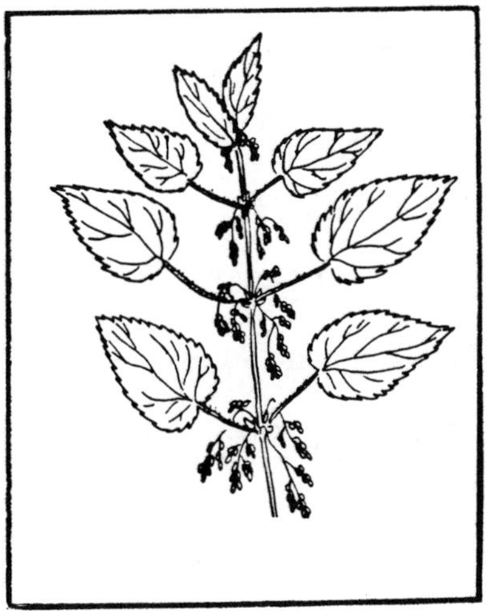

Common name — Stinging nettle, Nettle

Botanical name — *Urtica lyalli*

Plant family — Urticaceae

Fiber length — 19-80 mm

Fiber type — Bast

Plant location in the U.S. — Throughout mainland U.S., except Florida, parts of Georgia, northwestern and southwestern states

Plant part used — Stems from perennial herb

Season harvested — Summer, fall

Fiber purchase — Not available

Pre-cooking preparation — Field ret by leaving stalks to stand until winter, or remove leaves, steam stems and strip fiber.

Cooking — For fresh stems, it is sometimes helpful to do two separate 2-hour cooks with lye, rinsing in between.

Beating method — Hand pounding, or Hollander beater

Sheet formation — Japanese or Western if beaten in the Hollander

Paper color —
1. Greenish
2. Cream (depending on cook)

Historical references — Specimen of nettle paper included in 1770 book about Jacob Schaffer's experiments with plant fibers for papermaking. Fragment of nettle textile found in late bronze age urn in Denmark. Cultivated for centuries in Scandinavia as a fiber plant.

Further notes — The fiber was used in Germany for weaving until quite recently. Papermakers in Nepal use the nettle fiber to weave the cloth screening used on their papermaking molds.

Common name

Botanical name

Plant family

Fiber length

Fiber type

Plant location in the U.S.

Plant part used

Season harvested

Fiber purchase

Pre-cooking preparation

Cooking

Beating method

Sheet formation

Paper color

Historical references

Further notes

Data sheet from

Liliaceae Press
1970 S. Davis St.
McMinnville, Oregon 97128

"PLANT FIBERS FOR PAPERMAKING"
by Lilian A. Bell

LEAF FIBERS

Length Range: 3 mm to 7 mm

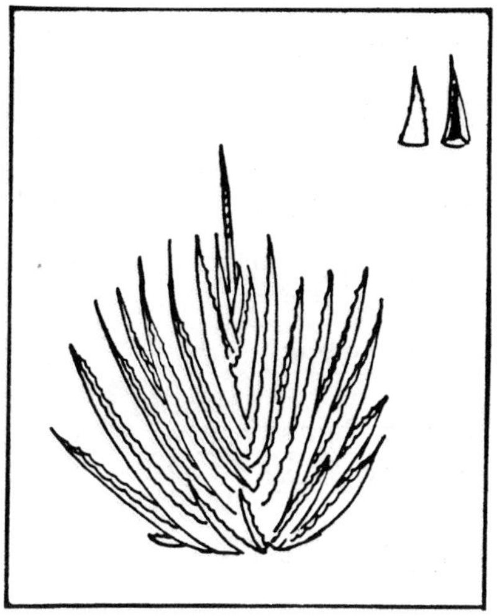

Common name	Sisal
Botanical name	*Agave sisalana*
Plant family	Agavaceae
Fiber length	3-7 mm
Fiber type	Leaf
Plant location in the U.S.	Florida, other species grow in Hawaii, California
Plant part used	Leaves from herbaceous plant
Season harvested	Summer
Fiber purchase	Available from weaving supply, hardware stores, and by mail order
Pre-cooking preparation	Scrape fresh leaves to remove outer skin. Both dry and fresh fibers should be pre-soaked for 24 hours in clear water prior to cooking.
Cooking	2 hours with lye
Beating method	Ball mill 4 hours, or Hollander
Sheet formation	Japanese, Western or Nepalese/Indian
Paper color	Cream
Historical references	Agave species used for cordage, sandals 1100-1425 A.D. by U.S. Pueblo Indians, also used by Aztecs in Mexico.
Further notes	Grown commercially in Kenya and Tanzania and sold worldwide for rope and papermaking as fiber or in dry pulp sheets. Used for tea bags and filter papers.

Common name	Bromeliad, Pita, Silk Grass
Botanical name	*Aechmea magdalenae*
Plant family	Bromeliaceae
Fiber length	1.5 to 6.1 mm
Fiber type	Leaf
Plant location in the U.S.	Other spp. in southern California
Plant part used	Leaves from perennial herb
Season harvested	
Fiber purchase	Not available
Pre-cooking preparation	Scrape leaves with shell, soak in clear water 24 hours prior to cooking.
Cooking	
Beating method	
Sheet formation	
Paper color	
Historical references	Fiber used for rope and twine in South America.
Further notes	

Common name	Pineapple
Botanical name	*Ananas comosus*
Plant family	Bromeliaceae
Fiber length	3-7 mm
Fiber type	Leaf
Plant location in the U.S.	Hawaii
Plant part used	Leaves from perennial herb
Season harvested	After the fruit crop has been harvested
Fiber purchase	Sometimes available
Pre-cooking preparation	Scrape fresh leaves to remove outer skin. Pre-soak in clear water for 24 hours prior to cooking.
Cooking	2 hours with lye
Beating method	Ball mill 3 hours, or Hollander
Sheet formation	Japanese
Paper color	Off-white
Historical references	Fiber used in the Philippines for Pina cloth.
Further notes	For a whiter paper, ferment the leaves after scraping. Pineapple produces a similar paper to Gampi, translucent, but not as lustrous.

Common name	Canna lily
Botanical name	*Canna spp.*
Plant family	Cannaceae
Fiber length	Not known
Fiber type	Leaf
Plant location in the U.S.	Southern states, Hawaii, California; a garden bulb elsewhere
Plant part used	Leaves from perennial herb
Season harvested	Summer, after flowers have faded
Fiber purchase	Not available
Pre-cooking preparation	Scrape leaves to remove outer skin, soak leaves in water 24 hours prior to cooking.
Cooking	2 hours with lye
Beating method	Blender
Sheet formation	Nepalese/Indian, Japanese
Paper color	Light tan brown
Historical references	Australians developing fiber from Canna for pulp, twine, and sacking.
Further notes	Plant matures in twelve weeks.

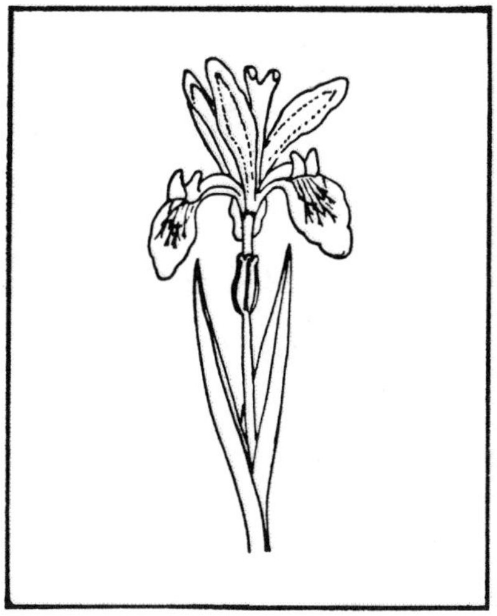

Common name	Iris
Botanical name	*Iris spp.*
Plant family	Iridaceae
Fiber length	Not known
Fiber type	Leaf
Plant location in the U.S.	Gardens throughout U.S.
Plant part used	Leaves from perennial herb
Season harvested	Summer, after flowers have faded
Fiber purchase	Not available
Pre-cooking preparation	Scrape fresh leaves and pre-soak in clear water 24 hours prior to cooking.
Cooking	2 hours with lye
Beating method	Ball mill 3 hours
Sheet formation	Japanese, Nepalese/Indian
Paper color	Light tan
Historical references	Iris douglasiana, Iris purdyi and Iris macrosiphon used by Californian and Oregon Indians for ropes, nets and canvas cloth. A single fiber was taken from the margin of each leaf.
Further notes	Dry, field retted leaves could be harvested in winter and cooked without prior scraping. Color of paper would most likely be different.

Easy to cut - Beat in blenders. Paper light green aging to tan. Did not scrape lvs.

Common name	Ti plant
Botanical name	*Cordyline terminalis*
Plant family	Liliaceae
Fiber length	0.8 to 8.1 mm
Fiber type	Leaf
Plant location in the U.S.	Hawaii, southern California
Plant part used	Leaves from shrub or tree
Season harvested	Summer
Fiber purchase	Not available
Pre-cooking preparation	Scrape leaves with shell, soak in clear water 24 hours prior to cooking.
Cooking	
Beating method	
Sheet formation	
Paper color	
Historical references	Fibers used by Hawaiians and Maori for clothing.
Further notes	C. australis used for papermaking in New Zealand.

Common name	Uki uki
Botanical name	*Dianella sandwicensis* *D. odorata*
Plant family	Liliaceae
Fiber length	Not known
Fiber type	Leaf
Plant location in the U.S.	Hawaii, other species in California
Plant part used	Leaf of perennial herb
Season harvested	
Fiber purchase	Not available
Pre-cooking preparation	Scrape leaves to remove outer skin; soak leaves in water 24 hours prior to cooking.
Cooking	
Beating method	
Sheet formation	
Paper color	
Historical references	Used by the Hawaiians for cordage.
Further notes	

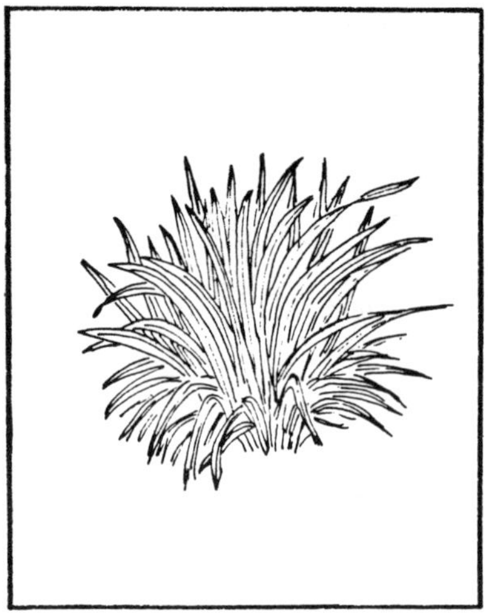

Common name	New Zealand "flax"
Botanical name	*Phormium tenax*
Plant family	Liliaceae
Fiber length	3-7 mm
Fiber type	Leaf
Plant location in the U.S.	California, Oregon
Plant part used	Leaves from perennial tall herb
Season harvested	Summer
Fiber purchase	Not available
Pre-cooking preparation	Scrape fresh leaves to remove outer skin. Pre-soak in clear water 24 hours prior to cooking.
Cooking	2 hours with lye
Beating method	Ball mill 4 hours, possibly blender
Sheet formation	Japanese, Nepalese/Indian
Paper color	Cream
Historical references	Used for centuries by the Maori in New Zealand for clothing and other fiber uses.
Further notes	

Common name	Sansevieria, Bowstring hemp
Botanical name	*Sansevieria trifasciata*
Plant family	Liliaceae
Fiber length	2 mm approximately
Fiber type	Leaf
Plant location in the U.S.	Hawaii, southern California
Plant part used	Leaves from perennial herb
Season harvested	Summer
Fiber purchase	Not available
Pre-cooking preparation	Scrape fresh leaves to remove outer skin. Pre-soak in water for 24 hours prior to cooking.
Cooking	2 hours with lye
Beating method	Ball mill 4 1/2 hours, or possibly Hollander
Sheet formation	Japanese, Nepalese/Indian
Paper color	Greenish cream
Historical references	Used in Africa and India for twine.
Further notes	S. zeylanica used for papermaking in Ceylon.

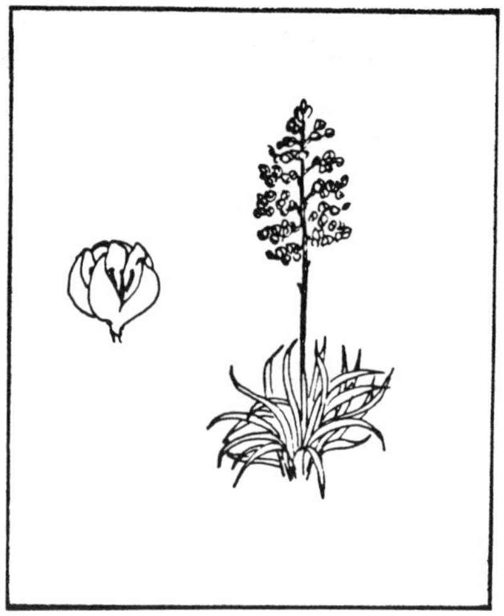

Common name	Yucca
Botanical name	*Yucca filamentosa*
Plant family	Liliaceae
Fiber length	1-5 mm
Fiber type	Leaf
Plant location in the U.S.	Throughout the U.S.
Plant part used	Leaves from perennial herb
Season harvested	Summer
Fiber purchase	Not available
Pre-cooking preparation	Scrape fresh leaves to remove outer skin. Pre-soak in clear water for 24 hours prior to cooking.
Cooking	2 hours with lye
Beating method	Ball mill 4 hours, or blender for more textured paper
Sheet formation	Japanese, Nepalese/Indian or Western
Paper color	Cream
Historical references	Yucca species used for blankets, brushes, sandals 1-450 A.D. by Southwest Indians in the U.S.
Further notes	Yucca baccata was used for making burlap. Stiff leaved Yuccas produce a stronger fiber than the soft leaved varieties.

Yucca filamentosa used - harvested wild in Clay Co., NC
Paper w/small yellow flecks. Blender beat.

Common name	Abaca, Manila hemp
Botanical name	*Musa textilis*
Plant family	Musaceae
Fiber length	3-7 mm
Fiber type	Leaf
Plant location in the U.S.	Other Musa spp. in Hawaii, Florida, California
Plant part used	Leaf stem of tall perennial herb

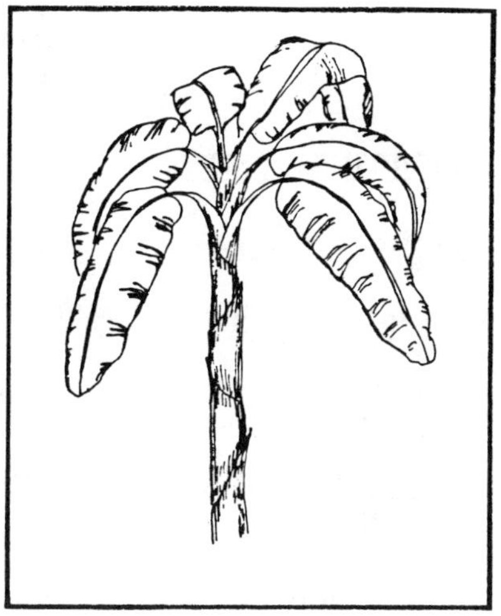

Season harvested	Summer
Fiber purchase	Available by mail order or from hardware stores
Pre-cooking preparation	Cut into useable pieces and soak in clear water 24 hours prior to using. In the Philippines the leaf stems are decorticated.
Cooking	2 hours or more with lye
Beating method	Ball mill 4 1/2 hours, or Hollander
Sheet formation	Nepalese/Indian, Japanese or Western
Paper color	1. Cream if dry fiber was used 2. Pale yellow if rope was used 3. Tan if fresh leaf stems were used
Historical references	Textiles were being made in the Philippines from abaca when Magellan visited in 1521, also used in Japan for cloth. 1837 was first use of old Manila ropes for paper in the United States.
Further notes	The plant is indigenous to the Philippine islands. In 17th century Japan paper was made from Musa sapientum (Basho) in Okinawa. Basho paper is currently being made in Japan by Mr. Katsu. Musa spp. was cultivated in the Indus Valley by 327 B.C.

Common name	Raffia, Piassava, Bass, Jupati palm
Botanical name	*Raphia ruffia*
Plant family	Palmaceae
Fiber length	1-2 mm
Fiber type	Leaf
Plant location in the U.S.	Southern Florida
Plant part used	Leaves and leaflets from palm
Season harvested	Not known
Fiber purchase	Available at weaving supply stores
Pre-cooking preparation	Pre-soak dry fiber for 24 hours in water prior to cooking.
Cooking	2 hours with lye
Beating method	Blender, or ball mill 2 hours
Sheet formation	Nepalese/Indian, Japanese, or Western
Paper color	Yellow gold
Historical references	A West African fiber that is used for hats, baskets, brushes and cloth.
Further notes	Has largest leaves of any known plant, up to 70 feet long. Wax is sometimes obtained from the leaves. The stems are also used for palm wine. Other palm leaf fibers that could perhaps be used for papermaking grow in California or possibly Florida: Acrocomia spp. (1-3 mm), Chamaerops spp. (1-3 mm), Arenga spp., Livistonia spp., Caryota spp. (fishtail palm) and Sabal palmetto.

Common name	Screw pine
Botanical name	*Pandanus odoratissimus*
Plant family	Pandanaceae
Fiber length	1.0 to 4.2 mm
Fiber type	Leaf
Plant location in the U.S.	Hawaii
Plant part used	Leaves from tree
Season harvested	Summer
Fiber purchase	Not available
Pre-cooking preparation	Scrape leaves with a shell, or cut leaves into strips, and soak for 24 hours in clear water prior to cooking.
Cooking	2 hours with lye
Beating method	Ball mill 2 hours
Sheet formation	Japanese
Paper color	Gold tan
Historical references	Used in all of Polynesia for a variety of purposes including mats and baskets. The dry seed pods are used as paintbrushes for painting the Tapa.
Further notes	

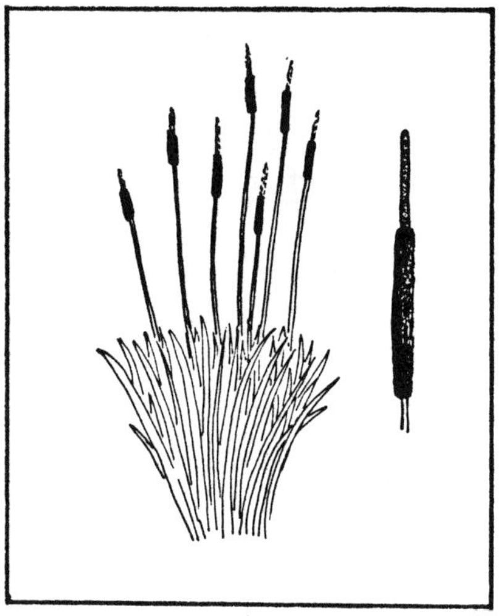

Common name	Cattail
Botanical name	*Typha latifolia*
Plant family	Typhaceae
Fiber length	0.8 to 4.6 mm
Fiber type	Leaf
Plant location in the U.S.	Throughout the mainland U.S.
Plant part used	Leaves from perennial herb
Season harvested	Fall or winter, or summer
Fiber purchase	Not available
Pre-cooking preparation	Pre-soaking in clear water for 24 hours till thoroughly wetted.
Cooking	2 hours in soda ash
Beating method	Ball mill 1 1/2 hours or blender
Sheet formation	Nepalese/Indian or Japanese
Paper color	1. Green brown from summer harvesting 2. Brown from winter harvesting
Historical references	Used for papermaking in Korea prior to 1596. Cattail paper included in 1765 book by Jacob Schaffer describing his plant fiber experiments. Used by Northwest Indians of the U.S. for string, mats and capes.
Further notes	

GRASS FIBERS

Length Range: 1.5 mm to 7 mm

Common name	Papyrus
Botanical name	*Cyperus papyrus*
Plant family	Cyperaceae
Fiber length	1-4 mm
Fiber type	Grass
Plant location in the U.S.	California, other Cyperus spp. grow in many locations in U.S. as weeds, other species available as house-plants

Plant part used	Stem of perennial herb
Season harvested	Summer. Dry brown stems can be harvested any time from houseplant.
Fiber purchase	Sometimes available
Pre-cooking preparation	Cut stems into useable pieces and soak for 24 hours in clear water prior to cooking.
Cooking	2 hours with soda ash
Beating method	Ball mill 1 1/2 hours, or blender, or hand pounding
Sheet formation	Nepalese/Indian, Japanese
Paper color	Light tan brown
Historical references	C. papyrus first used for pounded paper in Egypt prior to 3,000 B.C.
Further notes	C. alternifolius is a houseplant commonly called the Umbrella Plant and makes a brown paper.

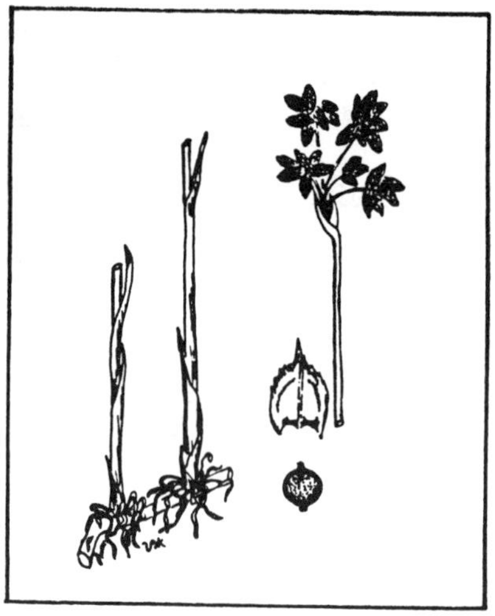

Common name	Tule, Hardstem bulrush
Botanical name	*Scirpus acutus*
Plant family	Cyperaceae
Fiber length	0.5 to 1.3 mm
Fiber type	Grass
Plant location in the U.S.	Throughout the mainland U.S.
Plant part used	Stems from perennial herb
Season harvested	Summer (fresh) or dry any season
Fiber purchase	Not available
Pre-cooking preparation	Split and cut stems into useable pieces, soak 24 hours in clear water prior to cooking.
Cooking	1 1/2 hours with lye
Beating method	Blender
Sheet formation	Nepalese/Indian or Japanese
Paper color	Beige/brown
Historical references	Used by native Americans for baskets, S. validus basketry fragment 1686 years old found in Lovelock Cave, Nevada.
Further notes	The tule can also be used to make papyrus. In 1869 experiments were done with tule fibers for paper-making in California.

Common name	Sea grass, Beach grass
Botanical name	*Ammophila arenaria*
Plant family	Graminae
Fiber length	0.7 mm approximately
Fiber type	Grass
Plant location in the U.S.	Atlantic and Pacific coasts of North America and shores of Great Lakes
Plant part used	Stems from perennial grass
Season harvested	Summer
Fiber purchase	Available from weaving supply stores where it is sold as Sea grass rope
Pre-cooking preparation	Cut stems and rope (no need to unbraid) into useable pieces and soak for 24 hours in clear water prior to cooking.
Cooking	2 hours with soda ash
Beating method	Ball mill 1 1/2 hours or blender
Sheet formation	Nepalese/Indian or Japanese
Paper color	Tan-brown
Historical references	Patent issued in 1838 in the U.S. for making beach grass paper. Also used for making chair seats, baskets and for papermaking in England.
Further notes	

Common name	Pampas grass
Botanical name	*Cortaderia selloana*
Plant family	Graminae
Fiber length	Not known
Fiber type	Grass
Plant location in the U.S.	Throughout the U.S. as a garden plant
Plant part used	Leaves of perennial grass
Season harvested	Fall
Fiber purchase	Not available
Pre-cooking preparation	Cut stems, leaves into useable pieces; soak in clear water for 24 hours prior to cooking.
Cooking	2 hours with lye
Beating method	Blender
Sheet formation	Nepalese/Indian
Paper color	Yellow
Historical references	Used in some parts of South America for papermaking.
Further notes	

Common name	Rice straw
Botanical name	*Oryza sativa*
Plant family	Graminae
Fiber length	0.7 to 3.5 mm
Fiber type	Grass
Plant location in the U.S.	California, Florida
Plant part used	Stem from annual grass
Season harvested	After the crop has been harvested (may be several crops per year in some climates)
Fiber purchase	Not available
Pre-cooking preparation	Cut stems into useable pieces and soak for 24 hours in clear water prior to cooking.
Cooking	2 hours with lye or soda ash
Beating method	Ball mill 1 1/2 hours or blender
Sheet formation	Nepalese/Indian or Japanese
Paper color	Yellow, greenish cream
Historical references	1521 earliest recorded use in China for paper.
Further notes	The Chinese have used rice straw papers for a variety of uses. Rice has been grown intensively for more than 7,000 years. The fiber is still used extensively in China and Southeast Asia for papermaking.

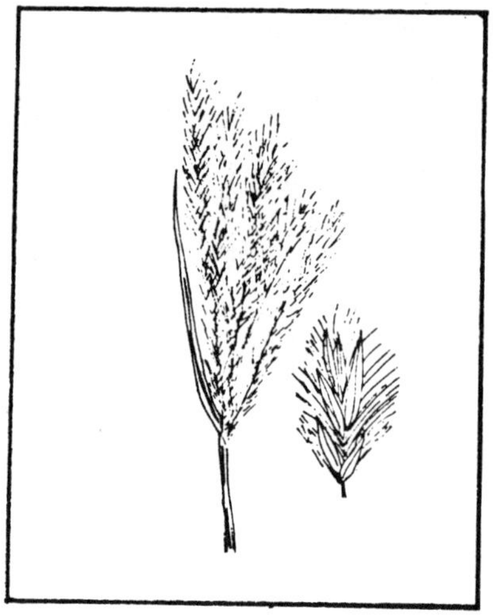

Common name	Feather grass
Botanical name	*Phragmites communis*
Plant family	Graminae
Fiber length	0.8 to 3.0 mm
Fiber type	Grass
Plant location in the U.S.	Throughout the U.S.
Plant part used	Leaves and stems from perennial grass
Season harvested	Summer
Fiber purchase	Not available
Pre-cooking preparation	Cut leaves and stems into useable pieces; soak 24 hours in clear water prior to cooking.
Cooking	2 hours with lye
Beating method	Blender
Sheet formation	Nepalese/Indian
Paper color	Khaki
Historical references	Used in Egypt, Romania, and Russia for papermaking. Used by native Americans for mats and baskets.
Further notes	

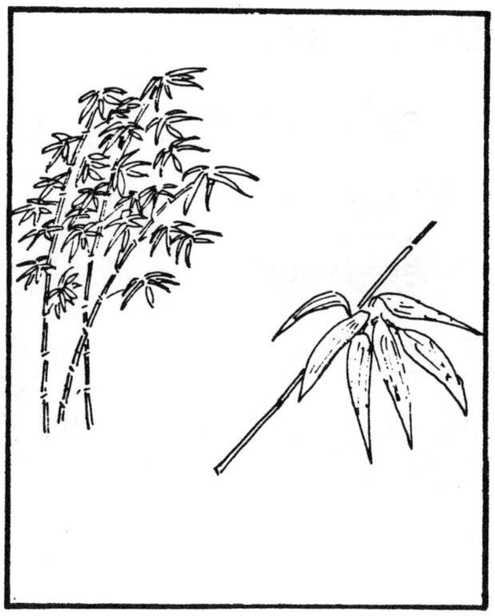

Common name	Bamboo, Clump bamboo
Botanical name	*Phyllostachys aurea*
Plant family	Graminae
Fiber length	1-5 mm
Fiber type	Grass
Plant location in the U.S.	Variety of species grown throughout U.S. for garden plants
Plant part used	Stem of tall woody grass
Season harvested	Anytime
Fiber purchase	Sometimes available in braided bundles
Pre-cooking preparation	Crush stems with hammer to aid chemical penetration and free large amounts of air that stems hold.
Cooking	2 hours + with lye, may benefit with pressure cooking first with soda ash , then with lye*
Beating method	Ball mill 4 hours
Sheet formation	Japanese or Nepalese/Indian
Paper color	Varies from yellow/gold to cream depending on thickness and cooking treatment
Historical references	Fiber used for papermaking by Chinese since the 10th century.
Further notes	Species used for papermaking commercially are: Melocanna baccifera (0.8 to 4.8 mm), Oxytenanthera nigrociliata (1-6 mm), Schizostachyum zollingeri; 2 tons of fresh bamboo on average yields 1 ton dry pulp.

*Bamboo contains 20 percent starch which normally stains pulp, so cook in two processes: (1) weak alkali cook to remove starch, (2) caustic soda cook for 3 hours, and then wash and rinse.

Common name	Sugar cane, Bagasse
Botanical name	*Saccharum officinarum*
Plant family	Graminae
Fiber length	1.5 to 2.7 mm
Fiber type	Grass
Plant location in the U.S.	Hawaii
Plant part used	Stem from tall perennial grass
Season harvested	After the crop has been harvested
Fiber purchase	Not generally available, possibly available in large quantities from sugar companies in the form of crushed damp stems
Pre-cooking preparation	Soak crushed stems in clear water 24 hours prior to cooking.
Cooking	2 hours with lye
Beating method	Ball mill 2 hours,or blender, or Hollander
Sheet formation	Nepalese/Indian, Japanese, Western
Paper color	1. Brown 2. White if bleached
Historical references	Bagasse paper was made in 1884 in Louisiana.
Further notes	The crop residue is used all over the world for papermaking. By 327 B.C. sugar cane had become a noteworthy crop in India.

Common name	Esparto grass
Botanical name	*Stipa tenacissima*
Plant family	Graminae
Fiber length	0.5 to 3.5 mm
Fiber type	Grass
Plant location in the U.S.	Other Stipa spp. in Colorado, Texas, Arizona, California and Midwest
Plant part used	Stems/leaves from perennial grass
Season harvested	Summer
Fiber purchase	Available in dry pulp sheets
Pre-cooking preparation	Cut stems/leaves into useable pieces, soak in clear water for 24 hours prior to cooking.
Cooking	
Beating method	
Sheet formation	
Paper color	
Historical references	Used for papermaking in the Mediterranean area. Also used for ropes, sails and mats.
Further notes	Stipa comata is one species growing in U.S., used in western states for cattle forage. Stipa vaseyi, Porcupine grass fibers, were used by U.S. Indians for brooms and brushes.

Common name	Wheat straw
Botanical name	*Triticum aestivum*
Plant family	Graminae
Fiber length	0.5 to 2.5 mm
Fiber type	Grass
Plant location in the U.S.	Major crop producing areas are in midwestern and western states
Plant part used	Stems from annual grass
Season harvested	Fall or late spring after the crops have been harvested
Fiber purchase	Sometimes available at weaving stores as braided straw. Straw bales can be purchased at country fruit stands or from farmers.
Pre-cooking preparation	Cut stems into useable pieces and soak 24 hours in clear water prior to cooking.
Cooking	2 hours in lye or soda ash
Beating method	Ball mill 1 1/2 hours or blender
Sheet formation	Nepalese/Indian or Japanese
Paper color	Green-tan
Historical references	Straw paper made by Jacob Schaffer in 1765. In England a patent was issued to Matthias Koops in 1800 for his straw paper. Earliest grains of domesticated wheat date to 7500 B.C. approximately from Syria.
Further notes	Straw also used for hats, baskets, and chair seats.

Common name	Cornhusks, Cornstalks
Botanical name	*Zea mays*
Plant family	Graminae
Fiber length	1.52 mm approximately
Fiber type	Grass
Plant location in the U.S.	Throughout U.S., major crop producing areas in midwestern and western states
Plant part used	Stalks and seed husks from annual grass
Season harvested	Fall, after crop has been harvested
Fiber purchase	Not available
Pre-cooking preparation	Split stalks in half, cut into useable pieces, separate husks, soak both in clear water 24 hours prior to cooking.
Cooking	2 hours with soda ash
Beating method	Ball mill 1 1/2 hours or blender
Sheet formation	Japanese or Nepalese/Indian
Paper color	Light greenish cream
Historical references	Used for paper by Jacob Schaffer in his 1766 book about his experiments with plant fibers. In 1802 a patent was issued in the U.S. for cornhusk paper.
Further notes	If fiber is overcooked, it can produce a sticky pulp that is very hard to form into paper. Corn was grown since pre-Columbian times by native Americans, possibly as early as 2,000 B.C.

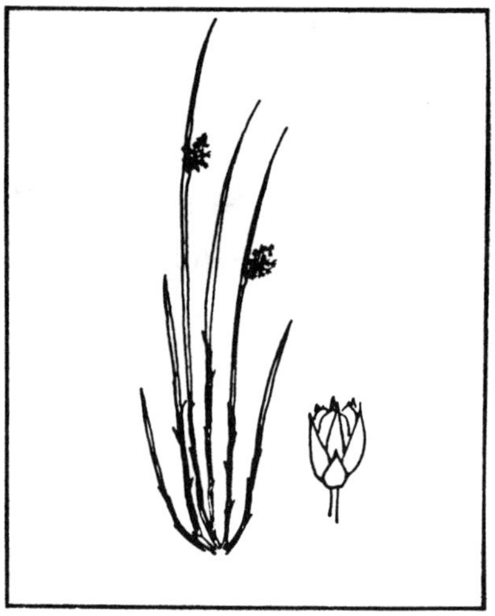

Common name	Common rush
Botanical name	*Juncus effusus*
Plant family	Juncaceae
Fiber length	0.90 mm approximately
Fiber type	Grass
Plant location in the U.S.	Throughout the U.S.
Plant part used	Stems of perennial herb
Season harvested	Summer, fall
Fiber purchase	Not available
Pre-cooking preparation	Split stems and cut into useable pieces; soak for 24 hours in clear water prior to cooking. (see note)
Cooking	2 hours with lye
Beating method	Blender
Sheet formation	Nepalese/Indian
Paper color	Off-white
Historical references	Used by American Indians for basketry and mats.
Further notes	Rush is used in combination with mulberry for the making of stencil paper in South Korea.

8 Dec. 2001 Unusually warm day (70°c). Collected ½ a large (30 gal.) trash bag full of Juncus... I used ½ the bag and once cut into ~1-2" segmts. it filled 3/4 of my stainless pot. I did not soak the stems for 24 hrs, Nor did I split them. Fibers were short and paper was green when poured and when 1st dry (very lite). One pour was enough to make one thin sheet of paper. The pulp went quite a long way - on the 2nd day we combined w/ Hollyhock - nice mix texturally & color.

APPENDIX

A selected list of fiber plants for those wishing to grow their own specimen plants or who are ambitious enough to attempt a plantation:

Bast fiber plants	Source
1. Asclepias spp. (Milkweed)	W
2. Broussonetia papyrifera (Paper mulberry)	W
3. Ficus carica (Edible fig)	L
4. Daphne odora (Daphne)	L
5. Dirca palustris (Leatherwood)	W
6. Mitsumata	E

Leaf fiber plants	
1. Ananas comosus (Pineapple)	S
2. Phormium tenax (New Zealand "Flax")	L
3. Yucca filamentosa (Yucca)	L

Grass fiber plants	
1. Cyperus papyrus (Papyrus)	T
2. Phyllostachys aurea (Bamboo)	L
3. Zea mays (Corn)	L

E - Echo Gardens/Don Jacobs, P.O. Box 1227, Decatur, Georgia 30031 (404) 294-6468, send $1 for plant list
L - Your local nursery
S - Supermarket, new plant can be grown by planting the leaves from the top of the fruit
T - William Tricker, P.O. Box 7843, Independence, Ohio 44131 (216) 524-3491
W - Woodlanders Inc., 1128 Colleton Avenue, Aiken, South Carolina 29801 (803) 648-7522

While most of the plants listed in this work are interesting from various aesthetic points of view, the majority are probably not suitable for archival papers needing a high degree of permanence and durability. As a guide for those readers wishing to make papers for these purposes, the following rating system of #1 POOR through #9 SUPERIOR can be used as a guide. The criteria in use here are (1) fiber length, (2) natural whiteness of fiber, and (3) final paper strength.

Plant Families

1. Agavaceae 7
2. Apocynaceae 7
3. Asclepiadaceae 9

4. Betulaceae 3
5. Bromeliaceae 7
6. Cannaceae 6

7. Compositae 3
8. Cupressaceae 3
9. Cyperaceae 5

10. Graminae 4
11. Iridaceae 4
12. Juncaceae 5

13. Leguminosae 5
14. Liliaceae 7
15. Linaceae 8

16. Malvaceae 6
17. Moraceae 9
18. Musaceae 7

19. Myrtaceae 3
20. Palmaceae 6
21. Pandanaceae 3

22. Rosaceae 3
23. Salicaceae 3
24. Taxodiaceae 3

25. Thymelaeaceae 8
26. Tiliaceae 4
27. Typhaceae 3

28. Ulmaceae 4
29. Urticaceae 9

In terms of the most promising fibers to work with using the above criteria, the following would be my personal choices:

Bast

1. Mulberry
2. Milkweed
3. Flax
4. Breadfruit
5. Gampi or Daphne
6. Stinging nettle
7. True hemp

Leaf

1. Yucca
2. Abaca
3. New Zealand flax
4. Sisal
5. Pineapple

Grass

1. Bamboo
2. Rush
3. Cornhusks/stalks

BIBLIOGRAPHY

AUSTIN, R., UEDA, K., LEVY, D., "Bamboo," Weatherhill, New York, 1975.

ARMITAGE, F. D., "An Atlas of Commoner Papermaking Fibers," Guildhall, London, England, 1957.

BAILEY, L. H., "The Standard Dictionary of Horticulture," 3 volumes, MacMillan, London, England, 1930.

BALL, Edward K., "Early Uses of California Plants," University of California, Berkeley, California, 1975.

BARRETT, Timothy, "Japanese Papermaking: Traditions, Tools and Techniques," with an appendix by WINIFRED LUTZ on Japanese fiber preparation, Weatherhill, Tokyo, Japan and New York, 1983.

BIRRELL, Verla, "The Textile Arts," Schocken, New York, 1973.

BRITT, K. W., "Handbook of Pulp and Paper Technology," Van Nostrand Reinhold, New York, 1964.

BUCHANAN, Rita, "A Weaver's Garden," Interweave Press, Loveland, Colorado, 1987.

COOK, Gordon, "Handbook of Textile Fibers," Volume 1, Natural Fibers, Merrow, Watford, England, 1968.

CORRELL, Donovan S., "Aquatic and Wetland Plants of the Southwestern United States," Government Documents, Washington, D.C., 1977.

CROCKETT, Lawrence J., "Wildly Successful Plants," MacMillan, New York, 1977.

DENSMORE, Frances, "How Indians Use Wild Plants for Food, Medicine, and Crafts," Dover, New York, 1974.

DODGE, Charles Richard, "The Leaf Fibers of the United States," Report #5, U.S. Department of Agriculture, Washington, D.C., 1893.

DODGE, Charles Richard, "Uncultivated Bast Fibers of the United States," Report #6, U.S. Department of Agriculture, Washington, D.C., 1894.

DODGE, Charles Richard, "Useful Fiber Plants of the World," Report #9, U.S. Department of Agriculture, Washington, D.C., 1897.

FORBES, R. J., "Studies in Ancient Technology," Volume 4, Brill, Leiden, Netherlands, 1964.

GILKEY, Helen, "Weeds of the Pacific Northwest," Oregon State University, Corvallis, Oregon, 1957.

GUNTER, Erna, "The Ethnobotany of Western Washington," University of Washington, Seattle, Washington, 1973.

HANAUSEK, T. F., "Microscopy of Technical Products," Wiley, New York, 1907.

HELLER, Jules, "Papermaking," Watson Gupthill, New York, 1978.

HILL, A. F., "Economic Botany," McGraw, New York, 1952.

HUGHES, Sukey, "Washi--the World of Japanese Paper," Kodansha International, New York, 1978.

HUNTER, Dard, "Papermaking: History and Technique of an Ancient Craft," Dover, New York, 1978.

HYLANDER, Clarence, "The World of Plant Life," MacMillan, New York, 1956.

JACQUES, H. E., "Plants We Eat and Wear," Dover, New York, 1975.

KENT, Kate Peck, "Montezuma Castle Archeology," Pt. 2 Textiles, Southwest Monuments Association, Technical Series, Volume 3, #2, Globe, Arizona, 1954.

KIRBY, R. H., "Vegetable Fibers," Interscience, New York, 1963.

KOOIJMAN, Simon, "Tapa in Polynesia," Bishop Museum, Honolulu, Hawaii, 1972.

KRAUSS, Beatrice, "Ethnobotany of Hawaii," University of Hawaii, Honolulu, Hawaii, 1974.

LONG, Paulette, Editor, "Paper: Art and Technology," World Print Council, San Francisco, California, 1979.

MASON, Otis Tufton, "Aboriginal American Indian Basketry," Peregrine Smith, Santa Barbara, California, 1976.

MAUERSBERGER, H. R., Editor, "Matthews Textile Fibers," Wiley, New York, 1947.

McCURRACH, James, "Palms of the World," Harper, New York, 1960.

MERRILL, Elmer, "Plant Life of the Pacific World," The Infantry Journal, Washington, D.C., 1945.

MURPHY, Edith Van Allen, "Indian Uses of Native Plants," South Mendocino County Historical Society, Fort Bragg, California, 1950.

NARITA, Kiyofusa, "A Life of Tsai Lung and Japanese Papermaking," Paper Museum, Tokyo, Japan, 1976.

NEAL, Marie, C., "In Gardens of Hawaii," Bishop Museum, Honolulu, Hawaii, 1965.

PARHAM, R. A. and KAUSTINEN, H. P., "Papermaking Materials: An Atlas of Electron Micrographs," Institute of Paper Chemistry, Appleton, Wisconsin, 1974.

PURSEGLOVE, J. W., "Tropical Crops," Halsted, London, 3 volumes, 1972-75.

RAGAB, Dr. Hassan, "Papyrus through History," Papyrus Institute, Cairo, Egypt, 1976.

ROTH, Ling, "The Maori Mantle," Ruth Bean, Bedford, England, 1979.

SARJEANT, Peter, "Handpapermaking Manual," Papermake, Covington, Virginia, 1975.

SAUNDERS, C., "Useful Wild Plants of the U.S. and Canada," McBride, New York, 1926.

SHURTLEFF, William, "A Book of Kudzu," Autumn Press, Brookline, Mass., 1977.

SIMMONDS, N. W., Editor, "Evolution of Crop Plants," Longman, New York, 1976.

SUNSET BOOKS, Editors, "New Western Garden Book," Lane, Menlo Park, California, 1979.

SUTERMEISTER, Edwin, "The Chemistry of Pulp and Papermaking," Wiley, New York, 1941.

TOALE, Bernard, "The Art of Papermaking," Davis Publications, Worcester, Massachusetts, 1983.

TRIER, Jesper, "Ancient Paper of Nepal," Jutland Archeological Society Publications, Volume X, Copenhagen, Denmark, 1972.

TSUEN-HSUIN, Tsien, "Written on Bamboo and Silk," University of Chicago Press, Chicago, Illinois, 1962.

UPHOF, J. C. Th., "Dictionary of Economic Plants," Hafner, New York, 1959.

U.S. DEPARTMENT OF AGRICULTURE, "Common Weeds of the United States," Dover, New York, 1971.

VON HAGEN, Victor, "Aztec and Maya Papermakers," Hacker, New York, 1977.

WEEKS, Lyman, "History of Papermaking in the United States," Lockwood, New York, 1916.

WINGATE, Dr. Isabel B., "Fairchild's Dictionary of Textiles," Fairchild, New York, 1979.

ZIM, H. and MARTIN, A., "Trees," Golden Press, New York, 1956.

CASTING REFERENCES

CUNNING, Sheril, "Handmade Paper," Ravensword Press, Escondido, California, 1987.

FAERBER, Ruth, "Casting and Hand Forming Bas-Relief Paperworks," pp.35-36, Post Conference Papers, 1st National Paper Conference, Hobart, Australia, 1987.

JOUKOWSKY, Martha, "A Complete Manual of Field Archaeology, pp.248-253, Prentice Hall, New Jersey, 1980

LUTZ, Winifred, "Casting to Acknowledge the Nature of Paper," pp.26-38, Proceedings, International Conference of Hand Papermakers, Carriage House Press, Brookline, Massachusetts, 1981. The article describes three-dimensional thin sheet casting with plant pulps containing mucilage on fabric molds held under tension to achieve their shape.

NEWMAN, Thelma, "Innovative Printmaking," Crown, New York, New York, 1977.

ROUKES, Nicholas, "Sculpture in Paper," Davis, Worcester, Massachusetts, 1993.

ROWLEY, Kathleen, "Pulp Spraying," pp.35-36, Proceedings, International Papermakers Conference, Marilyn Wold Press, Kailua-Kona, Hawaii, I988.

WILLIAMS, Arthur, "Sculpture: Technique, Form and Content," Davis, Worcester, Massachusetts, 1989.Contains chapters about cast and molded paper, working with plaster, and body casting.

The publications of organizations listed on page 17 include information on paper casting from time to time.

BOTANICAL NAME INDEX

COMMON NAME INDEX

signal. LB secretion was stimulated with either ATP (100 μmol/L) or ionomycin (5 μmol/L) or phorbol-12-myristate-13-acetate (300 nmol/L) together with ATP. When indicated, the cells were treated with the membrane permeable calcium chelator BAPTA-AM (20 μmol/L) for 30 min at 37°C before imaging. Phospholipase D (PLD) was inhibited by incubating the cells with C2 ceramide (150 μmol/L), 1–2 h; 25 μmol/L) in bath solution). In the actin–green fluorescent protein (GFP) experiments, the cells were infected with 5–10 μL/mL purified adenoviruses (which correspond to 5 × 10^5 plaque-forming units) immediately after seeding (see also *Adenovirus vectors and cell transfection* below). When imaging of actin–GFP and labeling of filamentous actin were combined, the actin–GFP-transfected cells were fixed with 4% paraformaldehyde in PBS for 10 min, permeabilized with PBS/Triton X-100 (0.1%) for 30 min, and incubated with Alexa Fluor 568 phalloidin for 20 min at the concentration recommended by the manufacturer.

Dark-field and Fluorescence Imaging with the RTM 2.5 Microscope

An upright Richardson RTM 2.5 microscope (Improvision, Tübingen, Germany) was used for combined dark-field and fluorescence imaging. For dark-field illumination a red filter in the condenser was used. Dark-field and fluorescence light were detected with a 3CCD color video camera (ExwaveHAD DXC-C33, Sony; supplied by Improvision, Germany). All images were acquired with the 63× Leica water immersion objective (numerical aperture = 0.90) and Volocity software. For fluorescence imaging the RTM 3.0 blue filter cube [490 nm dichroic, 400–500 nm excitation filter, 510 long-pass (LP) emission filter] was used for all fluorescence experiments. For the actin–GFP fluorescence measurements, a square region of interest was defined around the fused LB on which fluorescence and dark-field light intensity were measured.

Imaging of Fixed Cells

Fixed cells transfected with actin–GFP and labeled with Alexa 568 phalloidin were imaged on a Zeiss Meta LSM with a 488 nm excitation laser/500–530-nm emission filter and a 543-nm excitation laser/LP 560-nm emission filter. We used a Zeiss planapochromate 63× oil immersion objective.

ß-actin–GFP and Lamp3–GFP Adenovirus Vectors and Cell Transfection

The cDNA of the GFP–actin fusion protein was amplified using the primer pair eGFPß-Actin-F1g (GGGGACAAGTTTGTACAAAAAAGCAGGCTTCGAAGGAGATAGAACCATGGTGAGCAAGGGCGAGGAGC) and eGFP-ßActin-R1g (GGGGACCACTTTGTACAAGAAAGCTGGGTAGTGGATCCCTAGAAGCATTTGCGG) from the plasmid pEGFP-actin (Clontech, Heidelberg, Germany) using standard PCR techniques. The PCR product was cloned into the entry vector pDONR221 (Invitrogen, Karlsruhe, Germany), using the BP Clonase II enzyme mix (Invitrogen, Germany), and propagated in *Escherichia coli* ccdB survival T1 Phage-resistant cells (Invitrogen, Germany). For generating the GFP entry vector, the multicloning site and open reading frame (ORF) of EFGP1 were amplified from the plasmid pEGFP1-N1 (Clontech, Germany), using the primers attB1_MCSgfp-F1 (GGGGACAAGTTTGTACAAAAAAGCAGGCTATGCTAGCGCTACCGGACTCAG) and attb2_gfp-R1 (GGGGACCACTTTGTACAAGAAAGCTGGGTATTTACTTGTACAGCTCGTCC) by standard PCR techniques. The PCR product was cloned into the entry vector pDONR221 (Invitrogen, Germany), using the BP Clonase II enzyme mix (Invitrogen, Germany), and propagated in *E. coli* ccdB survival T1 Phage-resistant cells (Invitrogen, Germany). The obtained plasmid was linearized, using XhoI and BamHI (both Fermentas, St. Leon-Rot, Germany).

Lamp3 cDNA was amplified from the full-length cDNA clone IRBPp993D126D2 obtained from the German resource center and primary database of the human genome project (RZPD, Berlin, Germany), using the primers lamp3-F3 (GGACTCAGATCT-CGAGATGCCTGGGCAGACCTCTGCA-GTAGCTG) and lamp3-R3 (GGCGACCG-GTGGATCGATTCTCTGGTATGCAGAT-GACTGACGC). The PCR product was introduced into the XhoI/BamHI linearized GFP entry vector by *in vitro* recombination, using the InFusion dry-down PCR cloning kit (Clontech, Germany). The recombination product was cloned and propagated in *E. coli* Fusion Blue (Clontech, Germany). The cDNAs encoding the eGFP–Actin and Lamp3–GFP fusion proteins were transferred from the entry vector into the adenoviral vector pAd/CMV/V5-DEST (Invitrogen, Germany) by *in vitro* recombination, using the LR Clonase enzyme mix (Invitrogen, Germany), according to the manufacturers' protocol. The expression plasmid was propagated in *E. coli* OmniMax cells (Invitrogen, Germany), confirmed by sequencing, and transfected into HEK293A cells (Invitrogen, Germany) after PacI linearization.

Transfected cells were cultivated at 7.5% CO_2 and 95% humidity in DMEM medium containing 4.5 g/L glucose, 10% FCS, 1% penicillin/streptomycin, and MEM nonessential amino acids (Gibco/Invitrogen, Karlsruhe, Germany) for 5 to 8 days. Afterwards, cells were lysed by repetitive freezing in liquid nitrogen and thawing at 37°C. Adenoviral particles were isolated using the Vivapure adenopack 20 virus isolation kit (Vivascience, Hannover, Germany), according to manufacturers' protocol. Infectiousness of virus particles was confirmed by HEK293A cell infection.

Image Analysis and Data Evaluation

The measurements and image analysis were performed either with Volocity software or with ImageJ. Excel and SigmaPlot 10.0 were used for statistics and graphs. Some images were corrected for brightness and contrast with Adobe Photoshop. Z-slices were digitally produced with Amira software from confocal XY section.

Results

LBs contain surfactant lipids and proteins tightly packed within an acidic pH.[9] Therefore, acidotropic dyes, such as LysoTracker Green DND-26 (LTG), almost exclusively accumulate in LBs.[22] Under dark-field illumination, these vesicles were surrounded by a ring of scattered light which exceeded the light intensity of other membranous structures (Fig. 1A). LBs contain lysosomal enzymes and were classified as "lysosome-related organelles."[23] In type II cells expressing Lamp3–GFP, a fusion protein between GFP and the lysosome-associated membrane protein 3 (Lamp3), the fluorescence and dark-field light signals overlapped (Fig. 1B). Thus, dark-field imaging displays the limiting membrane of LBs or their compound organelles. The reason for this light-enhanced dark-field signal at the border of LBs is yet unclear but consistent with dark-field microscopy on giant liposomes[24] and largely abolished by treatment with detergent (Triton-X, Fig. 1C), suggesting a role of lipid organization or composition. This "scattered light" signal may include a combination of reflection, refraction, and diffraction.

In contrast to the less-intense signal from the interior of the LB, secreted surfactant appears in dark-field illumination as an amorphous bright structure without a limiting circular ring (Fig. 1D). Hydration of lipid layers with restricted water content before secretion[25] could account for this effect. Consistently, surfactant becomes also visible by dark-field illumination inside the LB after fusion but before secretion (see below).

In combined dark-field and FM 1–43 fluorescence ($F_{FM1\text{–}43}$) measurements on cells

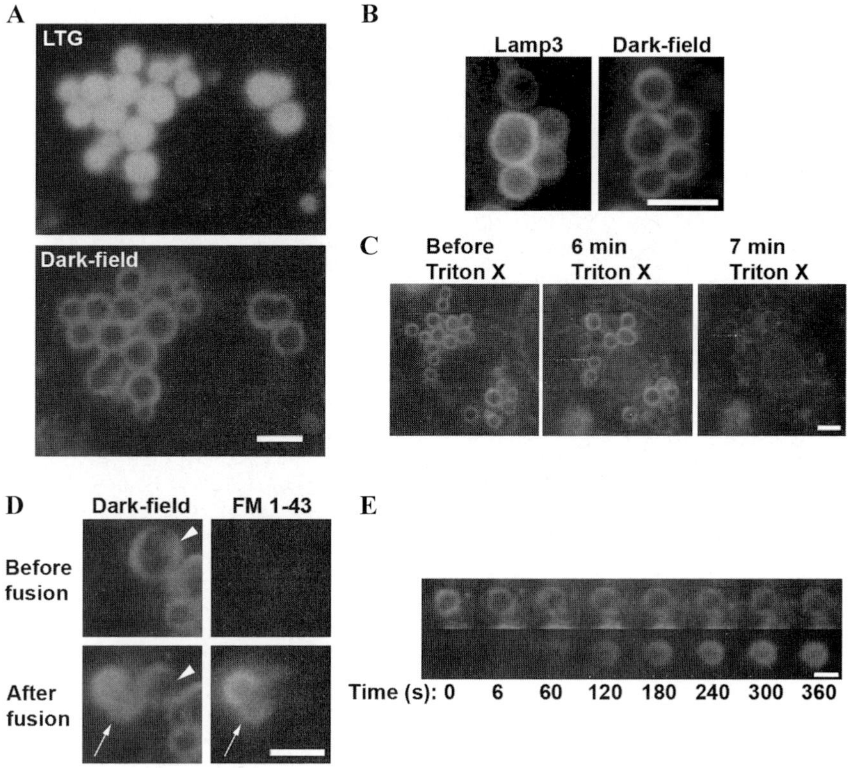

Figure 1. Dark-field and fluorescence imaging of lamellar bodies (LBs) with the Richardson RTM 2.5 microscope. (**A**) Fluorescence (above) and dark-field (below) image of a single type II cell. LBs before fusion (green) are stained with the fluorescent dye LysoTracker Green DND-26. In dark field, LBs are surrounded by a bright ring of scattered red light. (**B**) Fluorescence (left) and dark-field (right) image of a part of a type II cell expressing lysosomal marker Lamp3–green fluorescent protein (GFP). Note that LB-limiting membrane and dark-field signals overlap. (**C**) Dark-field image of one type II cell before, 6 min, and 7 min after application of Triton X (1 μmol/L). Note sequential disappearance of scattered light. (**D**) Dark-field (left) and F_{FM1-43} (right) images of surfactant before (top) and after (bottom) fusion of a single LB (arrowhead) with the plasma membrane. Dark-field illumination reveals surfactant as a bright structure but without a limiting ring (arrow). In the fluorescence image, extruded surfactant is stained with FM 1–43 from the bath solution. Note the shrinkage of the fused LB in the dark-field images during squeeze-out of surfactant. (**E**) Temporal sequence of dark-field (top) and F_{FM1-43} (bottom) images of an LB in a cell stimulated with ATP. FM 1–43 (0.3 μmol/L) was present in the bath solution and stained LB content (surfactant) by passing through the fusion pore. Time = 0 indicates last image before fusion, as evidenced by scattered light intensity decrease (SLID). Scale bars: 4 μm for A–D and 2 μm for E.

stimulated with ATP or ionomycin, we observed that in single LBs the onset of F_{FM1-43} increase was always preceded by a prominent, scattered, light-intensity decrease (SLID, Fig. 1E). Uptake of the dye FM 1–43[26] from the bath solution into surfactant through the fusion pore (from the bath solution) is a very sensitive and specific way to measure single LB fusion with the plasma membrane. The delay between the onset of SLID and F_{FM1-43} intensity change ranged between a few hundred milliseconds and several seconds, which is, on average, longer than the delay between the change in cell membrane capacitance and F_{FM1-43} (< 1 second[27]). It is therefore intriguing to speculate, however beyond the scope of this

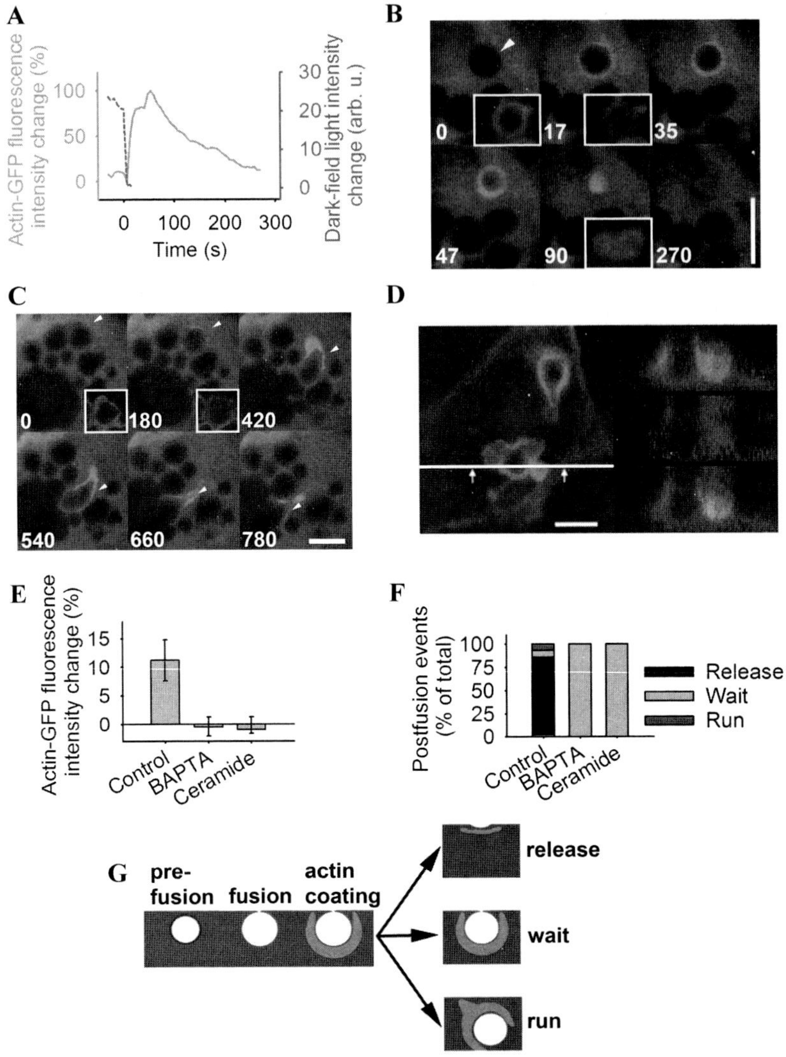

Figure 2. Actin coating and subsequent actions in actin–GFP-transfected cells. (**A**) Time course of changes (maximum change = 100%) of SLID and actin–GFP fluorescence intensities around a single fused LB shown in (**B**). Last frame before SLID was arbitrarily set as time = 0. Image acquisition rate = 10/min. (**B**) Content release: Fluorescence images of a part of a cell as a function of time after fusion of a single LB (arrowhead) with the plasma membrane (time 0 = immediately before fusion). The three inserts show the simultaneous dark-field images (red) indicating fusion of this LB (times 0 and 17 s) and extrusion of surfactant (time 90 s). Images between 17 to 90 s show actin ring formation around the fused vesicle and contraction of the ring as surfactant was squeezed out. The image at 270 s shows the disappearance of the actin coat when surfactant was extruded from the cell. (**C**) "Run": The fused LB (arrowhead) moved on top of its actin tail for a distance of 17 μm following fusion and actin coating. The two inserts show the simultaneous dark-field images indicating (transient) fusion of this LB. Note that in contrast to (**B**), the interior of the fused LB was never visible in dark field, indicating lack of extrusion. (**D**) "Wait": Confocal images of actin accumulation around a fused LB that neither released nor ran. Left: Horizontal section through the cell; note subcortical actin accumulation and strong actin coating of at least two LBs. Right: Z-stacks through the LB (left) at the vertical plane (white line) between the white arrows. Top, actin–GFP; middle, Alexa 568 phalloidin immunostaining; bottom, overlay. Scale bars represent 5 μm. (**E**) Actin coat formation around fused vesicles after treatment with BAPTA-AM (20 μmol/L for 30 min) or in the presence of

article, that the onset of SLID may reflect stages earlier than fusion pore formation (e.g., hemifusion). Alternatively, long delays between LB–plasma membrane fusion and F_{FM1-43} increase in the presence of compound exocytosis may reflect long diffusion distances.[11] In any case, the reliable and selective uptake of FM 1-43 into the LB lumen clearly indicates that SLID and fusion pore formation are tightly coupled events.

In less than 10 s after the onset of SLID, an increase of perivesicular β-actin–GFP fluorescence initiated around the fused LB (Fig. 2A) but was entirely absent in nonfused LBs. Actin coating was from polymerization of actin monomers, as evidenced by staining of actin coats with Alexa 568 phalloidin (Fig. 2D).

The further fate of the fused and coated vesicles varied considerably (Fig. 2G):

1 Most LBs (>80%) reshrunk after coat formation and expelled their contents while the actin coat flattened out and disappeared (kiss-coat-and-release, Fig. 2F).
2 Some LBs (<10%) were dislocated inside the cell after coating (kiss-coat-and-run, Fig. 2F).
3 Some LBs (<10%) remained coated for up to 20 min of observation without any further action (kiss-coat-and-wait, Fig. 2F).

Actin coating was inhibited by cytoplasmic Ca^{2+} chelation using BAPTA-AM (Fig. 2E). Since we have shown earlier that LB fusion with the plasma membrane is only abolished when BAPTA-AM is used in the absence of bath Ca^{2+},[28] we used here a bath Ca^{2+} concentration sufficient to maintain fusion. Inhibition of actin coating under these conditions indicates that free Ca^{2+} ions are required. Under this condition, "release" or "run" was never observed ($n = 43$, Fig. 2F) and all fused LBs remained in the waiting position, indicating that Ca^{2+} and actin coating are required for generation of force needed to expel contents or translocate the vesicle. In nonfused vesicles, global $[Ca^{2+}]_c$ elevations (by ATP or ionomycin) never induced actin coating ($n = 100$).

Actin coating was also completely inhibited by the phosphatidylcholine-specific phospholipase D inhibitor C2 ceramide ($n = 15$, Fig. 2E). This suggests involvement of the diacylglycerol-, protein kinase Cβ-, cell division cycle 42 protein- (Cdc42), and neuronal Wiskott-Aldrich Syndrome protein- (N-WASP) dependent pathway described by Yu and Bement.[21] As with BAPTA, fused vesicles of C2 ceramide-treated cells did not "release" or "run" but remained in their "waiting" position (Fig. 2F).

The actin coat remained as long as the vesicle resided in its waiting position, in agreement with a stabilizing function of actin.[19] Following release, the actin coat disappeared at the site of fusion (Fig. 2B). When run occurred, the coated vesicle was pushed forward in the direction opposing actin accumulation at the tip of a dynamic actin tail (Fig. 2C).[29]

Discussion

Here we demonstrate that actin coating in the postfusion phase of LB exocytosis is a required step for either content (surfactant) release (kiss-coat-and-release) or vesicle retrieval (kiss-coat-and-run) in type II pneumocytes. The

←

C2 ceramide (25 μmol/L). Bars represent median change of perivesicular actin–GFP fluorescence 60 s after fusion of the vesicle with the plasma membrane [$n = 29$ (control), 43 (BAPTA-AM), 15 (C2 ceramide); error bars represent interquartile ranges]. **(F)** Relative number of "release," "run," and "wait" events as percentage of all fusion events under the different experimental conditions; absolute numbers of events as in **(E)**. **(G)** Schematic illustration of postfusion events.

waiting condition was favored when the free activity of Ca^{2+} ions was decreased and actin coat formation inhibited. Under experimental conditions, this can also be caused by Ca^{2+} indicators, which are Ca^{2+} buffers (in fact, cells loaded with fluo-4 show a reduced ability to release, P. Miklavc, unpublished observation). Whether or not kiss-coat-and-wait is a physiologically relevant condition, however, cannot be assessed with certainty. In this situation, actin coating could even act to prevent full-fusion pore expansion by "stabilizing" the vesicle, as proposed by Sokac and Bement.[19] Our data do not allow a conclusion if resealing of the fusion pore may still occur after actin coating. Late closures are unlikely, however, because partial content release before vesicle retrieval was never observed and earlier capacitance measurements failed to reveal late-fusion pore closures.[27]

The function of occasional "kiss-coat-and-run" events could be to induce compartment mixing, that is, the exchange of lipids and proteins [such as soluble *N*-ethylmaleimide-sensitive factor attachment protein receptor (SNARE) proteins] between LB membrane and plasma membrane, for successive LB–LB fusions: Uptake of plasma membrane components into the fused vesicle might enable its fusion with other vesicles before these prefused vesicles fuse with the plasma membrane, as in pancreatic acinar cells.[30] We know from combined patch clamp and fluorescence measurements that this special form of compound exocytosis exists in type II cells,[27] which is also supported by dark-field observations of LBs fused with each other. This mechanism could help to secrete sufficient amounts of surfactant through a limited apical membrane area.

In kiss-coat-and-run events, the retrieved LB moved on top of an actin tail, similar to endocytic vesicles.[29] Hence, the actin coat exerts a force resulting in propulsion of the LB. It is not yet clear which factors determine if this force is directed to push the LB forward instead of releasing its contents. It is likely that this decision is made earlier on the basis of full or transient fusion.

Actin-powered exocytosis, or kiss-coat-and-release, is probably an important mechanism in type II cells because in the absence of actin coating, release was never observed. Since actin coating occurred in the fused LB selectively, this form of exocytosis may depend on two prerequisites, as suggested by Yu and Bement[21] for cortical granules in oocytes: an elevation of $[Ca^{2+}]_c$ and mixing between vesicle and plasma membrane.

Inhibition of actin coating by C2 ceramide in type II cells suggests that, as in oocytes, elevation of diacylglycerol at the membrane under the influence of a phosphatidylcholine-specific PLD may be involved.

Actin-powered exocytosis may be an important mechanism in the lung to accelerate the delivery of LB material to the air–liquid interface where it is converted to a surfactant film for reduction of surface tension and facilitation of inspiration.[31,32] This process will most likely occur during a deep breath, which causes an elevation of $[Ca^{2+}]_c$ by Ca^{2+} entry from the extracellular space and release from stores.[33–35]

Acknowledgments

This work was supported by the Fonds zur Förderung der Wissenschaftlichen Forschung (FWF), grants P15742 and P15743, the Deutsche Forschungsgemeinschaft (DFG), grant D1402, and the 6th framework of the European Union, Pulmo-Net. We thank Dr. Thomas Haller and Dr. Manfred Frick for critical reading of the manuscript. Technical assistance by Melanie Timmler, Tatiana Felder, and Christiane Buchholz is gratefully acknowledged.

Conflicts of Interest

The authors declare no conflicts of interest.

References

1. Jahn, R., T. Lang & T.C. Südhof. 2003. Membrane fusion. *Cell* **112:** 519–533.

2. Breckenridge, L.J. & W. Almers. 1987. Final steps in exocytosis observed in a cell with giant secretory granules. *Proc. Natl. Acad. Sci. USA* **84:** 1945–1949.
3. Zimmerberg, J. & L.V. Chernomordik. 1999. Membrane fusion. *Adv. Drug Deliv. Rev.* **38:** 197–205.
4. Alvarez de Toledo, G., R. Fernandez-Chacon & J.M. Fernandez. 1993. Release of secretory products during transient vesicle fusion. *Nature* **363:** 554–558.
5. Chow, R.H., L. von Ruden & E. Neher. 1992. Delay in vesicle fusion revealed by electrochemical monitoring of single secretory events in adrenal chromaffin cells. *Nature* **356:** 60–63.
6. Fesce, R., F. Grohowaz, F. Valtorta, *et al.* 1994. Neurotransmitter release: fusion or 'kiss-and-run'? *Trends Cell Biol.* **4:** 1–4.
7. Chernomordik, L.V., J. Zimmerberg & M.M. Kozlov. 2006. Membranes of the world unite! *J. Cell Biol.* **175:** 201–207.
8. Michael, D.J., H. Cai, W. Xiong, *et al.* 2006. Mechanisms of peptide hormone secretion. *Trends Endocrinol. Metab.* **17:** 408–415.
9. Dietl, P. & T. Haller. 2005. Exocytosis of lung surfactant: from the secretory vesicle to the air-liquid interface. *Annu. Rev. Physiol.* **67:** 595–621.
10. Perez-Gil, J. 2008. Structure of pulmonary surfactant membranes and films: The role of proteins and lipid-protein interactions. *Biochim. Biophys. Acta,* **1778**(7–8): 1676–1695.
11. Haller, T., P. Dietl, K. Pfaller, *et al.* 2001. Fusion pore expansion is a slow, discontinuous, and Ca^{2+}-dependent process regulating secretion from alveolar type II cells. *J. Cell Biol.* **155:** 279–289.
12. Singer, W., M. Frick, T. Haller, *et al.* 2003. Mechanical forces impeding exocytotic surfactant release revealed by optical tweezers. *Biophys. J.* **84:** 1344–1351.
13. Tsilibary, E.C. & M.C. Williams. 1983. Actin in peripheral rat lung: S1 labeling and structural changes induced by cytochalasin. *J. Histochem. Cytochem.* **31:** 1289–1297.
14. van Weeren, L., A.M. de Graaff, J.D. Jamieson, *et al.* 2004. Rab3D and actin reveal distinct lamellar body subpopulations in alveolar epithelial type II cells. *Am. J. Respir. Cell Mol. Biol.* **30:** 288–295.
15. Becker, K.A. & N.H. Hart. 1999. Reorganization of filamentous actin and myosin-II in zebrafish eggs correlates temporally and spatially with cortical granule exocytosis. *J. Cell Sci.* **112**(Pt 1): 97–110.
16. Nemoto, T., T. Kojima, A. Oshima, *et al.* 2004. Stabilization of exocytosis by dynamic F-actin coating of zymogen granules in pancreatic acini. *J. Biol. Chem.* **279:** 37544–37550.
17. Eitzen, G. 2003. Actin remodeling to facilitate membrane fusion. *Biochim. Biophys. Acta* **1641:** 175–181.
18. Gasman, S., S. Chasserot-Golaz, M. Malacombe, *et al.* 2004. Regulated exocytosis in neuroendocrine cells: a role for subplasmalemmal Cdc42/N-WASP-induced actin filaments. *Mol. Biol. Cell* **15:** 520–531.
19. Sokac, A.M. & W.M. Bement. 2006. Kiss-and-coat and compartment mixing: coupling exocytosis to signal generation and local actin assembly. *Mol. Biol. Cell* **17:** 1495–1502.
20. Malacombe, M., M.F. Bader & S. Gasman. 2006. Exocytosis in neuroendocrine cells: new tasks for actin. *Biochim. Biophys. Acta* **1763:** 1175–1183.
21. Yu, H.Y. & W.M. Bement. 2007. Control of local actin assembly by membrane fusion-dependent compartment mixing. *Nat. Cell Biol.* **9:** 149–159.
22. Haller, T., J. Ortmayr, F. Friedrich, *et al.* 1998. Dynamics of surfactant release in alveolar type II cells. *Proc. Natl. Acad. Sci. USA* **95:** 1579–1584.
23. Weaver, T.E., C.L. Na & M. Stahlman. 2002. Biogenesis of lamellar bodies, lysosome-related organelles involved in storage and secretion of pulmonary surfactant. *Semin. Cell Dev. Biol.* **13:** 263–270.
24. Nomura, F., T. Inaba, S. Ishikawa, *et al.* 2004. Microscopic observations reveal that fusogenic peptides induce liposome shrinkage prior to membrane fusion. *Proc. Natl. Acad. Sci. USA* **101:** 3420–3425.
25. Grathwohl, C., G.E. Newman, P.J. Phizackerley, *et al.* 1979. Structural studies on lamellated osmiophilic bodies isolated from pig lung. ^{31}P NMR results and water content. *Biochim. Biophys. Acta* **552:** 509–518.
26. Betz, W.J. & G.S. Bewick. 1992. Optical analysis of synaptic vesicle recycling at the frog neuromuscular junction. *Science* **255:** 200–203.
27. Mair, N., T. Haller & P. Dietl. 1999. Exocytosis in alveolar type II cells revealed by cell capacitance and fluorescence measurements. *Am. J. Physiol.* **276:** L376–L382.
28. Frick, M., S. Eschertzhuber, T. Haller, *et al.* 2001. Secretion in alveolar type II cells at the interface of constitutive and regulated exocytosis. *Am. J. Respir. Cell Mol. Biol.* **25:** 306–315.
29. Merrifield, C.J., S.E. Moss, C. Ballestrem, *et al.* 1999. Endocytic vesicles move at the tips of actin tails in cultured mast cells. *Nat. Cell Biol.* **1:** 72–74.
30. Thorn, P. & I. Parker. 2005. Two phases of zymogen granule lifetime in mouse pancreas: ghost granules linger after exocytosis of contents. *J. Physiol.* **563:** 433–442.
31. Bertocchi, C., A. Ravasio, S. Bernet, *et al.* 2005. Optical measurement of surface tension in a miniaturized air-liquid interface and its application in lung physiology. *Biophys. J.* **89:** 1353–1361.
32. Haller, T., P. Dietl, H. Stockner, *et al.* 2004. Tracing surfactant transformation from cellular release to insertion into an air-liquid interface. *Am. J. Physiol. Lung Cell Mol. Physiol.* **286:** L1009–L1015.

33. Ashino, Y., X. Ying, L.G. Dobbs, *et al.* 2000. $[Ca^{2+}]_i$ oscillations regulate type II cell exocytosis in the pulmonary alveolus. *Am. J. Physiol. Lung. Cell Mol. Physiol.* **279:** L5–L13.
34. Frick, M., C. Bertocchi, P. Jennings, *et al.* 2004. Ca^{2+} entry is essential for cell strain-induced lamellar body fusion in isolated rat type II pneumocytes. *Am. J. Physiol. Lung Cell Mol. Physiol.* **286:** L210–L220.
35. Wirtz, H.R. & L.G. Dobbs. 1990. Calcium mobilization and exocytosis after one mechanical stretch of lung epithelial cells. *Science* **250:** 1266–1269.

Exocytosis of Insulin

In Vivo Maturation of Mouse Endocrine Pancreas

Aldo Rozzo,[a,b] Tiziana Meneghel-Rozzo,[a] Saša Lipovšek Delakorda,[c] Shi-Bing Yang,[a,d] and Marjan Rupnik[a,c]

[a]*European Neuroscience Institute Göttingen, Göttingen, Germany*

[b]*Department of Medicine, Children's Hospital Boston, Harvard Medical School, Boston, Massachusetts, USA*

[c]*Faculty of Medicine University of Maribor, Maribor, Slovenia*

[d]*Howard Hughes Medical Institute, San Francisco, California, USA*

The aim of this study was to define when an insulin-positive cell becomes functional *in vivo* and starts to exocytose insulin in a regulated nutrient-dependent manner. Insulin-positive cells appear in embryonic life (midgestation) and complete their maturation, presumably around birth. In order to work with embryonic and newborn endocrine pancreas, we used organotypic slices. The mouse embryonic pancreas slices show high basal insulin release that is not further elevated by high glucose levels. Despite the presence of functional voltage-activated ion channels, the cells are not electrically active in the presence of secretagogues. At birth, the high basal insulin release drops and, after postnatal day 2, the insulin-positive cells show both adult-like bursting electrical activity and hormone release induced by high glucose levels. These properties allowed us to define them as beta cells. Despite the apparent stability of the transcription factor profile reported in insulin-positive cells during late-embryonic life, functional beta cells appear only 2 days after birth.

Key words: **insulin release; embryo; newborn; beta-cell maturation; developing pancreas**

Introduction

In mouse embryos, endocrine pancreatic precursor cells first appear in the pancreatic bud at the 14-somite stage [about embryonic day (E8.5), midgestation[1,2]]. At E10.5, these precursor cells further differentiate into insulin-positive cells in the ductal epithelium.[3] The processes of cell proliferation, migration, and maturation[4] persist throughout the embryonic stage, and, just before birth, these precursor cells form the islets of Langerhans and still maintain a high proliferation rate for about 2 weeks after birth.[3,5,6] Mammalian embryos have relatively higher plasma insulin levels compared to postnatal stages.[7] Right after birth, the neonates lose maternal energy supplies and the whole body, including the pancreas, has to adapt quickly to gain the control of energy homeostasis.

However, in contrast to the morphogenesis of islets of Langerhans, functional development of the pancreatic beta cell is still largely unknown. A mature pancreatic beta cell is defined as an insulin-secreting cell, and the secretion is tightly regulated upon stimulation (primarily glucose). The glucose responsiveness in rat beta cells matures progressively in postnatal life[8] as a consequence of increasing glucokinase activity, but a

Address for correspondence: Dr. Marjan Rupnik, Institute of Physiology, Faculty of Medicine, University of Maribor, Slomškov trg 15, 2000 Maribor, Slovenia. Voice: +386 2 330 5854; fax: +386 2 330 5853. marjan.rupnik@uni-mb.si

Mechanisms of Exocytosis: Ann. N.Y. Acad. Sci. 1152: 53–62 (2009).
doi: 10.1111/j.1749-6632.2008.04003.x

detailed functional analysis of the development of excitability and secretory competence in the mouse is lacking.

In this study, we used organotypic cultured slices from mouse pancreas, a novel preparation, to study the functional development of pancreatic beta cells.[9] The main reason for using this approach is the high yield of endocrine tissue from young animals when the pancreatic tissue is scarce and fragile. In fact, the insulin-positive cell clusters appear as small elongated shapes close to the ducts in late-embryonic life.[9] We found that embryonic beta cells have a higher basal insulin secretion rate but very little glucose responsiveness. This fact can be at least explained by its depolarized membrane potential. At postnatal day 2, the basal insulin secretion is significantly reduced to the adult level, coinciding with the gradually hyperpolarized membrane potentials. This change is primarily a result of delayed expression of ATP-sensitive potassium channels (K_{ATP}).

Materials and Methods

Experimental Animals

Naval Medical Research Institute (NMRI) mice were provided by Göttingen University (ZTE) and used between embryonic day 17 (E17) and postnatal day 7 (P7). The adults (pregnant) were killed by cervical dislocation and decapitation and embryos and newborns by decapitation. All animal procedures were conducted according to the National Institutes of Health guidelines and approved by the local animal care authorities. Blood samples were collected immediately after decapitation from the carotid artery to measure glucose (One-Touch Ultra, LifeScan, Milpitas, CA) and to extract plasma for insulin measurement (EDTA-prepared microtubes, Sarstedt, Nuernbrecht, Germany; ultrasensitive mouse insulin ELISA, Mercodia, Uppsala, Sweden) according to the manufacturer's instructions.

Slices

Whole pancreas slices were prepared as previously reported.[9] Briefly, low-melting point agarose (1.9% in extracellular solution at 37°C; Cambrex Corporation, East Rutherford, NJ) was injected under the organ capsule to stabilize the tissue. Blown-up pancreas was embedded in the same agarose and cut into 100-μm thick slices, using a vibratome (VT1000S, Leica Microsystems AG, Wetzlar, Germany) in ice-cold solution bubbled with 95% O_2 and 5% CO_2. To wash out the pharmacologically active compounds released by the cutting procedure, the slices were stored in cold, bubbled, extracellular solution for at least 1 h and then used for experiments. Between E17 and P3 both the animals and pancreata increased their weight more than 300%[10] and, accordingly, the number of slices prepared from each animal increased from more than 10 to about 20, with five to 15 insulin-positive cell clusters or islets each. This gave us the possibility of using tissue from the same animal for electrophysiological hormone-release and immunostaining experiments.

Immunostaining

The slices were used fresh or after overnight culture in a humid incubator (37°C with 5% CO_2). For overnight culture, slices were placed in 24-well plates, free floating in about 1 mL of incubation medium. Both fresh and cultured slices were fixed with 4% paraformaldehyde in PBS (pH 7.4) for 1 h. The slices were than incubated in the blocking solution (reducing unspecific binding and increasing the penetration of antibodies) for 1 h at room temperature or overnight at 4°C with two primary antibodies: anti-insulin (prepared in mouse, 1:1000 dilution, Sigma, St Louis, MO) and anti-S100 (in rabbit, 1:200, DAKO, Carpinteria, CA). After washing, the slices were incubated with secondary antibodies anti-mouse Alexa 488 and anti-rabbit Alexa 647 (in goat, 1:500; Molecular Probes, Carlsbad, CA) for 1 h in blocking solution. All the incubations were at room temperature unless otherwise indicated. Finally, slices were mounted using a solution of 95% glycerol in PBS, and fluorescence immunostaining was acquired using a confocal microscope (Leica TCS SP2 AOBS; Leica Microsystems, Mannheim, Germany) using 488 nm (Ar) and 633 nm (He-Ne) lasers for sequential excitation. The images were processed by Leica confocal software.

Transmission Electron Microscopy

For the electron microscopy, the tissue was embedded in epoxy resin (Agar Scientific Ltd., Essex, UK). The tissue was then dissected and fixed in 2.45% glutaraldehyde and 2.45% paraformaldehyde in 0.1 mol/L sodium cacodylate buffer (pH 7.4) at room temperature for 3 h and at 4°C overnight. Small fragments of tissue were washed in 0.1 mol/L sodium cacodylate buffer at room temperature for 3 h and postfixed in 2% OsO_4 at room temperature for 2 h. Then, the tissue was washed in 0.1 mol/L sodium cacodylate buffer (pH 7.4) at room temperature for 3 h and dehydrated using a graded ethanol series and embedded in epoxy resin. The samples were sectioned on an Ultracut E Ultramikrotom (Reichert, Depew, NY). Semithin sections were stained with 0.5% toluidine blue in aqueous solution. For electron microscopy, ultrathin sections (70 nm) were obtained and transferred to copper grids (300 mesh). Ultrathin sections were stained with uranyl acetate and lead citrate and examined in a Zeiss EM 902 transmission electron microscope.

Insulin Release

After a washing step, the slices were incubated at 37°C in culture medium for at least 10 min. Single slices were incubated sequentially in low (5 mmol/L) and high (15 mmol/L) glucose in 0.2 mL solution in a humid incubator (37°C, 5% CO_2) and the media prewarmed. After 30 min of incubation a sample of medium was collected to measure the insulin released and the remaining medium was discarded. Each slice was then homogenized and the insulin content extracted by sonication in acid ethanol (25 min, 200 μL[11]) to normalize the release to the hormone cellular content in the slice.

Electrophysiology

The slices were superfused (1.5 mL/min) with warm, bubbled, extracellular solution (35°C; 95% O_2, 5% CO_2; pH 7.4) in a recording chamber (400 μL). A platinum frame with nylon fibers kept the slices flat on the bottom of the recording chamber. An upright microscope (Eclipse E600FN, Nikon, Tokyo, Japan) with a 60× water objective was used. Borosilicate capillaries (GC150F-15; Harvard Apparatus Ltd, Edenbridge, UK) were pulled to resistances of 3–6 MΩ by a horizontal puller (P-97 Sutter Instruments, Novato, CA).

Recordings were performed in whole-cell configuration with a standard patch-clamp amplifier (SWAM IIC, Celica, Ljubljana, Slovenia; or EPC9/10, HEKA Electronik, Lambrecht, Germany). Data were acquired with WinWCP 3.2.9 software (John Dempster, University of Strathclyde, Glasgow, UK) via an A/D converter (PCI-6035E, National Instruments, Austin, TX). The same software was used to analyze data and export them to Sigmaplot (version 7.0, SPSS Inc., Chicago, IL). Putative beta cells were identified by the characteristic voltage-activated Na^+ current inactivation properties.[9,12] ATP concentration inside the pipette was 0, 2, or 5 mmol/L as indicated.

Solutions and Chemicals

Extracellular solution (in mmol/L) was as follows: NaCl 125, KCl 2.5, $MgCl_2$ 1, $CaCl_2$ 2 or 10, NaH_2PO_4 1.25, $NaHCO_3$ 26, Na-pyruvate 2, ascorbic acid 0.25, *myo*-inositol 3, lactic acid 6, and glucose between 3 and 15, pH 7.4 (with carbogen at 35°C). K-intracellular (in mmol/L) solution was as follows: KCl 150, HEPES 10, $MgCl_2$ 2, EGTA 0.05, Na_2-ATP as indicated, pH 7.2 (with KOH). Cs-intracellular

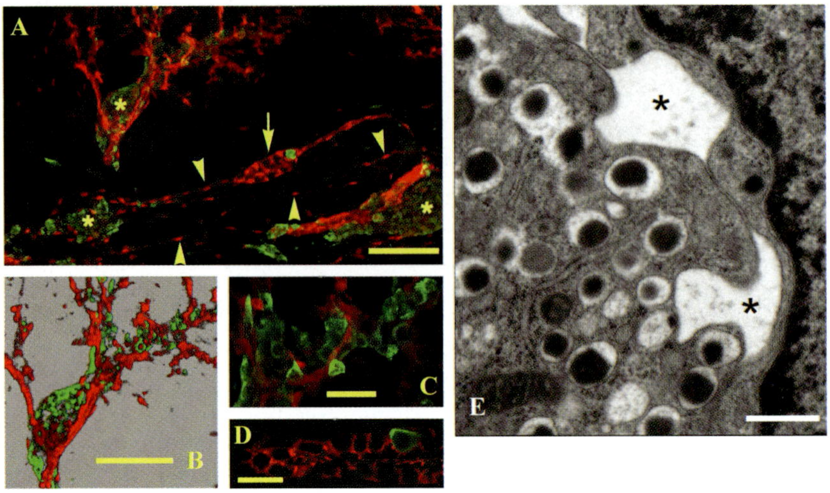

Figure 1. Relation between insulin-positive cells and ducts in postnatal day 0 (P0) mice. (**A**) Confocal image of double staining of 100-μm thick pancreas slice from P0 with anti-insulin (green) and anti-S-100 (Schwann cells and ducts, red) antibodies. Many S-100 positive cells are evident in the lower part of the picture and can be identified as Schwann cells (arrowheads) along fibers, around the ganglion (arrows), and the islets (asterisks). The solitary insulin-positive cell in the ganglion seems to be part of a neuroinsular complex. The islet in the upper left area of the picture is located at the bifurcation of a duct and is magnified and rendered in three dimension in (**B**). The anatomical relation between ducts and beta cells appears strict at P0 with numerous, isolated, insulin-positive cells along the ducts, the putative source of endocrine precursors. At higher magnification (**C**) the two cell populations appear mixed and focal insulin accumulations in the cytoplasm are visible. (**D**) Insulin-positive cell (green) embedded in the pancreatic duct (S-100, red). (**E**) Transmission electromicrograpy showing the dilatations of the extracellular space and absence of docked insulin granules in P0 beta cells. Scale bars: (**A**) 250 μm; (**B**) 100 μm; (**C**) 50 μm; (**D**) 25 μm; (**E**) 500 nm.

(in mmol/L): CsCl 140, HEPES 10, $MgCl_2$ 2, EGTA 10, ATP-Na_2 2 or 5, TEA-Cl 20, pH 7.2 [with tetraethylammonium (TEA)-OH]. Incubation medium was CMRL-1066 (Invitrogen Co., Paisley, UK) without L-glutamine and supplemented with 10% fetal bovine serum, penicillin 100 U/mL, streptomycin 100 μg/mL, and ciprofloxacin 4 μg/mL. Blocking solution consisted of 3% normal goat serum, 3% BSA, and 0.3% Triton in PBS buffer. Extraction solution consisted of 71% EtOH and 0.5% HCl in water (acid/ethanol[11]). All chemicals were from Sigma unless otherwise indicated.

Statistics

Statistics are given as means ± SEM. The statistical significance for the comparison was assessed using paired or unpaired Student's t-test.

Results

Morphology

We first used immunostaining to determine the morphological development of pancreatic beta cells in organotypic slices. Double staining of anti-insulin and anti-S-100 antibodies during the first postnatal days showed the presence of many insulin-positive cells associated with ducts in the terminal region of the pancreatic ductal system (Fig. 1B). In embryonic and newborn pancreata, Schwann cells were found around islets, ganglia, and neuronal fibers (Fig. 1A). Unexpectedly ductal structures with typical morphology were also stained.

As already reported, the insulin signal distribution in embryo and P0 beta cells is condensed to a perinuclear area (Fig. 1C).

More rarely, the anti-GluT2 (Fig. 1D) stained apical parts of the ductal tree, which harbored a small number of isolated insulin-positive cells, confirming the ongoing neogenesis of endocrine cells from precursors during the first few postnatal days.[13,14] Larger dilatations of extracellular space observed between endocrine cells at P0 may represent disconnections from cell replication (Fig. 1E; Susan Bonner-Weir, Joslin Diabetes Center, personal communication).

Insulin versus Glucose Levels

The plasma glucose and insulin levels are shown in Figure 2A. Before birth (E17–18) the plasma level of insulin was high. In contrast, the plasma glucose level in embryos was relatively low, although it was largely controlled by the mother. The plasma glucose levels increased just before birth (E19), despite an elevated insulin level (Fig. 2A). After birth, the glucose level rose above 4 mmol/L (Fig. 2A) and then progressively reached a plateau after P2. In the same period, the insulin level decreased. This result indicated the establishment of an adult-like regulatory mechanism for insulin release.

Hormone Release from Slices

Insulin release was tested using static incubation (Fig. 2B) in basal (5 mmol/L, black hatched) and stimulatory (15 mmol/L, hatched) glucose concentrations. The basal insulin release was high in embryonic life and dropped after birth from 6.7 ± 0.9 to 2.6 ± 0.7% of its cellular content (E19 vs. P0). Despite this difference, the response to high glucose levels was negligible in both ages. This high basal release in the embryo was stable for hours and was inhibited when incubated at room temperature (data not shown).[15,16] From P2 onward, 15 mmol/L glucose resulted in an additional insulin release of 2.3 ± 0.8% on top of its basal release (3.4 ± 0.5%).

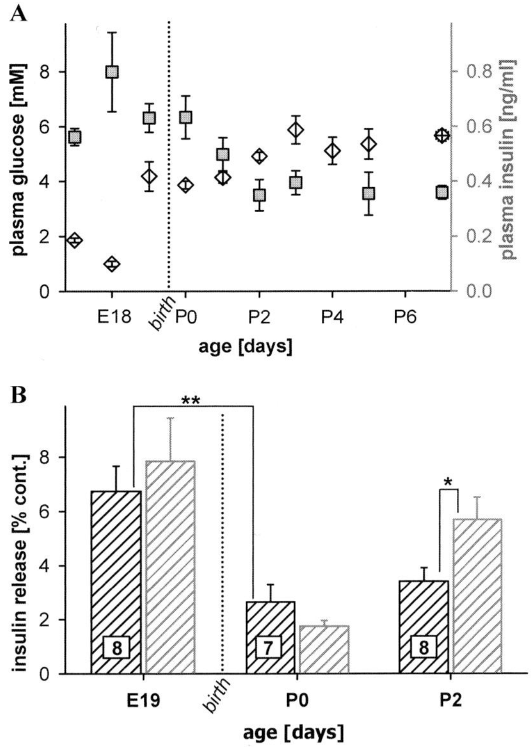

Figure 2. Plasma insulin and glucose levels and insulin release in embryo and newborn mice. (**A**) Age profile of the glucose level in plasma (diamonds) and the insulin level in plasma (squares) in 141 (minimum 10 animals/age) and 59 (minimum five animals/age) mice, respectively. The glucose level starts to rise just before birth and reaches a plateau 3 days after birth. Insulin, on the other hand, is high before birth and starts to decline after; it reaches a plateau in about 3 days. The insulin drop between P0 and P2 is statistically relevant ($P < 0.03$); glucose levels rise the day before birth [$P < 0.01$, embryonic day (E)18 vs. E19]. (**B**) Insulin release in organotypic pancreas slices induced by low (5 mmol/L, basal release, black hatched bars) and high glucose (15 mmol/L, stimulated release, grey hatched bars) levels via sequential static incubation of 23 slices (minimum seven slices/age). The basal release dropped immediately after birth; responsiveness to glucose started only at P2. Values in boxes are means ± SEM; (**$P < 0.02$, *$P < 0.05$; paired *t*-test comparing the responses of each slice, unpaired between ages).

Electrical Activity

Electrical activity was recorded in whole-cell mode in pancreas slices. Until P0 the cells were depolarized (-47 ± 5 mV, $n = 13$) and

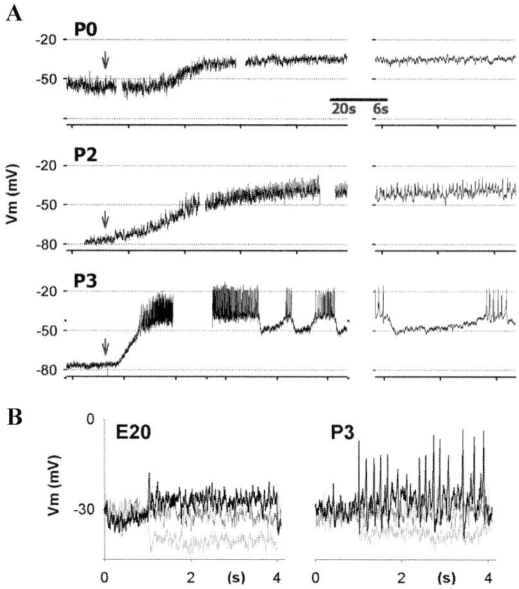

Figure 3. Electrical activity induced by application of tolbutamide. Three representative traces of the responses induced by tolbutamide (100 μmol/L, the arrows mark the start of application) in newborn beta cells at different ages. (**A**) At P0 (top trace), beta cells have a relatively depolarized resting membrane potential and tolbutamide only slightly depolarizes the membrane. After P1 (middle and bottom trace) the resting membrane potential was more hyperpolarized and the tolbutamide elicited much stronger depolarization, and from P3 onward, even bursting activity can be observed sometimes upon tolbutamide application. Right panels show expanded traces. (**B**) Electrical activity after the injection of different current pulses in E20 (left trace) and P3 (right trace) beta cells.

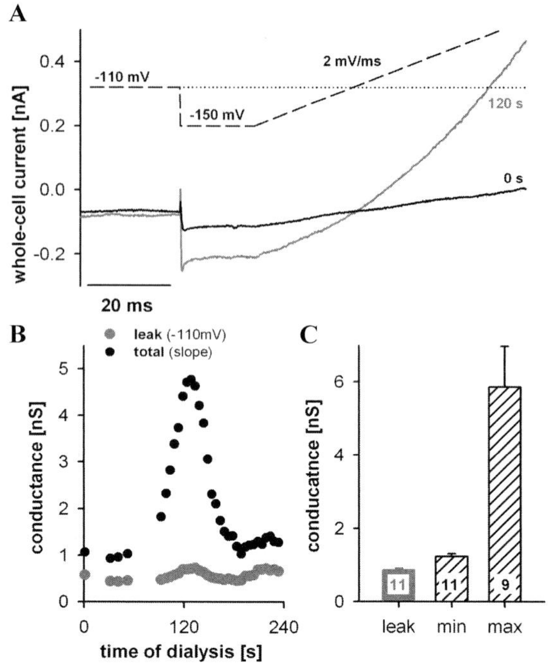

Figure 4. K_{ATP} currents in P0 insulin-positive cells. (**A**) The plot shows a representative cell where the initial low K^+ conductance (black, 0 s) increases after cell dialysis with zero mmol/L ATP (gray, 120 s). In (**B**) the panel shows the time course of K_{ATP} currents upon zero ATP dialysis. The zero ATP dialysis opens the K_{ATP} in about 2 min and then inactivates them. (**C**) The histograms show the statistics for 11 cells; note the increase in conductance from dialysis (two cells were not included in "max" because the recording was too short).

application of the K_{ATP} modulator, tolbutamide (100 μmol/L), induced only a mild depolarization (Fig. 3A). At P2, the cells showed more hyperpolarized potential (−83 ± 7 mV, $n = 6$) and the K_{ATP} block induced a sustained depolarization and firing activity. However, only from P3 did the cells start to show adult-like bursting activity with plateau phases between clusters of Ca^{2+} firing.[17] The injection of a current (+10 pA steps) also showed absence of firing in insulin-positive cells in animals younger than P0; however, similar current injection induced action potential-like electrical activity in insulin-positive cells of older animals (>P2; Fig. 3B).

K_{ATP} Channels

On pancreatic beta-cell membrane, the K_{ATP} channel is the key molecule linking metabolism to the electrical activity. We used voltage ramps between −150 and −100 mV to estimate the K_{ATP} conductance (Fig. 4AB). Immediately after establishment of the whole-cell recording, the K_{ATP} currents developed within 2 min as a result of washing out of intracellular ATP (Fig. 4B). We found that the K_{ATP} channels at P0 were closed under resting conditions (0.4 ± 0.1 nS) but opened after a few minutes of zero ATP dialysis (peak conductance 5.9 ± 1.1 nS; Fig. 4BC). In beta cells from P2–3 animals, more K_{ATP} channels were opened under resting conditions (0.6 ± 0.1 nS, $P < 0.05$);

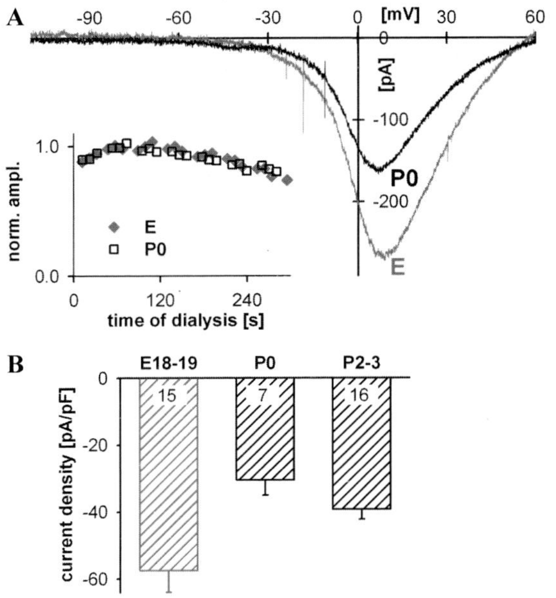

Figure 5. Development of Ca^{2+} currents in insulin-positive cells. (**A**) Two representative traces of Ca^{2+} currents induced in embryo (E, gray) and newborn (P0, black) applying a ramp from −110 to +60 mV with a slope of 0.5 mV/ms. Note that the P0 cell has not only a smaller current but also a larger surface (+0.7 pF) according to its increase in cell size. The dialysis effect on current amplitude is shown in the insert. (**B**) The Ca^{2+} current density is compared between different ages. Numbers on bars indicate the number of cells tested.

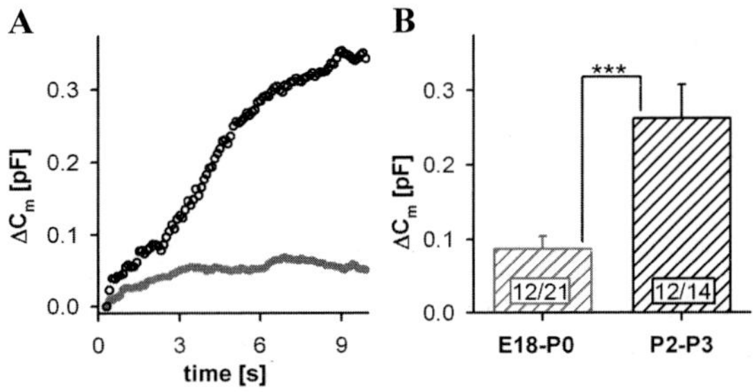

Figure 6. Increase in capacitance triggered by activation of voltage-activated Ca^{2+} channels. A train of depolarizations from −60 to +10 mV was applied (10 Hz, 40 ms per pulse). (**A**) Two representative traces in P0 (gray) and P3 (black) insulin-positive cells. (**B**) Statistically significant difference in capacitance change ($P < 0.002$) and responsiveness of the insulin-positive cells. At P0 12/21 cells responded to a depolarization train (>20 fF), while 12/14 cells responded at P2. The ATP concentration inside the pipette was 5 mmol/L to exclude any response limitation from ATP washout.

peak conductance after zero ATP washout was not significantly different from P0 (not shown).

Ca^{2+} Channels

The pancreatic beta cell has calcium channel-mediated action potentials.[17] The difference in membrane activity between E19 and P2 could be partially a result of different calcium channel expression on the membrane. We used Cs^+-based solution to record the Ca^{2+} currents from both embryonic and newborn beta cells (Fig. 5A). We found the normalized Ca^{2+} peak current amplitudes were higher in embryonic cells and dropped significantly at birth (from -57.6 ± 6.4 to -30.6 ± 4.4 pA/pF, respectively; Fig. 5B).

In E20, a relatively rare condition (approximately 1%) of delayed birth from a small litter size (four or five pups vs. 14 and 18 pups) was observed; the weight of these embryos (2.09 ± 0.03 g) was comparable to P2 (2.03 ± 0.15 g) and was 58% higher than E19 (1.32 ± 0.05 g). Despite the increased body size, the insulin-positive cells maintained their embryonic features, suggesting a critical role for delivery or the start of independent life in triggering the functional maturation of beta cells (not shown).

Single Cell Exocytotic Activity

The application of a train of 100 depolarizing pulses from a holding potential at −60 mV to +10 mV (40-ms duration, 10 Hz) was used to open voltage-activated Ca^{2+} channels and test the secretory capacity in a single beta cell (Fig. 6A). This protocol was applied between 20 and 30 s after the establishment of the whole-cell patch-clamp mode. The number of responsive cells increased from 57 to 86% between E18–P0 and P2–P3, and the capacitance change amplitude increased from less than 100 fF to about 300 fF (Fig. 6B).

Discussion

Insulin-positive cells at birth are not responsive to stimulatory levels of glucose but show a high basal insulin release (Fig. 2). This observation can account for both low levels of insulin immunostaining close to the plasma membrane[9] and the minuscule capacitance change induced by the depolarization train (Fig. 6). This suggests that insulin takes over the glycemic control only postnatally.

At birth, basal insulin release drops, and this event seems to be correlated to the drop in voltage-activated Ca^{2+} conductance. Two days after birth, the insulin and glucose levels are comparable with those in adult animals, suggesting that the glucose levels take over the control of insulin release. At this age, the cells appear similar to adult beta cells, considering both the secretory capacity and excitability. This last transition may be associated with an increase in resting K_{ATP} conductance that reduces the resting membrane potential.

The idea of nonmature hormonal release at birth is not surprising as insulin does not seem to be vital during the first days after birth. Recently, a number of knockout (KO) animals were produced eliminating genes involved in islet-cell differentiation. Some of these KOs, such as Nkx2.2,[18] Nkx6.1,[19] Pax4,[20] and Arx-Pax4[21] almost completely lack insulin-producing cells, but despite this they survive through gestation and at birth are almost indistinguishable (or slightly smaller) from their wild-type and heterozygous littermates. By the second day postpartum all of these KOs have severe problems related to glucose homeostasis. The phenotype of the aforementioned mutants and the high insulin plasma level observed and reported in wild-type embryos[7] would suggest a predominant trophic and hemodynamic action for insulin during embryonic life.

The interruption of the glucose supply forces the animal to maintain glycemia compatible with the active behavior of feeding. The fact that the Ca^{2+} current density does not increase significantly between nonresponsive P0 and mature-like beta cells at P2–3 (Fig. 5) suggests that the insensitivity of P0 to high glucose is not a result of Ca^{2+} channel properties and the redundant Ca^{2+} conductance in embryos is apparently causing the high basal release. It would be possible to speculate about the nonresponsiveness to high glucose levels in embryos, assuming saturation of the system already at basal glucose concentration (close to 7% in 30 min compared to less than 3% in adults). This would also explain the depletion of insulin-positive granules from the cell periphery (Fig. 1E). Furthermore, the lower response in capacitance suggests a rapid exhaustion of secretory vesicles in embryonic and P0 beta cells, reaching a plateau phase in a few seconds, blunting the glucose-stimulated insulin release.

S-100 has not yet been reported as a ductal marker in pancreas. This might stem from the fact that this feature seems to be absent in animals older than 3 weeks (data not shown). The high expression of this protein family in pancreatic ductal adenocarcinomas[22] suggests a correlation between S-100 and the high proliferation rate present in embryo and newborn ducts as in tumors.

The strict anatomical relationship between insulin-positive cells and ductal cells has already been shown with citokeratin-20[9] and with GluT2.[13,14] Using thicker slices allowed us to reconstruct larger portions of the tissue and obtain evidence that many insulin-positive cells are still not aggregated at birth and are putative precursor cells that are destined to migrate and/or aggregate into new islets (Fig. 1B).

In adults the K_{ATP} activity is known to be crucial to maintain the resting membrane potential,[23] and the total K_{ATP} conductance was estimated to be about two times higher than in P0–P3. In beta cells, both perinatal and adult, dialysis with 2 mmol/L ATP slightly hyperpolarizes the cell, suggesting that, in P0, the resting ATP is higher than 2 mmol/L. Until P0, the resting K_{ATP} conductance is close to 6% of the maximum; in P3 it reaches 12%, and in adults it was reported to be close to 20%.[24] We do not have enough evidence to explain what causes the difference between these two different stages, but putative cytosolic factor(s) controlling ATP sensitivity may still be identified.

The knowledge on the maturation of functional beta cells in wild-type animals is vital for understanding the basic physiology of these important cells. This information is also key for evaluating the results of the numerous manipulations devoted to producing new or modified insulin-releasing cells that could improve cell replacement therapy in diabetic patients. The process of beta-cell neogenesis, naturally occurring between the late-embryonic and the early postnatal stages, appears to be a relevant model for studying regeneration. A detailed characterization of beta-cell physiology during the early postnatal period will also provide a solid base to evaluate the insulin-positive cells in many model mice with perinatal lethal phenotype (KO of Pax6, Pax4, Arx genes) and newly differentiated insulin-positive cells from embryonic[25] or adult stem cells or from transdifferentiated cells both *in vitro* and *in vivo*. In addition, further studies are required to assess the epigenetic influence on the maturation of insulin-positive cells.

Acknowledgments

We thank Marion Niebeling, Heiko Röhse, and Rudi Mlakar for excellent technical support. Part of the work was financed by the Growbeta European Union grant (5th framework, No. QLG1-CT-2001-02233). The European Neuroscience Institute Göttingen (ENI-G) is jointly funded by the Göttingen University Medical School, the Max-Planck-Society, and Schering AG. MR is supported from the Slovenian Research Agency (J3-7186-2334).

Conflicts of Interest

The authors declare no conflicts of interest.

References

1. Edlund, H. 2002. Pancreatic organogenesis–developmental mechanisms and implications for therapy. *Nat. Rev. Genet.* **3:** 524–32.
2. Lammert, E., O. Cleaver & D. Melton. 2003. Role of endothelial cells in early pancreas and liver development. *Mech. Dev.* **120:** 59–64.
3. Habener, J.F., D.M. Kemp & M.K. Thomas. 2005. Minireview: transcriptional regulation in pancreatic development. *Endocrinology* **146:** 1025–1034.
4. Bonner-Weir, S. *et al.* 2004. The pancreatic ductal epithelium serves as a potential pool of progenitor cells. *Pediatr. Diabetes* **5**(Suppl 2): 16–22.
5. Murtaugh, L.C. & D.A. Melton. 2003. Genes, signals, and lineages in pancreas development. *Annu. Rev. Cell Dev. Biol.* **19:** 71–89.
6. Georgia, S. & A. Bhushan. 2004. Beta cell replication is the primary mechanism for maintaining postnatal beta cell mass. *J. Clin. Invest.* **114:** 963–968.
7. Alvarez, C. *et al.* 1997. Contrasted impact of maternal rat food restriction on the fetal endocrine pancreas. *Endocrinology* **138:** 2267–2273.
8. Taniguchi, S., K. Tanigawa & I. Miwa. 2000. Immaturity of glucose-induced insulin secretion in fetal rat islets is due to low glucokinase activity. *Horm. Metab. Res.* **32:** 97–102.
9. Meneghel-Rozzo, T. *et al.* 2004. In vivo and in vitro development of mouse pancreatic beta-cells in organotypic slices. *Cell Tissue Res.* **316:** 295–303.
10. Montanya, E. *et al.* 2000. Linear correlation between beta-cell mass and body weight throughout the lifespan in Lewis rats: role of beta-cell hyperplasia and hypertrophy. *Diabetes* **49:** 1341–1346.
11. Detimary, P., J.C. Jonas & J.C. Henquin. 1995. Possible links between glucose-induced changes in the energy state of pancreatic B cells and insulin release. Unmasking by decreasing a stable pool of adenine nucleotides in mouse islets. *J. Clin. Invest.* **96:** 1738–1745.
12. Gopel, S. *et al.* 1999. Voltage-gated and resting membrane currents recorded from B-cells in intact mouse pancreatic islets. *J. Physiol.* **521**(Pt 3): 717–728.
13. Pang, K., C. Mukonoweshuro & G.G. Wong. 1994. Beta cells arise from glucose transporter type 2 (Glut2)-expressing epithelial cells of the developing rat pancreas. *Proc. Natl. Acad. Sci. USA* **91:** 9559–9563.
14. Bonner-Weir, S. & A. Sharma. 2002. Pancreatic stem cells. *J. Pathol.* **197:** 519–526.
15. Henquin, J.C. *et al.* 1975. Quinine-induced modifications of insulin release and glucose metabolism by isolated pancreatic islets. *FEBS Lett.* **57:** 280–284.
16. Atwater, I. *et al.* 1984. Cooling dissociates glucose-induced insulin release from electrical activity and

cation fluxes in rodent pancreatic islets. *J. Physiol.* **348:** 615–627.

17. Dean, P.M. & E.K. Matthews. 1970. Glucose-induced electrical activity in pancreatic islet cells. *J. Physiol.* **210:** 255–264.
18. Sussel, L. *et al.* 1998. Mice lacking the homeodomain transcription factor Nkx2.2 have diabetes due to arrested differentiation of pancreatic beta cells. *Development* **125:** 2213–2221.
19. Naya, F.J. *et al.* 1997. Diabetes, defective pancreatic morphogenesis, and abnormal enteroendocrine differentiation in BETA2/neuroD-deficient mice. *Genes Dev.* **11:** 2323–2334.
20. Sosa-Pineda, B. *et al.* 1997. The Pax4 gene is essential for differentiation of insulin-producing beta cells in the mammalian pancreas. *Nature* **386:** 399–402.
21. Collombat, P. *et al.* 2005. The simultaneous loss of Arx and Pax4 genes promotes a somatostatin-producing cell fate specification at the expense of the {alpha}- and {beta}-cell lineages in the mouse endocrine pancreas. *Development* **132:** 2969–2980.
22. Nakamura, T. *et al.* 2004. Genome-wide cDNA microarray analysis of gene expression profiles in pancreatic cancers using populations of tumor cells and normal ductal epithelial cells selected for purity by laser microdissection. *Oncogene* **23:** 2385–2400.
23. Smith, P.A., F.M. Ashcroft & P. Rorsman. 1990. Simultaneous recordings of glucose dependent electrical activity and ATP-regulated K(+)-currents in isolated mouse pancreatic beta-cells. *FEBS Lett.* **261:** 187–190.
24. Speier, S. *et al.* 2005. KATP-channels in beta-cells in tissue slices are directly modulated by millimolar ATP. *Mol Cell Endocrinol.* **230:** 51–58.
25. Blyszczuk, P. *et al.* 2004. Embryonic stem cells differentiate into insulin-producing cells without selection of nestin-expressing cells. *Int. J. Dev. Biol.* **48:** 1095–1104.

Compound Exocytosis in Pituitary Cells

Nina Vardjan,[a,b] Jernej Jorgačevski,[b] Matjaž Stenovec,[a,b] Marko Kreft,[b,a] and Robert Zorec[b,a]

[a]*Celica Biomedical Center, Ljubljana, Slovenia*

[b]*Laboratory of Neuroendocrinology-Molecular Cell Physiology, Faculty of Medicine, University of Ljubljana, Ljubljana, Slovenia*

Neurotransmitter and hormone release from vesicles involves fusion between the vesicle and the plasma membranes, a process termed exocytosis. Recently we reported that most of the spontaneous and stimulated exocytotic events in pituitary lactotrophs are transient and repetitive, appearing in bursts lasting more than 100 s. However, whether this is also the case in compound vesicle-to-vesicle exocytosis is unknown. Here we investigated compound exocytotic events in resting and stimulated lactotrophs by using optical and cell-attached patch-clamp capacitance measurements. Elementary compound exocytotic events were characterized by multiple-amplitude on-steps in synaptopHluorin fluorescence and in membrane capacitance signals. Multiple-amplitude on-steps appeared either as a relatively large upward step, indicating that vesicles were fused with each other prior to fusion of the vesicle membrane with the plasma membrane (multivesicular exocytosis), or as a time-dependent stepwise signal increase, indicating sequential fusion of two or more vesicles with the plasma membrane (sequential exocytosis). In the majority of membrane capacitance recordings (>90%), multiple-amplitude on-steps terminated as multiple-amplitude off-steps. These complex amplitude events were repetitive, indicating that transient fusion pore openings reflect repetitive interactions of a single vesicle or vesicles in a cluster with the plasma membrane. Out of many mechanisms, these interactions may enable the diffusion of fusion proteins from the plasma membrane to the membrane of the primary fused vesicles, consequently enabling vesicle-to-vesicle fusion. The incidence of compound exocytotic events increased by 33% after stimulation, which is consistent with the enhanced efficiency of hormone secretion after the stimulus.

Key words: **compound exocytosis; neuroendocrine; fusion pore; peptidergic release; optical imaging; pore conductance**

Introduction

The principal event of exocytotic secretion, the opening of the fusion pore that connects the secretory vesicle lumen to the extracellular space, has been elucidated in increasing detail as a result of the development of sensitive electrophysiological[1–3] and optical techniques using different fluorescence markers.[4–6] After formation fusion pore either opens irreversibly and the secretory vesicle membrane is recovered slowly by endocytosis (full fusion exocytosis)[7] or it closes in the process known as transient ("kiss-and-run") exocytosis.[8–10] Fusion pores may also flicker between open and closed states repeatedly and randomly, and a flicker burst may end with the pore shut or irreversibly open.[11–13]

Exocytotic events in secretory cells may involve more than one vesicle (i.e., compound exocytosis).[14,15] Before exocytosis, vesicles may fuse with each other homotypically, forming large mulitivesicular structures, which then fuse with the plasma membrane via a single fusion pore (multivesicular exocytosis). Alternatively, initial fusion of a secretory vesicle (primary vesicle) with the plasma membrane is followed by

Address for correspondence: Robert Zorec, Ph. D., Professor, University of Ljubljana, Faculty of Medicine, Institute of Pathophysiology, LN-MCP, Zaloška 4, 1000 Ljubljana, Slovenia. Voice: +386 1 543 70 20; fax: +386 1 543 70 36. robert.zorec@mf.uni-lj.si

Mechanisms of Exocytosis: Ann. N.Y. Acad. Sci. 1152: 63–75 (2009).
doi: 10.1111/j.1749-6632.2008.04008.x

the fusion of a second vesicle with the already exocytosed primary vesicle (sequential exocytosis) (reviewed in Ref. 16).

In eukaryotes intracellular membrane fusion requires SNARE (soluble *N*-ethylmaleimide-sensitive factor attachment protein receptor) proteins. Molecular recognition between target-membrane SNARE (t-/Q-SNARE) and vesicle-associated SNARE (v-/R-SNARE) proteins is needed to form complexes bridging the target and vesicle membrane and to induce the formation of a fusion pore (reviewed in Ref. 17). In some cells t-/Q-SNARE proteins [e.g., synaptosome-associated protein of 23 and 25 kDa (SNAP-23 and SNAP-25)] are also present on vesicles.[18–20] It has been proposed that flickering fusion events between a single vesicle and the plasma membrane may enable the relocation of plasma membrane t-/Q-SNARE fusion proteins into the vesicle membrane.[21] This would allow t-/Q-SNARE and v-/R-SNARE protein recognition and vesicle-to-vesicle fusion, which may lead to the formation of multivesicular structures in the vicinity of the plasma membrane and consequently to multivesicular exocytosis. Similarly, in sequential exocytosis, t-/Q-SNARE fusion proteins could move from the plasma membrane into the membrane of the primary fused vesicles to make them competent for fusion with the oncoming vesicles.[19,21,22]

Pituitary lactotrophs, which possess several distinct properties that render them favorable preparations for study of peptidergic vesicle fusion, engage single-vesicle and compound exocytosis.[12,13,23,24] Repetitive transient fusion events with the flicker-burst duration of more than 100 s represent the principal mode of single-vesicle exocytosis in lactotrophs.[12,13] In this study we investigated fusion pore dynamics of compound-exocytotic events in lactotrophs by optophysiological methods with pHluorine-tagged v-/R-SNARE synaptobrevin 2 [i.e., synaptopHluorin (spH)][5] and green fluorescent protein (GFP)-tagged atrial natriuretic peptide [ANP-emd (ANP fused with the emerald-GFP)][25] and by high-resolution electrophysiological measurements of membrane capacitance.

Materials and Methods

Cell Preparation and Transfection

Lactotroph-enriched cell cultures were prepared from adult male Wistar rat anterior pituitaries as described.[26] The animals were euthanized in accordance with the International Guiding Principles for Biomedical Research Involving Animals developed by the Council for International Organizations of Medical Sciences and the Directive on Conditions for Issue of License for Animal Experiments for Scientific Research Purposes (Official Gazette of the Republic of Slovenia 40/85 and 22/87). Cells were plated on glass coverslips coated with poly-L-lysine and maintained in high-glucose Dulbecco's modified Eagle's medium (Invitrogen, Carlsbad, CA) supplemented with 10% newborn calf serum and L-glutamine in an atmosphere of humidified air (92%) and CO_2 (8%). One-day-old cultures were transfected with spH in a pCI neo vector (Promega, Madison, WI) or ANP-emd[25] using Lipofectamine Plus (Invitrogen) as described.[5] The spH plasmid construct was a generous gift from G. Miesenböck and ANP-emd was a kind gift from E.S. Levitan. All experiments were carried out at room temperature 2 to 4 days after transfection. All chemicals were from Sigma-Aldrich (St. Louis, MO).

Time-lapse Confocal Microscopy

Laser-scanning fluorescence images of lactotrophs were acquired with an inverted confocal microscope (Zeiss LSM 510, Jena, Germany). Coverslips with transfected cells were mounted in standard saline solution [10 mmol/L HEPES/NaOH (pH 7.2), 10 mmol/L D-glucose, 130 mmol/L NaCl, 8 mmol/L $CaCl_2$, 1 mmol/L $MgCl_2$, and 5 mmol/L KCl] in a perfusion-stimulation chamber on the microscope stage. In experiments with lactotrophs transfected with the ANP-emd-plasmid construct, the standard saline solution contained 4 μmol/L FM 4–64 (Invitrogen). Fluorescent probes were exited with a 488-nm laser line and fluorescence emission was collected through a plan apochromatic objective (63×, 1.4 NA) with a 505–530-nm band-pass emission filter (spH, ANP-emd) and a 585-nm long-pass emission filter (FM 4–64). Time-lapse fluorescence images (512 × 512 pixels) were obtained every ca. 0.5 s for up to 2 min. The thickness of optical sections was approximately 2 μm. Images were quantitatively analyzed by the LSM 510 and MATLAB (Math Works, Natick, MA) software.

Image and Data Analysis

Cells were visualized for individual exocytotic events by rapidly replaying the image sequences in forward and reverse directions. The onset of exocytosis was defined as the first frame showing a significant increase in spH or decrease in ANP-emd fluorescence. A circular region of interest (ROI) of 15 pixels in diameter was centered on the spot where the exocytotic event occurred, and average changes in the fluorescence intensities of spH, ANP-emd, and FM 4–64 were monitored over time. Single-exponential ($F = F_0 + c * \exp(-t/\tau)$) or double-exponential ($F = F_0 + c_1 * \exp(-t/\tau_1) + c_2 * \exp(-t/\tau_2)$) decay functions were fitted to the time-dependent spH fluorescence changes using SigmaPlot (Systat Software, San Jose, CA). F is fluorescence at time t, F_0 is baseline fluorescence, c is fluorescence amplitude of $F_{(t=0)}$-F_0, c_1 and c_2 are the fluorescence amplitudes of individual exponential components of a double-exponential function, and τ is the time constant of the individual exponential component. The goodness of the exponential fits was judged from the calculated coefficient of determination, R^2. The linear function fits ($y = a + b * x$) in scatter plot diagrams were obtained with SigmaPlot. Images in Figures 1 and 2 were contrasted with linear interpolation and filtered with a Gaussian filter (radius of 0.5 pixels). Images of a single exocytotic vesicle (ROI) in Figures 1 and 2 were cropped (43 × 43 pixels) from the original image. Cropped images were resampled into images of 189 × 189 pixels, using bilinear interpolation. Values are reported as means ± SEM. Statistical significance was evaluated by Student's t-test. Unless stated otherwise, the level of significance was at $P < 0.05$.

Electrophysiology

The cell-attached patch-clamp measurements of membrane capacitance (C_m) were performed at room temperature with a dual-phase, lock-in, patch-clamp amplifier (SWAM IIC, Celica, Ljubljana, Slovenia) as described.[13,27] Briefly, lactotrophs were voltage clamped at 0 mV to which a 1.6 kHz sine wave (111 mV r.m.s.) was applied. The in-phase (real part of admittance trace; *Re*) and out-of-phase (imaginary part of admittance trace; *Im*) outputs of the lock-in amplifier were low-pass filtered (100 Hz, -3 dB), digitalized by A/D converter (CED 1401, Cambridge, UK), and stored on a personal computer (PC). Phase was adjusted to null the changes in *Re* in response to manually generated 10 fF calibration steps. We used fire-polished thick-wall pipettes (2 to 5 MΩ) heavily coated with silicone elastomer Sylgard and low surface level of bath solution (200 μL per coverslip) to reduce the noise level. The same standard solution was used in the bath solution and the pipettes: 10 mmol/L HEPES, 10 mmol/L D-glucose, 130 mmol/L NaCl, 8 mmol/L $CaCl_2$, 1 mmol/L $MgCl_2$, and 5 mmol/L KCl. The pH was adjusted to 7.2 using NaOH, and the osmolality was 310 ± 3 mosmol/kg (Osmomat030; Gonotech GmbH, Berlin, Germany). Calculation of C_m step amplitudes were manually made from *Im* signals (C_m signals) using the cursor option in the WCP software (Dr. J. Dempster, University of Strathclyde, UK), IgorPro (WaveMetrics, Portland, OR), or CellAn (Celica) written for MATLAB on a PC-compatible computer. The appearance of a step was ascertained by progressive filtering. The probability of step occurrence was calculated per temporal period

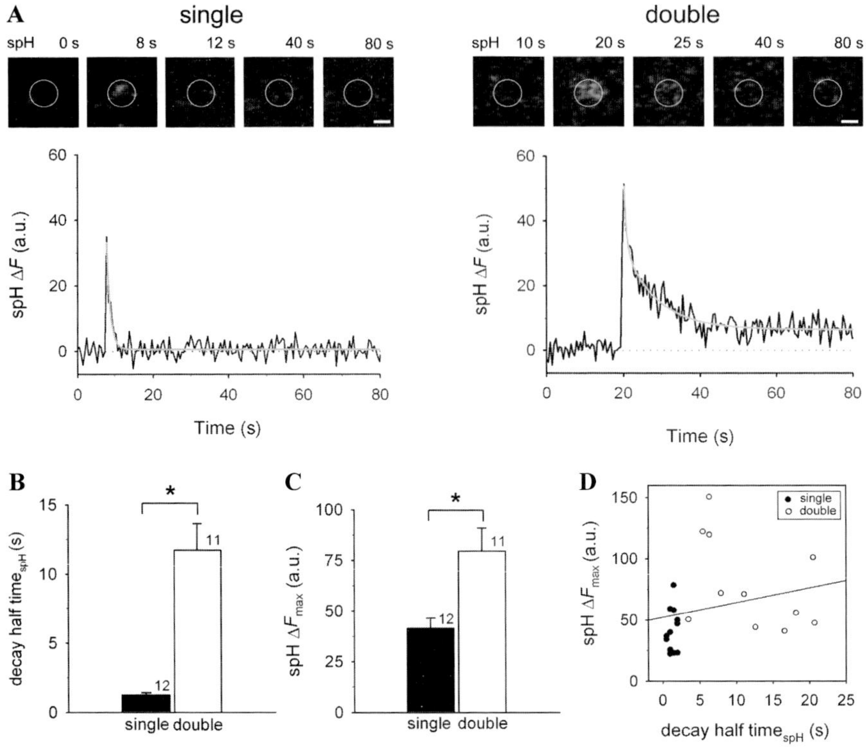

Figure 1. Differences in the kinetics and amplitudes of the synaptopHluorin (spH) fluorescence signal in spontaneous transient exocytotic events. (**A**) Representative sequential images of a single vesicle and changes in spH fluorescence at the vesicle site. Fluorescence signals are plotted as the change in fluorescence (ΔF) in arbitrary units (a.u.). Exocytosis resulted in a rapid increase in spH fluorescence that was followed by either (A, left) a rapid single-exponential decline with τ of 1 s or (A, right) a slow double-exponential decline in spH fluorescence with a fast component (τ_1) of 0.7 s and a slow component (τ_2) of 9.4 s. White circles indicate regions of interest (ROIs); gray lines show exponential fits; scale bar = 0.5 μm. (**B**) Mean decay half-times of spH fluorescence in single-exponential (1.27 ± 0.15 s; mean ± SEM) and double-exponential (11.72 ± 1.92 s) transient exocytotic events; $^*P < 0.001$. (**C**) Peak amplitudes of spH fluorescence increase (ΔF_{max}), separately for single-exponential (41.5 ± 5.1 a.u.) and double-exponential (79.6 ± 11.4 a.u.) transient exocytotic events; $^*P < 0.005$. (**D**) Relation between the peak amplitude of spH fluorescence increase and the half-time of fluorescence decay in two types of transient spontaneous events. The line has the format: $\Delta F_{max} = (51 \pm 11) + (1.3 \pm 1.1) * \text{half time}_{spH}$ [seconds]). The correlation coefficient (r) between the two parameters ($r = 0.24$; $P = 0.27$; $n = 23$) was not significantly different from 0.

of 10 ms. The probability of occurrence of two simultaneous independent steps at the same time was calculated as the multiplied probability for each step.

Results

Differences in the amplitudes and kinetics of the spH signal indicate two types of exocytotic events in resting lactotrophs.

To study fusion pore behavior of compound-exocytotic events, time-dependent fluorescence intensity changes of a fusion marker spH[5] were monitored by confocal microscopy in resting lactotrophs expressing spH. Fusion pore formation in vesicles carrying spH was indicated by an abrupt punctuate brightening in spH fluorescence from efflux of protons from the vesicle lumen through the pore.[5,13] After initial brightening, spH fluorescence either dimmed,

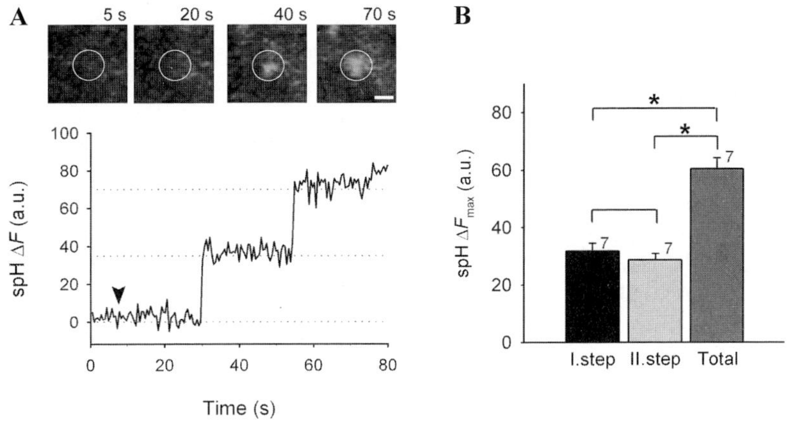

Figure 2. Staircase increase in spH fluorescence after stimulation with KCl. (**A**) Representative sequential images of a single vesicle and the changes in spH fluorescence increase at the vesicle site after stimulation with 100 mmol/L KCl solution (arrowhead). Fluorescence signal is plotted as time-dependent changes in fluorescence (ΔF) in a.u. Note that the signal resulted in a two-step increase in spH fluorescence (likely from sequential compound exocytosis of two vesicles); each fluorescence increment persisted for some time before the next step-like increase occurred. The thin dotted lines indicate levels of spH fluorescence; white circles indicate ROIs; scale bar = 0.5 μm. (**B**) Mean peak amplitudes of spH fluorescence increase (ΔF_{max}), separately displayed for the first step (31.73 ± 2.74 a.u.; mean ± SEM), the second step (28.71 ± 2.27 a.u.), and the total two-step increase in spH fluorescence (60.44 ± 3.75 a.u). Note that ΔF_{max} of the first and the second step are twofold lower than the ΔF_{max} of the total two-step increase ($^*P < 0.002$, $P = 0.66$; $n = 7$), indicating sequential exocytosis of two vesicles.

indicating fusion pore closure and reacidification of the vesicle lumen ($n = 23$, 64%, transient events; Fig. 1A), or spH fluorescence persisted bright throughout the monitoring ($n = 13$, 36%, persistent events, data not shown, see Ref. 13), indicating that the vesicle lumen remained in contact with the extracellular space throughout the monitoring.

Based on the kinetics of the spH fluorescence signal we could distinguish two types of transient exocytotic events in resting lactotrophs ($n = 23$). In 52% ($n = 12$) of spH-positive transient events, the appearance of green fluorescent puncta was followed by a fast decline that was best fitted by a single-exponential function with an average (± SEM) time constant (τ) of 1.2 ± 0.2 s (single-exponential events; for representative event see Fig. 1A, left). However, in 48% ($n = 11$) of spontaneous transient events, the spH fluorescence decline was slower and best fitted by a double-exponential function with an average fast component (τ_1) of 1.9 ± 0.6 s and an average slow component (τ_2) of 29.6 ± 5.3 s (double-exponential events; for representative event see Fig. 1A, right).

Analysis of spH fluorescence decay showed that the decay half-time was nine times faster in single- compared to double-exponential events (average half-time, 1.3 ± 0.2 s versus 11.7 ± 1.9 s) (Fig. 1B). Moreover, the mean peak amplitude of spH fluorescence was about twofold smaller in single-exponential events than in double-exponential events (41.5 ± 5.1 arbitrary units (a.u.) versus 79.6 ± 11.4 a.u., $P < 0.005$; Fig. 1C). When we plotted decay half-time of spH fluorescence in transient events against the peak amplitude, single-exponential exocytotic events were clustered tightly whereas double-exponential events largely scattered with no apparent correlation between the two measures (Fig. 1D). These findings are consistent with the existence of distinct populations of transient events in resting lactotrophs with different reacidification kinetics. However, twofold higher amplitudes (Fig. 1C) and double-exponential kinetics (Fig. 1A, right) of spH fluorescence signal observed in double-exponential events may also suggest transient compound exocytosis. In this case two vesicles, which were fused together before exocytosis, underwent multivesicular fusion via a single fusion pore opening at the plasma membrane, as indicated by the twofold higher amplitudes. Double-exponential kinetics of the spH decline implies that the vesicles are retrieved from the plasma membrane in a sequential manner [scheme in Fig. 6B(b)].

Persistent events observed in lactotrophs expressing spH are thought to represent fast, repetitive, transient fusion pore openings observed with cell-attached C_m recordings.[13] The effect of fusion pore closure in persistent events is poorly reflected in optical measurements using spH construct because of the reacidification, which is rather slow in comparison with the fusion pore closure observed in capacitance measurements.[13] We excluded persistent events from analysis because we could not follow the reacidification kinetics of these events.[13] However, the amplitudes of spH fluorescence in persistent events ranged from 28 a.u. to 78 a.u., suggesting that vesicles in persistent events and with higher amplitudes also underwent multivesicular compound exocytosis.

Stepwise Increase in spH Fluorescence in Stimulated Lactotrophs

To explore the fusion pore behavior of evoked compound events[23,24] we stimulated lactotrophs expressing spH with 100 mmol/L KCl solution. In contrast to resting conditions, 94% (n = 33) of all stimulated events were persistent,[13] with amplitudes ranging from 27 a.u. to 96 a.u. Two events were transient with double-exponential decay kinetics of the spH fluorescence signal (data not shown). Transient events with double-exponential kinetics and persistent events with large amplitudes likely indicate multivesicular exocytosis, as suggested before for spontaneous events.

In seven (21%) out of 33 persistent events, an abrupt punctuate brightening in spH fluorescence was followed by a second spH brightening increase, which was overlapping the first one (Fig. 2A, upper panel). In these events the increase in spH fluorescence consisted of two successive steps (Fig. 2A, lower panel) whereby mean peak amplitudes of the first and the second step were similar (31.7 ± 2.7 a.u. and 28.7 ± 2.3 a.u., respectively) and the total mean peak amplitude of spH fluorescence increase was 60.4 ± 3.8 a.u. (n = 7; Fig. 2B). The sequential two-step spH fluorescence increase is likely a consequence of a sequential fusion of two vesicles whereby a vesicle after fusion with the plasma membrane undergoes secondary fusion with a deeper lying vesicle (sequential compound exocytosis; scheme in Fig. 6A).

Stepwise Decrease in ANP-emd Fluorescence Intensity in Stimulated Lactotrophs

To further confirm compound exocytosis in stimulated lactotrophs and to optically demonstrate release of cargo from a compound vesicle fused with the plasma membrane, we performed additional studies in lactotrophs transfected with the plasmid encoding ANP-emd.[25] Measurements of ANP-emd discharge through the fusion pore reflect the hormone release from a vesicle, as it is associated with the fluorescence intensity decay.[12] Simultaneously with the measurements of ANP-emd release, single- and multiple-vesicle exocytotic events were monitored by labeling the vesicles with FM 4–64 dye, which stains the vesicle membrane and the vesicle matrix following fusion pore opening.[23,28] In two out of 22 cells stimulated by K^+-enriched solution, sequential decay in the ANP-emd signal was accompanied by a simultaneous FM 4–64 increase. Figure 3 (bottom right) shows two time-dependent sequential fluorescence step decreases in the ANP-emd signal, suggesting sequential fusion of two discrete vesicles in a small space beneath the plasma membrane whereby the second

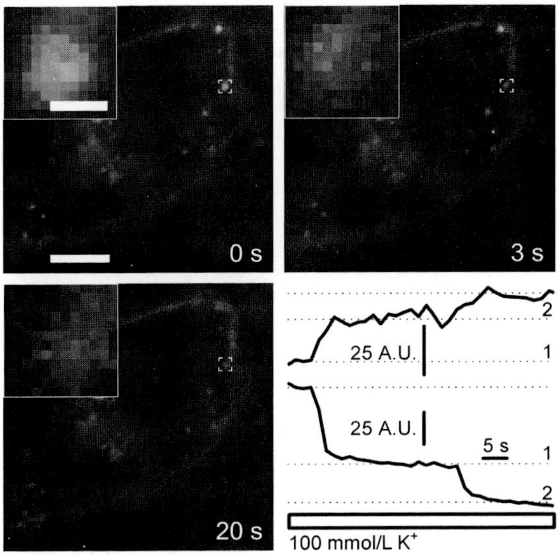

Figure 3. Sequential peptide discharge reveals compound-exocytotic events in a stimulated lactotroph. Confocal images (top and bottom left) displaying fluorescent peptide release from two, adjacent, likely interconnected vesicles in a lactotroph following stimulation with 100 mmol/L K^+. Scale bar: 5 μm. Insets show the magnified view of atrial natriuretic peptide fused with the emerald green fluorescent protein (ANP-emd) (green) and FM 4–64 (red) fluorescence at various times. Scale bar: 0.5 μm. Bottom right shows the time course of fluorescence intensity changes obtained in boxed vesicles. Note the two recognizable steps (1 and 2) in FM 4–64 (top trace) and distinctive steps in ANP-emd (bottom trace) fluorescence during stimulation by 100 mmol/L $[K^+]_o$ that likely indicate sequential discharge of fluorescent cargo followed by simultaneous vesicle loading by FM 4–64 in a compound-exocytotic event. The thin dotted lines indicate discrete levels of FM 4–64 and ANP-emd fluorescence.

vesicle apparently fused with the first one, which had already formed a fusion pore for the time period of several seconds.[29,30]

Spontaneous Discrete Multiple Amplitude Steps in C_m

The dynamics of the fusion pore can be investigated at the level of a single vesicle directly by cell-attached patch-clamp measurements of membrane capacitance (C_m).[31] Discrete steps in C_m indicate the occurrence of elementary single- and multiple-vesicle fusion events, which can respectively be distinguished by their amplitude and their time course.[21] While discrete upward (on-steps) in C_m are interpreted as vesicle fusion with the plasma membrane, discrete downward (off-steps) are attributed to vesicle fission.[31] An on-step shortly followed by an off-step of similar amplitude indicates transient fusion of the same vesicle.[31] In this study, we focused exclusively on more complex fusion events composed of multiple-amplitude on- and off-steps in C_m, indicating putative compound vesicle fusion.[2,21]

Spontaneous discrete changes in C_m were observed in 32 (37%) out of 87 C_m recordings. Six C_m recordings (19% of all recordings with C_m steps) exhibited events with multiple amplitudes. More than 90% of multiple-amplitude on- and off-steps were repetitive; some of them occurred in regular intervals (Fig. 4), as in single-amplitude events, termed *pulsing fusion pore*.[12]

Figure 4A shows a representative epoch of a cell-attached patch-clamp recording with spontaneous, multiple, amplitude steps in C_m. Every third event (starting with the first one on the left) displayed a second upward step, suggesting that a second fusion pore opened between the already fused vesicle and the second vesicle (sequential compound exocytosis).[16] The two sequential on-steps were followed by a larger amplitude off-step to the baseline, indicating the closure of the fusion pore. It is improbable that the large off-step represents a synchronous closure of two independent vesicles in a small patch of membrane (probability $P < 10^{-6}$).

A different recording, with three step sizes, is shown in Figure 4B (lower trace; *Re* shows no consistent projections from *Im*, proportional to C_m). The event marked with an asterisk is enlarged in Figure 4C; three consecutive on-steps are followed by an abrupt fall to baseline. The probability that the event highlighted in Figure 4C is from fusion of three independent vesicles is very low ($P < 10^{-9}$). In a few recordings, we observed an abrupt opening to a high level, followed by sequential downward steps

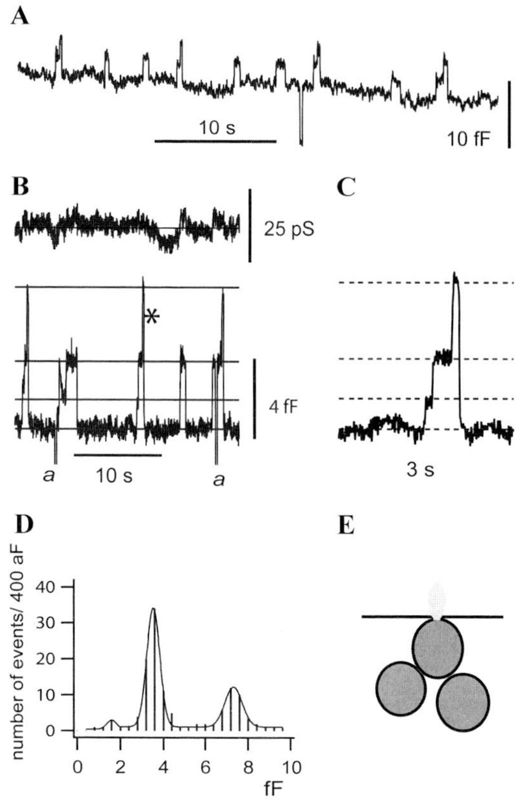

Figure 4. Multiamplitude membrane capacitance (C_m) steps reveal compound fusion pore openings in cell-attached patches before stimulation. **(A)** Rhythmic fusion pore openings with multiple amplitudes. Note that events with double amplitudes are pulsing as regularly as the events with lower amplitude. **(B)** A different epoch of recording from panel A. Note that changes in membrane capacitance (bottom) do not project to real part of admittance trace (*Re*) (top). Calibration pulses, used to set the phase of the lock-in amplifier outputs, are indicated by *a*. **(C)** Fusion pore event marked by an asterisk in panel B is shown at higher time resolution. The probability that this event is a result of the three independent vesicles is $P < 10^{-9}$. **(D)** Histogram of amplitudes of fusion events ($n = 123$) from experiment of panel A. Three Gaussian distributions were fitted to the data with parameters: 1.59 ± 0.31 fF; 3.52 ± 0.42 fF, and 7.32 ± 0.57 fF (mean ± SEM). Dashed lines in panel B and C are means determined from the amplitude histogram of panel D. **(E)** A possible interpretation of multiple amplitude fusion events consisting of a cluster of vesicles that are fused between each other when the amplitude of fusion events is maximal. Intermediate amplitudes of panel D are probably a result of a fusion of only one or two of these vesicles, respectively, with the plasma membrane.

to baseline (not shown), indicating multivesicular compound exocytosis.[16] A histogram of all step sizes observed in the recording is shown in Figure 4D, together with a cartoon illustrating one possible subcellular structural arrangement of granules and fusion sites (Fig. 4E).

Multiple Amplitude C_m Steps Recorded in Stimulated Cells

To see whether stimulation affects the incidence of events with multiple amplitudes,[13] we monitored C_m after depolarization with 100 mmol/L KCl. In eight (40%) out of 20 recordings, discrete C_m steps were recorded. In two (25%) out of eight recordings with C_m steps, repetitive multiple-amplitude on- and off-steps were observed (Fig. 5A), consistent with previous reports.[12,13] A histogram of all step amplitudes for representative events is shown in Figure 5B.

Under stimulated conditions the incidence of recordings exhibiting multiple-amplitude C_m steps increased by 33% in comparison to the spontaneous conditions, consistent with the results of single-amplitude events.[13] Some of the events with multiple amplitudes exhibited greater complexity than those observed under spontaneous conditions (Fig. 4 versus Fig. 5).

In general, two types of transient events with multiple amplitudes were observed before and after stimulation (Fig. 6). The first type is composed of multiple upward steps, indicating sequential exocytosis, followed by a single or multiple downward step(s) to the baseline (Fig. 6A). The second type of complex event is composed of a single, large-amplitude upward step, which indicates multivesicular exocytosis, followed by a single off-step or multiple off-steps to the baseline (Fig. 6B). It is unlikely that the events with multiple upward steps and the single amplitude off-step to the baseline [Fig. 6A(a)] and the events with the single large-amplitude on-step followed by the multiple off-steps to the

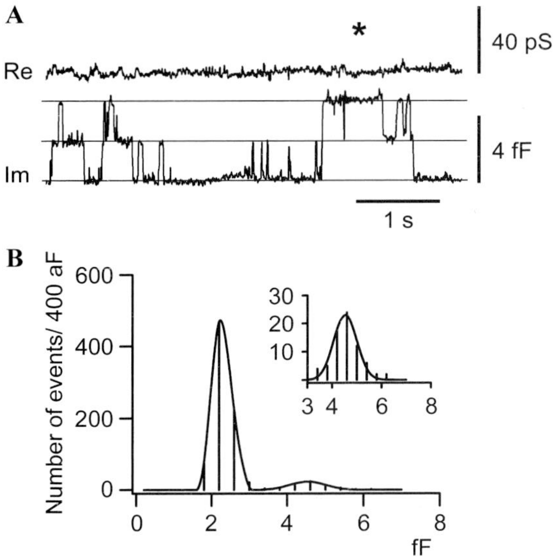

Figure 5. Compound fusion pore openings in a cell-attached patch after cell stimulation. (**A**) Repetitive fusion pore openings with two amplitude levels indicate fusion of at least two different vesicles. Note that the event marked with an asterisk begins with double level amplitude, likely because of the fusion of two "prefused" vesicles with the plasma membrane. The probability that two independent vesicles fuse at the same time in the whole recording is less than $P < 10^{-4}$. Lines represent basal line (closed fusion pores) and two mean levels determined from the amplitude histogram of panel B. (**B**) Histogram of amplitudes of fusion events ($n = 867$) from experiment of panel A. Double Gaussian distribution was fitted to the data with parameters: 2.28 ± 0.01 fF and 4.53 ± 0.03 fF (mean ± SEM). Inset shows the distribution of events with amplitudes from 3–8 fF on a larger scale of the ordinate.

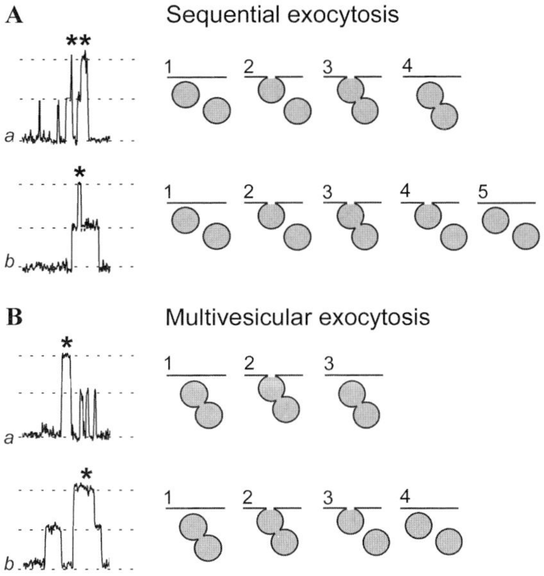

Figure 6. Different types of fusion events with multiple amplitudes. Cell-attached recordings revealed sequential and multivesicular exocytotic events. (**A**, left) Exocytotic events with ensuing on-steps, indicating sequential exocytosis, were followed by a multiple-amplitude off-step (*, probability that the double-amplitude event represents the fusion of two independent vesicles is $P < 10^{-6}$ for spontaneous and $P < 10^{-4}$ for stimulated events) or ensuing off-steps to the baseline. (**B**, left) Exocytotic events with a multiple-amplitude on-step, indicating multivesicular exocytosis, were followed by a multiple-amplitude off-step; (a) *, the probability that the double-amplitude event represents the fusion of two independent vesicles is $P < 10^{-9}$ for spontaneous and $P < 10^{-4}$ for stimulated events or ensuing off-steps to the baseline; (b) *, the probability that the double-amplitude event represents the fusion of two independent vesicles is $P < 10^{-6}$ for spontaneous and $P < 10^{-4}$ for stimulated events. (**A-B**, right) Diagrams of possible interpretations of fusion events with multiple amplitudes observed with cell-attached recordings in resting and stimulated lactotrophs.

baseline [Fig. 6B(b)] represent simultaneous fusion of the similarly sized vesicles, since the probability for independent fusion of two vesicles is $P < 10^{-6}$ in spontaneous and $P < 10^{-4}$ in stimulated conditions.

Discussion

Pituitary lactotrophs release hormone prolactin from secretory vesicles and may engage single-vesicle[12,13,23] and compound exocytosis.[24] Fusion pore behavior studies in resting and stimulated lactotrophs have shown that the fusion pore in single-vesicle exocytotic events tends to repetitively open and close.[12,13] In this study we used optical and electrophysiological approaches to investigate the fusion pore dynamics in compound-exocytotic events.

Amplitude and Signal Decay Time Differences in spH Fluorescence Puncta May Reflect Transient Multivesicular Exocytosis in Resting Lactotrophs

To monitor fusion pore dynamics of compound-exocytotic events in resting lactotrophs, we expressed a pH-sensitive fusion marker spH[5] in lactotroph vesicles and monitored exocytosis by confocal microscopy.[13] In resting lactotrophs two types of exocytotic events could be distinguished: transient (>70% of all events; Fig. 1) and persistent events (data not shown).[13]

Transient exocytotic events differed in the decay time and in the amplitude of the spH fluorescence signal (Fig. 1). In 48% of transient events, the decline in spH fluorescence was slow with complex double-exponential kinetics (Fig. 1A, right). The rest of the transient events (Fig. 1A, left) exhibited 90% faster single-exponential kinetics (Fig. 1B). The amplitude of the spH fluorescence in transient double-exponential events was twofold higher compared to single-exponential events (Fig. 1C), which is consistent with compound fusion,[32] as in large dense-core vesicles of stimulated lactotrophs,[23,24] eosinophils,[33] exocrine pancreatic acinar cells,[29,34] pancreatic β-cells,[35–37] PC12 cells,[20,30] and neutrophils.[38]

Transient large-amplitude events with complex double exponential kinetics (Fig. 1A, right) likely indicate the prefused multivesicular structures beneath the plasma membrane.[24] Such multivesicular structures then transiently fuse with the plasma membrane (multivesicular exocytosis; scheme in Fig. 6B). Consistent with our results in pituitary cells transient compound-exocytotic events were observed previously in mast cells[21] and neutrophils.[38] Moreover, double-exponential kinetics of the spH signal provide further evidence that in these events more than one vesicle was involved not only in the process of vesicle fusion but also in the process of vesicle fission and vesicle lumen reacidification, which affects the spH fluorescence. Presumably, after the fusion with the plasma membrane the "prefused" vesicles exhibit sequential fusion pore closures [scheme in Fig. 6B(b)].

Sequential Compound Exocytosis in Stimulated Lactotrophs

Electron microscopy studies revealed that stimulation of lactotrophs by extracellular elevation of K^+ results in the formation of multivesicular structures near the plasma membrane, which are sharing a single fusion pore opening with the plasma membrane (multivesicular exocytosis).[24] However, the time course of fusion pore openings of individual compound-exocytotic events is poorly studied. We exposed lactotrophs expressing spH to KCl in order to reveal the fusion pore behavior of these multivesicular exocytotic events.

Following KCl stimulation the majority of exocytotic events were persistent with the duration of greater than 100 s (94%). Some of the persistent events exhibited several times higher amplitudes in spH fluorescence than the other smaller amplitude events, indicating multivesicular exocytosis,[24] as in resting lactotrophs (data not shown). In 21% of persistent events the increase in spH was stepwise (Fig. 2). Similarly, stepwise release of hormone ANP-emd accompanied by simultaneous FM 4–64 uptake into the vesicles was observed in lactotrophs after stimulation (Fig. 3), consistent with sequential exocytosis.[20,30,36,37,39]

In stepwise exocytotic events, the primary fused vesicles remained in a stable form for the time period of several seconds before the fusion of the secondary vesicles (Figs. 2–3). During this time, the vesicle lumen remained in contact with the cell exterior, likely because of the prolonged fusion pore open state,[29,30,40] as also indicated by the persistence of the spH fluorescence signal (Fig. 2). Prolonged fusion pore lifetime is possibly adapted to permit sequential exocytosis, whereby the lingering primary vesicle acts as a transfer structure through which the contents of a secondary vesicle can be released.[29,30,40] However, the prolonged fusion pore open time is not necessary for complete hormone release,[41] which appears in a few seconds (Fig. 3). Therefore, it is possible that the fusion pores open and shut during this time (Figs. 2–3), which cannot be resolved by the spH construct.[13]

Repetitive Transient Compound-exocytotic Events

To learn more about the nature of the fusion pore in compound-exocytotic events observed in optical studies (Figs. 1–3), we performed C_m recordings in resting and in stimulated lactotrophs. Multiple-amplitude on- and off-steps, indicating compound fusion,[21,38] were observed in 19% of spontaneous (Fig. 4) and in 25% of stimulated C_m recordings (Fig. 5). Some fusion events displayed a more complex structure after stimulation (Fig. 4 vs. Fig. 5).

Multiple on-steps in C_m (Fig. 4) were interpreted as sequential compound exocytosis and single large-amplitude on-steps (Fig. 5) as multivesicular compound exocytosis (Fig. 6; reviewed in Ref. 16). In most of the compound events, on-steps were followed by off-steps shortly after the fusion (Figs. 4–5), indicating retrieval of vesicles and transient compound exocytosis.[21,38] This is consistent with transient compound events observed in optical studies (Fig. 1A, right). The retrieval of vesicles was either multivesicular or sequential, as indicated by single or multiple off-steps to the baseline [Fig. 6(a) vs. Fig. 6(b)].

The multiple-amplitude on- and off-events (Fig. 6) may also indicate that several vesicles were independently docked at the plasma membrane. However, since the simultaneous opening and closure of the fusion pore of two independent vesicles in the same patch of membrane [see Fig. 6A(a) and Fig. 6B(b)] is highly improbable ($P < 10^{-6}$ for spontaneous and $P < 10^{-4}$ for stimulated events, respectively), it is very unlikely that the records represent events of independently docked vesicles at the plasma membrane. They were rather arranged in series, with only one vesicle docked to the plasma membrane and the others docked to it.[21,38] But we cannot exclude the possibility that some of the multivesicular compound-exocytotic events with single large-amplitude on- and off-steps to the baseline [Fig. 6B(a)] were misinterpreted since compound exocytosis of two small vesicles cannot be distinguished from exocytosis of a single larger vesicle by C_m measurements.

More than 90% of transient compound-exocytotic events appeared repetitively in flicker bursts, as in single-vesicle exocytotic events (Figs. 4–5).[12,13,42] Some of the spontaneous multiple-amplitude C_m signals showed an unexpected type of behavior in which fusion pores opened and closed rapidly with notable regularity (Fig. 4A).[12] Fast repetitive on- and off-steps (Figs. 4–5) may not be resolved using the spH construct because of reacidification, which is, compared to the fusion pore closure observed as an off-step in C_m, rather slow.[13,43] Therefore, it is most likely that the persistence of the spH fluorescence signal in multistep persistent events (Fig. 2) is a consequence of fast, repetitive, transient fusion pore openings (Figs. 4–5)[13] and not a prolonged fusion pore open state.[29,30] This indicates that transient events were underestimated in optical studies. Moreover, with the time resolution of our optical imaging system (0.5 s/image), it is impossible to resolve the rapid multiple on-steps in C_m, which can appear in less than 1 s (Figs. 4–5). Thus some of the large-amplitude events observed in lactotrophs expressing spH (e.g., Fig. 1A, right) may actually undergo sequential (Fig. 6A) and not multivesicular (Fig. 6B) transient exocytosis, as suggested.

Conclusions

The results show compelling evidence of compound-exocytotic events (vesicle-to-vesicle) in pituitary lactotrophs, which also were transient and repetitive as were single-exocytotic events.[12,13] Such flickering activity of the fusion pore may facilitate the exchange of fusion proteins between the plasma and the vesicle membrane and the vesicle-to-vesicle fusion,[18–20] resulting in either sequential or multivesicular compound exocytosis.[21] The frequency of compound fusion observed by C_m measurements is increased by 33% after stimulation, consistent with previous results.[24] The increased occurrence of compound exocytosis is considered to play an important physiological role in increasing the efficiency of the hormone prolactin released into the cell exterior after the stimulus.[13]

Acknowledgments

We thank Ms. Sonja Grilc for continuous support. This work was supported by grants P3 310 381, Z3 7476 from the Ministry of Higher Education, Sciences and Technology of the Republic of Slovenia, NS36665 from the National Institutes of Health, European Commission (EC) grant #QLG3 2001-2004, EC support DECG, CLG3-CT-2001-02004, and EC GROWBETA QLG1-CT-2001-02233.

Conflicts of Interest

The authors declare no conflicts of interest.

References

1. Breckenridge, L. & W. Almers. 1987. Currents through the fusion pore that forms during exocytosis of a secretory vesicle. *Nature* **328:** 814–817.
2. Lollike, K., N. Borregaard & M. Lindau. 1995. The exocytotic fusion pore of small granules has a conductance similar to an ion channel. *J. Cell Biol.* **129:** 99–104.
3. Kreft, M. & R. Zorec. 1997. Cell-attached measurements of attofarad capacitance steps in rat melanotrophs. *Pflugers Arch.* **434:** 212–214.
4. Betz, W. & G. Bewick. 1992. Optical analysis of synaptic vesicle recycling at the frog neuromuscular junction. *Science* **255:** 200–203.
5. Miesenböck, G., D. De Angelis & J. Rothman. 1998. Visualizing secretion and synaptic transmission with pH-sensitive green fluorescent proteins. *Nature* **394:** 192–195.
6. Shaner, N., P. Steinbach & R. Tsien. 2005. A guide to choosing fluorescent proteins. *Nat. Methods* **2:** 905–909.
7. Heuser, J. & T. Reese. 1973. Evidence for recycling of synaptic vesicle membrane during transmitter release at the frog neuromuscular junction. *J. Cell Biol.* **57:** 315–344.
8. Ceccarelli, B., W. Hurlbut & A. Mauro. 1973. Turnover of transmitter and synaptic vesicles at the frog neuromuscular junction. *J. Cell Biol.* **57:** 499–524.
9. Alvarez de Toledo, G., R. Fernández-Chacón & J. Fernández. 1993. Release of secretory products during transient vesicle fusion. *Nature* **363:** 554–558.
10. Fesce, R. *et al.* 1994. Neurotransmitter release: fusion or 'kiss-and-run'? *Trends Cell Biol.* **4:** 1–4.
11. Fernandez, J., E. Neher & B. Gomperts. 1984. Capacitance measurements reveal stepwise fusion events in degranulating mast cells. *Nature* **312:** 453–455.
12. Stenovec, M. *et al.* 2004. Slow spontaneous secretion from single large dense-core vesicles monitored in neuroendocrine cells. *FASEB J.* **18:** 1270–1272.
13. Vardjan, N. *et al.* 2007. Subnanometer fusion pores in spontaneous exocytosis of peptidergic vesicles. *J. Neurosci.* **27:** 4737–4746.
14. Ichikawa, A. 1965. Fine structural changes in response to hormonal stimulation of the perfused canine pancreas. *J. Cell Biol.* **24:** 369–385.
15. Röhlich, P., P. Anderson & B. Uvnäs. 1971. Mast cell degranulation as a process of sequential exocytoses. *Acta. Biol. Acad. Sci. Hung.* **22:** 197–213.
16. Pickett, J. & J. Edwardson. 2006. Compound exocytosis: mechanisms and functional significance. *Traffic* **7:** 109–116.
17. Jahn, R. & R. Scheller. 2006. SNAREs–engines for membrane fusion. *Nat. Rev. Mol. Cell Biol.* **7:** 631–643.
18. Tagaya, M. *et al.* 1996. SNAP-25 is present on chromaffin granules and acts as a SNAP receptor. *FEBS Lett.* **394:** 83–86.
19. Castle, J., Z. Guo & L. Liu. 2002. Function of the t-SNARE SNAP-23 and secretory carrier membrane proteins (SCAMPs) in exocytosis in mast cells. *Mol. Immunol.* **38:** 1337–1340.
20. Kishimoto, T. *et al.* 2006. Vacuolar sequential exocytosis of large dense-core vesicles in adrenal medulla. *EMBO J.* **25:** 673–682.
21. Alvarez de Toledo, G. & J. Fernandez. 1990. Compound versus multigranular exocytosis in peritoneal mast cells. *J. Gen. Physiol.* **95:** 397–409.
22. Pickett, J., P. Thorn & J. Edwardson. 2005. The plasma membrane Q-SNARE syntaxin 2 enters the zymogen granule membrane during exocytosis in the pancreatic acinar cell. *J. Biol. Chem.* **280:** 1506–1511.
23. Angleson, J. *et al.* 1999. Regulation of dense core release from neuroendocrine cells revealed by imaging single exocytic events. *Nat. Neurosci.* **2:** 440–446.

24. Cochilla, A., J. Angleson & W. Betz. 2000. Differential regulation of granule-to-granule and granule-to-plasma membrane fusion during secretion from rat pituitary lactotrophs. *J. Cell Biol.* **150:** 839–848.
25. Han, W. *et al.* 1999. Neuropeptide release by efficient recruitment of diffusing cytoplasmic secretory vesicles. *Proc. Natl. Acad. Sci. USA* **96:** 14577–14582.
26. Ben-Tabou, S., E. Keller & I. Nussinovitch. 1994. Mechanosensitivity of voltage-gated calcium currents in rat anterior pituitary cells. *J. Physiol.* **476:** 29–39.
27. Zorec, R., S. Sikdar & W. Mason. 1991. Increased cytosolic calcium stimulates exocytosis in bovine lactotrophs. Direct evidence from changes in membrane capacitance. *J. Gen. Physiol.* **97:** 473–497.
28. Stenovec, M. *et al.* 2005. Concentration-dependent staining of lactotroph vesicles by FM 4–64. *Biophys J.* **88:** 2607–2613.
29. Thorn, P., K. Fogarty & I. Parker. 2004. Zymogen granule exocytosis is characterized by long fusion pore openings and preservation of vesicle lipid identity. *Proc. Natl. Acad. Sci. USA* **101:** 6774–6779.
30. Kishimoto, T. *et al.* 2005. Sequential compound exocytosis of large dense-core vesicles in PC12 cells studied with TEPIQ (two-photon extracellular polar-tracer imaging-based quantification) analysis. *J. Physiol.* **568:** 905–915.
31. Neher, E. & A. Marty. 1982. Discrete changes of cell membrane capacitance observed under conditions of enhanced secretion in bovine adrenal chromaffin cells. *Proc. Natl. Acad. Sci. USA* **79:** 6712–6716.
32. Pelkmans, L. & M. Zerial. 2005. Kinase-regulated quantal assemblies and kiss-and-run recycling of caveolae. *Nature* **436:** 128–133.
33. Hafez, I., A. Stolpe & M. Lindau. 2003. Compound exocytosis and cumulative fusion in eosinophils. *J. Biol. Chem.* **278:** 44921–44928.
34. Nemoto, T. *et al.* 2001. Sequential-replenishment mechanism of exocytosis in pancreatic acini. *Nat. Cell Biol.* **3:** 253–258.
35. Bokvist, K. *et al.* 2000. Compound exocytosis in voltage-clamped mouse pancreatic beta-cells revealed by carbon fibre amperometry. *Pflugers Arch.* **439:** 634–645.
36. Leung, Y. *et al.* 2002. Visualization of sequential exocytosis in rat pancreatic islet beta cells. *Biochem. Biophys. Res. Commun.* **292:** 980–986.
37. Kwan, E. & H. Gaisano. 2005. Glucagon-like peptide 1 regulates sequential and compound exocytosis in pancreatic islet beta-cells. *Diabetes* **54:** 2734–2743.
38. Lollike, K. *et al.* 2002. Compound exocytosis of granules in human neutrophils. *J. Leukoc. Biol.* **71:** 973–980.
39. Tran, V. *et al.* 2007. Characterization of sequential exocytosis in a human neuroendocrine cell line using evanescent wave microscopy and "virtual trajectory" analysis. *Eur. Biophys. J.* **37:** 55–69.
40. Scepek, S. & M. Lindau. 1993. Focal exocytosis by eosinophils–compound exocytosis and cumulative fusion. *EMBO J.* **12:** 1811–1817.
41. Thorn, P. & I. Parker. 2005. Two phases of zymogen granule lifetime in mouse pancreas: ghost granules linger after exocytosis of contents. *J. Physiol.* **563:** 433–442.
42. Vardjan, N. *et al.* 2007. Elementary properties of spontaneous fusion of peptidergic vesicles: fusion pore gating. *J. Physiol.* **585:** 655–661.
43. Atluri, P. & T. Ryan. 2006. The kinetics of synaptic vesicle reacidification at hippocampal nerve terminals. *J. Neurosci.* **26:** 2313–2320.

The Functions of Munc18-1 in Regulated Exocytosis

Robert D. Burgoyne, Jeff W. Barclay, Leo F. Ciufo, Margaret E. Graham, Mark T.W. Handley, and Alan Morgan

The Physiological Laboratory, School of Biomedical Sciences, University of Liverpool, Liverpool, United Kingdom

The activation of regulated exocytosis occurs by a rise in cytosolic Ca^{2+} concentration. Synaptotagmins act as the Ca^{2+} sensors, whereas the machinery that allows fusion of secretory vesicles with the plasma membrane consists of the soluble *N*-ethylmaleimide-sensitive factor attachment protein receptor (SNARE) proteins, including syntaxin 1, SNAP-25, and VAMP. Within the pathway leading to exocytosis, there is an essential requirement for a member of the conserved Sec1/Munc18 (SM) protein family, which in neurotransmitter and neurohormone release in mammalian cells is Munc18-1. The exact role of Munc18-1 and the steps within exocytosis in which it acts have been intensively investigated. Current evidence suggests that Munc18-1 acts via distinct modes of interactions with syntaxin 1 and the other SNARE proteins and influences all of the steps leading to exocytosis, including vesicle recruitment, tethering, docking, priming, and membrane fusion.

Key words: **amperometry; chromaffin cells; exocytosis; Munc18; neurotransmission; SM proteins; SNARE proteins; syntaxin**

Introduction

The controlled release of neurotransmitters and hormones is brought about by the process of regulated exocytosis in which a rise in cytosolic Ca^{2+} concentration triggers the fusion of a secretory vesicle with the plasma membrane and the release of its contents.[1–3] Complementary genetic and biochemical approaches have identified the basic membrane fusion machinery for exocytosis and established that this is conserved from yeast to man.[4] The core element of this conserved machinery is a set of proteins that assemble to form the soluble *N*-ethylmaleimide-sensitive factor attachment protein receptor (SNARE) complex. In neurons and neuroendocrine cells this comprises VAMP/synaptobrevin on the vesicle and syntaxin 1 and SNAP-25 on the plasma membrane.[5] All three neuronal SNARE proteins are proteolytic substrates for the clostridial neurotoxins, which thereby inhibit neurotransmitter release,[6] and genetic ablation of SNAREs in various organisms blocks exocytosis.[7] While the SNAREs have received much attention, several other proteins are crucial components of the exocytotic machinery that act in part through their interactions with the SNARE proteins or control the assembly/disassembly of the SNARE complex. Among these other proteins are the synaptotagmins, which appear to be the Ca^{2+} sensors for regulated exocytosis,[8] the Sec1/Munc18 (SM) proteins[9] and the Rab proteins.[10]

Evidence suggests that the SNAREs are intimately involved with membrane fusion during exocytosis, and reconstitution experiments using artificial liposomes suggest that formation of the SNARE complex is likely to be the driving force for membrane fusion.[11] The

Address for correspondence: Robert D. Burgoyne, The Physiological Laboratory, School of Biomedical Sciences, University of Liverpool, Crown Street, Liverpool L69 3BX, UK. Voice: +44-151-794-5305; fax: +44-151-794-5337. burgoyne@liv.ac.uk

Mechanisms of Exocytosis: Ann. N.Y. Acad. Sci. 1152: 76–86 (2009).
doi: 10.1111/j.1749-6632.2008.03987.x

in vitro fusion mediated by these minimal components is, however, orders of magnitude slower than fusion during regulated exocytosis, suggesting the involvement of additional proteins or other factors in exocytosis. Regulated exocytosis is also tightly controlled both in time and space and its modification is part of the mechanisms underlying forms of synaptic plasticity.[12] The Rabs comprise a large family of small GTPases that have long been implicated in controlling the fidelity of secretory vesicle targeting.[10] Rab proteins are thought to function by regulating both the transport of vesicles via the cytoskeleton and the assembly of tethering proteins between the vesicle and target membranes. A more crucially important family of proteins are the SM proteins. These derive their name from the yeast Sec1[13] and mammalian Munc18 proteins,[14] with Munc18 being named after UNC-18 in *Caenorhabditis elegans*.[15] The first evidence for a role of SM proteins in membrane fusion came from observations of defective secretion in yeast mutants and uncoordinated movement in *C. elegans unc18* mutants. Subsequent studies have established an essential requirement for SM genes in vesicle fusion throughout the cell in a variety of organisms, but their exact roles have been a matter of debate.

There are three mammalian Munc18 genes involved in exocytosis, with Munc18-1[14] (otherwise known as n-sec1[16]) being the neuronal isoform. Genetic manipulations in the mouse have demonstrated that Munc18-1 is essential for neurotransmitter[17] and hormone[18] release. Munc18-1 null mutant mice show no evidence of any neurotransmission,[17] and there is also a severe impairment of dense-core granule exocytosis in their adrenal chromaffin cells.[18] Neurotransmission is also substantially reduced in null mutants of the Munc18-1 orthologues UNC-18[19] and Rop[20] in *C. elegans* and *Drosophila,* respectively. Much still remains to be learned about the functions of the Sec1/Munc18-1 proteins in the steps leading to membrane fusion, which could be probed through analysis of specific mutations in the protein. In *C. elegans,* a mutation (R476H) of a highly conserved residue in UNC-18 does not have a general deleterious effect on the organism but disrupts one specific behavioral function, namely sperm transfer.[21] In addition, a nonsynonymous single nucleotide polymorphism in mice in the Munc18-1 gene has been linked to changes in alcohol preference,[22] suggesting that single amino acid changes in Munc18-1 can be linked to differing behaviors.

Modes of Interaction of Munc18-1 with Syntaxin

Munc18-1 was originally found to bind with high affinity to syntaxin 1A *in vitro,* precluding the binding of SNAP-25 and VAMP to syntaxin 1A.[23] This was taken as suggesting a possible inhibitory role for Munc18-1. In fact, in some[24] but not all systems[25] overexpression of Munc18-1 orthologues is inhibitory. A purely inhibitory role for Munc18-1 is not consistent, however, with its demonstrated requirement for normal function as described above. In the characterized Munc18-1/syntaxin 1A complex, syntaxin is in the so-called closed conformation[26,27] in which its N-terminal helices are folded back onto the C-terminal SNARE motif.[28] Release of syntaxin into an open conformation would be required to allow syntaxin to associate into the SNARE complex. It is not clear how this dissociation of Munc18-1 could be triggered nor has it been known whether the controlled transition between closed and open conformations is an essential aspect of Munc18-1/syntaxin function. It has been suggested that SM proteins are required as syntaxin chaperones that through binding to the closed form of syntaxin allow it to be trafficked correctly to the plasma membrane,[29,30] but this does not appear to be an essential role in neurons and neuroendocrine cells nor does mode 1 binding (Fig. 1) appear to be required.[19,31–33] Syntaxin with mutations that create a constitutively open form of the protein has been

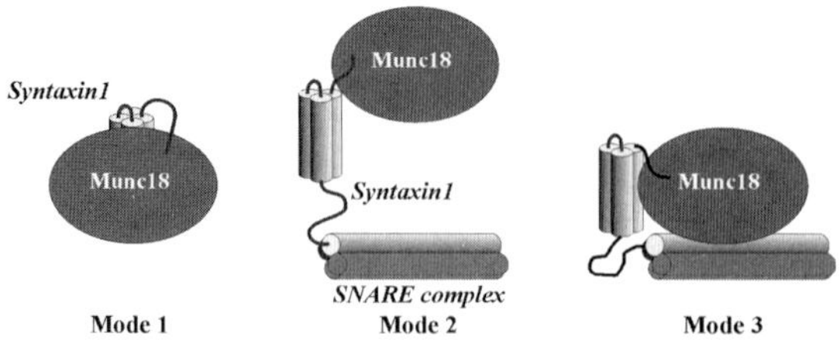

Figure 1. Schematic representation of the three known modes of Munc18-1 binding to syntaxin. Mode 1 is a well-characterized interaction of Munc18-1 with syntaxin 1 in its closed confirmation.[28,43] Mode 2 involves interaction of Munc18-1 with the extreme N-terminus of syntaxin 1. Mode 3 involves interaction of Munc18-1 with the assembled soluble *N*-ethylmaleimide-sensitive factor attachment protein receptor (SNARE) complex. The latter mode requires mode 2 to be achievable, but it is not known if both modes occur at the same time.

shown to fully support neurotransmission in *C. elegans* and can partially bypass a requirement for UNC-13[34] but, surprisingly, not the requirement for UNC-18.[19] This finding suggests that the role of Munc18-1 is not simply to assist in the provision of syntaxin in an open conformation and may be consistent with the existence of syntaxin-independent functions of UNC-18.

The interaction with a closed form of syntaxin has not been described for SM proteins other than Munc18-1. Indeed, the interaction of various SM proteins with their cognate syntaxins had appeared to be through distinct mechanisms (Fig. 1). For some time it was believed that Munc18-1 could only interact with monomeric syntaxin in its closed conformation (mode-1 binding).[26,28] In contrast, the yeast SM proteins Sly1p and Vps45p bind to the extreme N-terminus of their syntaxins (mode 2 binding)[35,36] and yeast Sec1 bind primarily to the assembled SNARE complex (mode 3 binding).[37,38] Evidence has recently accumulated indicating that multiple forms of binding can be used,[39] and Munc18-1, in particular,[40–44] may be able to participate in all three modes.[45] The exact functional significance of each mode of binding and the mechanisms that regulate transition between them are unknown. In addition, structural information is available for mode-1[28] and mode 2 binding[35,36,43,46] but as yet nothing is know about the structural basis of mode 3 binding.

Analysis of Munc18-1 Mutants in the Study of Its Role in Exocytosis

Our studies of the role of Munc18-1 in exocytosis in adrenal chromaffin cells[47–52] have been based on examination of the effects of expression of different Munc18-1 mutants (Table 1). These were modeled on known mutations from other species, such as *Drosophila* (Class A) or *C. elegans* (Class B), that result in single amino acid changes or, alternatively, novel mutations in residues in Munc18-1, including potentially phosphorylated amino acids (Class C). The effect of these constructs on exocytosis was assessed using carbon-fiber amperometry to monitor single fusion/release events from transfected adrenal chromaffin cells.[53] The resulting data have supported the idea that Munc18-1 has both early and late roles in the pathway leading to exocytosis.

The mutations that we studied are distributed over the Munc18-1 structure (Fig. 2) where they could potentially affect a range of protein interactions made by Munc18-1. Importantly, overexpression of wild-type Munc18-1 in our experiments had no discernable effects on exocytosis. In contrast, expression of certain of the mutant constructs resulted in two types of effects (Table 1). First, alterations were observed in the frequency of exocytotic fusion events, consistent with a stimulatory effect on the early prefusion stages of vesicle recruitment/docking (Fig. 3A–D). Second, alterations were seen in the kinetics and quantal size of single exocytotic release events (amperometric spikes) (Fig. 3E).

In an early study we used amperometric analysis of single exocytosis events in chromaffin cells overexpressing the mutant Munc18-1 (R39C). This has a reduced syntaxin binding affinity in an assay specific for mode 1 binding, and, indeed, R39C makes an important

TABLE 1. Effect of Mutations in Munc18-1 on Mode 1 Syntaxin Binding and Exocytosis in Adrenal Chromaffin Cells

Mutant class	Nature of mutation	Effect of mutation on syntaxin binding	Effect on exocytosis in in chromaffin cells	Reference
Class A (*C. elegans*)	*unc-18 (b403)* (E549K)	~90% inhibition	Unknown (not expressed)	49
Class A (*C. elegans*)	*unc-18 (md1401)* (I133V)	No effect	Increased quantal size	49
Class B (*D. melanogaster*)	rop^{G11} (D45N; D34N in Munc18-1)	~40% inhibition	No effect	49
Class B (*D. melanogaster*)	rop^{F3} (R50C; R39C in Munc18-1)	~60% inhibition	Decreased quantal size	47
Class B (*D. melanogaster*)	rop^{G17} (P253S; P242S in Munc18-1)	No effect	Increased quantal size	49, 52
Class B (*D. melanogaster*)	Rop^{A19} (H302Y; H293Y in Munc18-1)	~20% inhibition	Unknown (not expressed)	49
Class C (Munc18-1)	yeast *sly1-20* (E532K; E466K in Munc18-1)	No effect	Increased frequency	49, 51
Class C (Munc18-1)	PKC phosphomimetic (S306E/S313E)	~90% inhibition	Decreased quantal size	48, 52
Class C (Munc18-1)	Cdk5 nonphosphorylatable (T574A)	~25% inhibition	Increased quantal size	50

The effect of expression of various mutant forms of Munc18-1 on *in vitro* binding (mode 1) to syntaxin and exocytosis monitored using carbon-fiber amperometry in adrenal chromaffin cells. Where relevant, the mutation in the original orthologue and in Munc18-1 is given. Abbreviations: *C. elegans, Caenorhabditis elegans;* D. melanogaster, Drosophila melanogaster; PKC, protein kinase C.

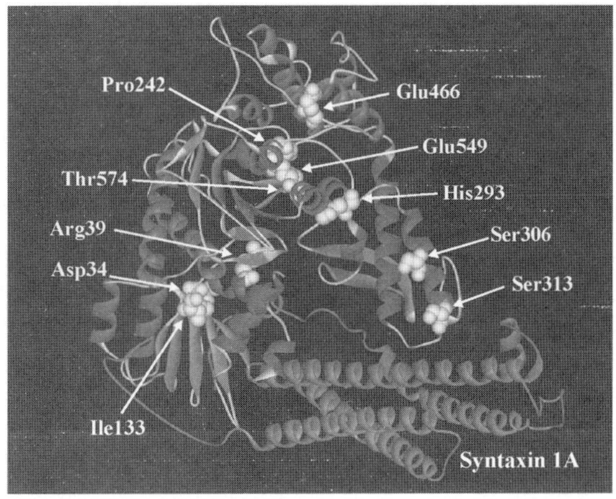

Figure 2. Mutations in Munc18-1 studied by the authors and their localization with the Munc18-1/syntaxin 1A structure. The structure that is shown is derived from the coordinates from the work of Misura *et al.*[28] The main amino acid chains are shown in ribbon format with syntaxin 1A in the lower part of the figure and the amino acids in which mutations have been studied are shown in space-fill format.

contact with the closed form of syntaxin in the structure of the Munc18-1/syntaxin 1A heterodimer.[28] This mutant revealed a potentially novel role for Munc18-1 in regulating the kinetics of very late-stage membrane fusion and thereby controlling quantal size.[47] These data also reinforced the generally held view that syntaxin binding is critical for the function of all SM proteins. Indeed, subsequent studies found that reciprocal mutations in syntaxin that reduce its binding affinity for Munc18-1 also resulted in changes in release kinetics and quantal size when overexpressed in chromaffin cells.[31] We interpreted the changes in release kinetics as being consistent with a late postdocking effect on fusion leading to changes in the kinetics of fusion pore expansion.[54] Additional analyses have shown significant effects of the expression of Munc18-1 mutants on the initial phase of release,[52] and it has recently been established that changes in the initial rate of release can indeed be directly accounted for by changes in fusion pore conductance.[55]

It is significant that some of the mutations that affected exocytosis had no effect on mode 1

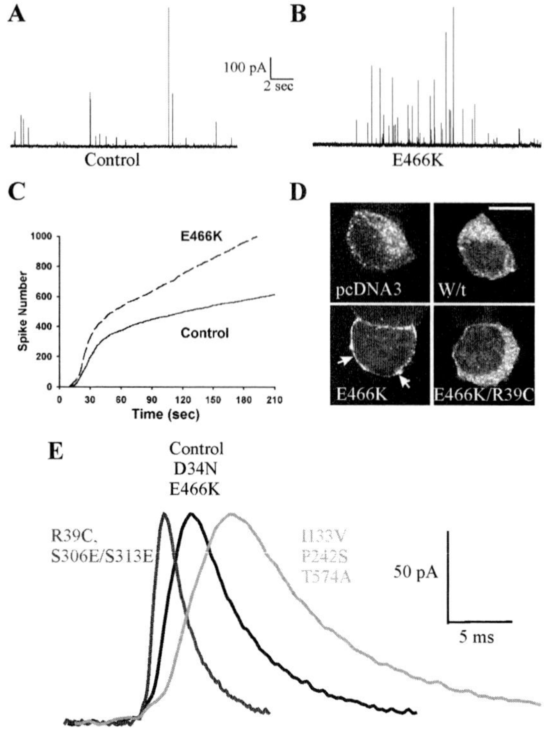

Figure 3. Examples of the modification of exocytosis in adrenal chromaffin cells following expression of Munc18-1 mutants. (**A**) and (**B**) amperometric traces during stimulation from chromaffin cells that are nontransfected (control) or transfected to express Munc18-1(E466K). (**C**) and (**A**) cumulative time course of amperometric spikes after stimulation for control and E466K-expressing cells. (**D**) localization of secretory vesicles expressing enhanced green fluorescent protein-Rab3A in PC12 cells transfected with control vector (pcDNA3), transfected to express wild-type Munc18-1, or constructs with the E466K or E466K/R39C mutations. Expression of only E466K resulted in redistribution of the secretory vesicle to the cell periphery (arrowhead). The scale bar represents 10 μm. (**E**) schematic examples of single amperometric traces to represent the average change in spike kinetics and area from chromaffin cells transfected with mutants that reduced (on the left) or increased (on the right) quantal size compared to a control spike in the center.

syntaxin binding[49,50] (Table 1) as expected from their position in the Munc18-1/closed syntaxin 1A complex. As these mutations had effects in our assay, and indeed those in classes A and B had already been shown to result in altered phenotypes in the organisms in which they were discovered, this suggests that some key functions of Munc18-1 proteins are independent of mode 1 syntaxin binding. Recent work on the yeast SM protein Sly1p has also implicated syntaxin-independent functions, and indeed a mutation in Sly1p that apparently prevented its interaction with its syntaxin did not prevent it from being functional in yeast.[56]

In addition to syntaxin, other Munc18-1 interacting proteins that were known include Doc2,[57] granuphilin,[58] Mint1, and Mint2.[59] The significance of these interactions for exocytosis and the molecular roles of these proteins in exocytosis are not well understood. The interaction with Doc2 has not been replicated by other labs,[49,60] and Doc2 may not be expressed in all secretory cells, including the adrenal chromaffin used in our studies.[61] The interaction with the Mint proteins may be more important and of wider significance as the Mints (also known as the X11 proteins) were found to modify the processing of the amyloid precursor protein.[62] Mints and Munc18-1 synergistically inhibit secretion of the β-amyloid peptides associated with Alzheimer's disease.[63,64] Interestingly, one Munc18-1 mutant with a syntaxin-independent functional effect (P242S) had an impairment in Mint binding.[49] It has also been suggested, but not shown directly, that a Munc18-1 mutant that cannot bind syntaxin modifies exocytosis via its interaction with Mint1.[60] The interaction of Munc18-1 with Mints remains to be explored in more detail. The interaction of Munc18-1 and the Rab 3/27 effector granuphilin appears to be important for the regulation of secretory vesicle docking at the plasma membrane.[65,66]

Interaction of Munc18-1 with Rab3 in Vesicle Docking

One mutation that we studied (E466K) was modeled on a yeast *sly1-20* mutant. A mutation in the Sly1p SM protein, which is involved in traffic from the Golgi complex, appeared to bypass the requirement for the Rab

GTPase Ypt1p.[67] It was later found, however, that its effect was to change the specificity of the transport step for Rab GTPases so that Ypt6p could replace Ypt1p.[68] The *sly1-20* mutation is in an exposed loop of domain III of Sly1p and we designed a mutation (E466K) within a similar exposed loop in domain III of Munc18-1in an attempt to mimic the *sly1-20* mutation.[49] Although sequences in this loop are not well conserved, peptide inhibition data have supported a functional role for this region of the Munc18 proteins.[69,70] Expression of Munc18-1(E466K) in adrenal chromaffin cells resulted in an increase in the number of exocytotic events[49] (Fig. 3A–C). This effect was dependent on the ability of Munc18-1(E466K) to undergo mode 1 binding to syntaxin as additional introduction of the R39C mutation that disrupts mode 1 binding prevented the effect.[51] Biochemical characterization of the E466K mutation suggested that it was a gain of function mutation that increased the direct binding of Munc18-1 to Rab3A.[51] Since vesicle docking stimulated by Rab3A was dependent on Munc18-1, a role of Rab3A interaction with Munc18-1in docking[71] would provide a satisfactory explanation for the effects of the mutant. Analysis of the distribution of secretory vesicles visualized by tagging them with enhanced green fluorescent protein (EGFP)-Rab3A indicated that expression of Munc18-1(E466K) increased the number of secretory vesicles at the cell periphery and this also required mode 1 syntaxin binding (Fig. 3D). These findings support the physiological significance of mode 1 syntaxin binding. Other recent work has also concluded that the mode 1 interaction by Munc18-1 is required for vesicle docking.[33] The data from E466K suggest that Munc18-1 can act as a linker between early Rab3-dependent docking (possibly involving Rab effectors, such as granuphilin) and later syntaxin-dependent steps in exocytosis. This interaction would, therefore, contribute to the ability of Munc18-1 to regulate both the early prefusion stages of vesicle recruitment/docking, leading to increased numbers of fusion events, as well as late stages of fusion pore expansion, leading to alterations in the kinetics of release and quantal size.

Physiological Regulation of Munc18-1/Syntaxin Interactions

The effect of expression of Munc18-1 R39C on exocytosis that we observed was very similar to one effect of protein kinase C (PKC) activation in these cells.[53] We, therefore, investigated a potential role for Munc18-1 as a mediator of the effect of PKC on the late stages of exocytosis.[48] To do this, we generated a Ser306-Glu/Ser313-Glu mutant based on the known PKC phosphorylation sites[72] and established that the mutant recombinant protein had a similarly reduced affinity for syntaxin as seen with the PKC-phosphorylated wild-type Munc18-1 *in vitro*. It was also subsequently shown that these phosphomimetic mutations reduce Munc18-1 interactions with syntaxin *in vivo*.[73] Overexpression of the phosphomimetic mutant in chromaffin cells caused a similar acceleration of exocytotic release kinetics and reduction in quantal size to that observed by PKC activation, suggesting that PKC regulates the late stages of exocytosis by phosphorylation of Munc18-1. In addition, activity-dependent phosphorylation of Munc18-1 on Ser313 in chromaffin cells and synaptosomes was demonstrated, consistent with the likely physiological relevance of phosphorylation on this amino acid.[74] In support of this conclusion, expression of the single phosphomimetic mutant S313D was sufficient to increase vesicle docking and to enhance the releasable vesicle pool in adrenal chromaffin cells.[75] An analysis of the role of PKC phosphorylation of Munc18-1 in synapses using a nonphosphorylatable mutant expressed on a Munc18-1 null background demonstrated that phosphorylation of Munc18-1 is necessary for phorbol ester-mediated potentiation of postsynaptic currents.[76] Surprisingly, it was found that phosphorylation of Munc18-1 was also essential for

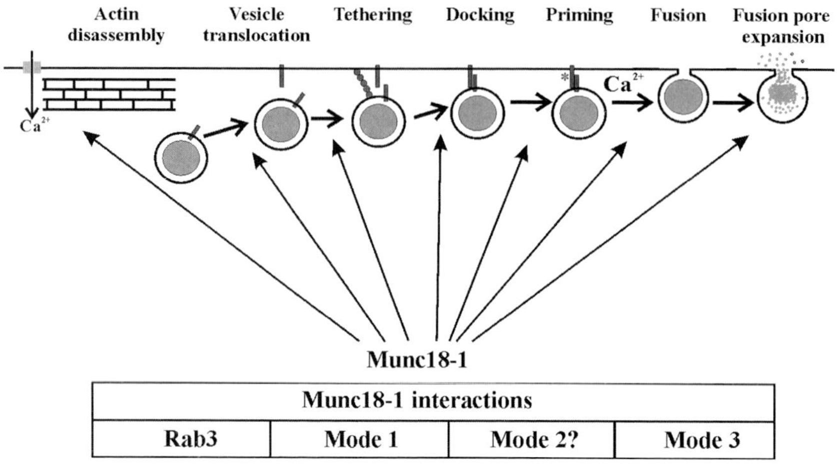

Figure 4. The multiple steps in regulated exocytosis and steps at which Munc18-1 has been suggested to act. The figure shows the functionally identifiable steps in exocytosis, which have all been suggested to be affected by Munc18-1. At the bottom is shown a progression in Munc18-1 protein interactions that may underlie progress through the exocytotic pathway.

a separate pathway involving direct activation of Munc13 by diacylglycerol, giving Munc18-1 phosphorylation a key role in synaptic plasticity.

Physiological regulation of syntaxin 1 interaction with Munc18-1 may also occur though nitrosylation of syntaxin, specifically on Cys145. This posttranslational modification reduces the affinity of mode 1 binding of syntaxin 1 to Munc18-1 and nitrosomimetic C145 mutants of syntaxin 1 (C145M and C145W) had a similarly reduced binding affinity. Furthermore, expression of the C145W mutant modified the kinetics of exocytosis in adrenal chromaffin cells,[77] suggesting that this modification could have a physiological role in switching off the mode 1 Munc18-1/syntaxin1 interaction and regulating exocytosis.

Where Does Munc18-1 Act in Exocytosis?

As described above, analysis of the effect of expression of Munc18-1 mutants suggests actions on both early and late stages of exocytosis. It had previously been debated whether the major function of Munc18 is in vesicle docking at the plasma membrane[9] or whether it also functions at the level of membrane fusion.[78] A role in docking is supported by observations that dense-core granule docking in the adrenal chromaffin cells of Munc18-1 knockout mice[18] and synaptic vesicle docking in *C. elegans* synapses[19] are impaired. In contrast, various observations suggest that SM proteins also function at a late, postdocking stage of the membrane fusion process. As noted above, yeast Sec1 will not bind efficiently to its cognate syntaxin in isolation but only when it is assembled into a binary or ternary SNARE complex.[37] This suggests that Sec1 functions at the heart of the membrane fusion process, acting at or after SNARE assembly. Indeed, recombinant Sec1 protein and Munc18-1 have been shown to accelerate SNARE-dependent membrane fusion between synthetic liposomes *in vitro*.[38,41] This effect required the mode 2 and mode 3 interactions with syntaxin and the SNARE complex. Evidence in support of a late postdocking role for SM proteins in mammalian exocytosis comes from analysis of synapses in the Munc18-1 knockout mouse. Here, there is no change in the proportion of morphologically docked synaptic vesicles, despite a total block in neurotransmitter release.[17] The work described above using Munc18-1 mutants has implicated Munc18-1 in late aspects of fusion and release, including in controlling fusion pore dynamics.[47] A simple explanation of these apparently contradictory findings is that SM proteins act in both early (docking) and late (fusion) stages in the exocytotic process. Study of an *in vitro* docking and fusion assay using giant artificial vesicles has shown two separable roles for Munc18-1 in adhesion and also in driving fusion.[79] Other work has implicated Munc18-1 in effects on actin disassembly, vesicle translocation, tethering, docking, and priming.[33,80–82] Increasingly, distinct roles for Munc18-1 are beginning to be distinguished through the analysis by various groups of the effects of specific Munc18-1 mutants. It now seems to be clear that Munc18-1 has roles in multiple steps in the pathway leading to exocytosis (Fig. 4).

Key Unanswered Questions

Various issues remain for a fuller understanding of the role of Munc18-1 and other SM proteins. It is not clear why the mode 1 interaction of Munc18-1 is physiologically important. This mode of interaction had not been observed for other SM proteins, leading to the idea that this may reflect some neuronal-specific function of Munc18-1, but more recent work has suggested its existence for other SM proteins.[43,83] It is unclear how Munc18-1 transitions through the distinct modes of interaction with syntaxin or to what extent these interaction modes overlap, although recent work suggests that they can.[43,84] Munc18-1 can associate with syntaxin 1 in the presence of SNAP-25 and VAMP on native membranes,[85] and there is some indication that the ability of Munc18-1 to undergo mode 2/3 interactions may be regulated by fatty acids.[44,86,87] The physiological significance of this regulation remains to be established. We need further information on the structural nature of the mode 3 interaction of Munc18-1 with the SNARE complex. Finally, a clear picture of the relationship between sequential modes of protein interaction with SNAREs and other proteins and the distinct steps of the exocytotic pathway will require considerably more experimental information.

Acknowledgments

Work described from the authors' laboratory has been supported by the Wellcome Trust, the Biotechnology and Biological Sciences Research Council, and the Medical Research Council.

Conflicts of Interest

The authors declare no conflicts of interest.

References

1. Burgoyne, R.D. & A. Morgan. 1995. Ca^{2+} and secretory vesicle dynamics. *Trends Neurosci.* **18:** 191–196.
2. Zucker, R.S. 1996. Exocytosis: A molecular and physiological perspective. *Neuron* **17:** 1049–1055.
3. Burgoyne, R.D. & A. Morgan. 2003. Secretory granule exocytosis. *Physiol. Rev.* **83:** 581–632.
4. Jahn, R., T. Lang & T.C. Sudhof. 2003. Membrane fusion. *Cell* **112:** 519–533.
5. Sutton, R.B. *et al.* 1998. Crystal structure of a SNARE complex involved in synaptic exocytosis at 2.4A resolution. *Nature* **395:** 347–353.
6. Montecucco, C. & G. Schiavo. 1994. Molecular mechanisms of action of bacterial protein toxins. *Mol. Microbiol.* **13:** 1–8.
7. Fernandez-Chacon, R. & T.C. Sudhof. 1999. Genetics of synaptic vesicle function: Toward the complete functional anatomy of an organelle. *Ann. Rev. Physiol.* **61:** 753–776.
8. Fernandez-Chacon, R. *et al.* 2001. Synaptotagmin I functions as a calcium regulator of release probability. *Nature* **410:** 41–49.
9. Toonen, R.F.G. & M. Verhage. 2003. Vesicle trafficking: pleasure and pain from SM genes. *Trends Cell Biol.* **13:** 177–186.
10. Zerial, M. & H. McBride. 2001. Rab proteins as membrane organizers. *Nat. Rev. Mol. Cell Biol.* **2:** 107–117.
11. Weber, T. *et al.* 1998. SNAREpins: minimal machinery for membrane fusion. *Cell* **92:** 759–772.

12. Zucker, R.S. 1999. Calcium- and activity-dependent synaptic plasticity. *Curr. Opin. Neurobiol.* **9:** 305–313.
13. Novick, P. & R. Schekman. 1979. Secretion and cell-surface growth are blocked in a temperature-sensitive mutant of saccharomyces cerevisiae. *Proc. Natl. Acad. Sci. USA* **76:** 1858–1862.
14. Hata, Y., C.A. Slaughter & T.C. Sudhof. 1993. Synaptic vesicle fusion complex contains unc-18 homologue bound to syntaxin. *Nature* **366:** 347–351.
15. Brenner, S. 1974. The genetics of caenorhabdits elegans. *Genetics* **77:** 71–94.
16. Pevsner, J., S.C. Hsu & R. Scheller. 1994. n-Sec1: A neural-specific syntaxin-binding protein. *Proc. Natl. Acad. Sci. USA* **91:** 1445–1449.
17. Verhage, M. *et al.* 2000. Synaptic assembly of the brain in the absence of neurotransmitter secretion. *Science* **287:** 864–869.
18. Voets, T. *et al.* 2001. Munc-18 promotes large dense-core vesicle docking. *Neuron* **31:** 581–591.
19. Weimer, R.M. *et al.* 2003. Defects in synaptix vesicle docking in unc-18 mutants. *Nature Neurosci.* **6:** 1023–1030.
20. Harrison, S.D. *et al.* 1994. Mutations in the drosophila rop gene suggest a function in general secretion and synaptic transmission. *Neuron* **13:** 555–566.
21. Schindelman, G. *et al.* 2006. Initiation of male sperm-transfer behavior in Caenorhabditis elegans requires input from the ventral nerve cord. *BMC Biol.* **4:** 26.
22. Fehr, C. *et al.* 2005. The syntaxin binding protein 1 gene (stxbp1) is a candidate for an ethanol preference drinking locus on mouse chromosome 2. *Alcohol. Clin. Exp. Res.* **29:** 708–720.
23. Pevsner, J. *et al.* 1994. Specificity and regulation of a synaptic vesicle docking complex. *Neuron* **13:** 353–361.
24. Wu, M.N. *et al.* 1999. Syntaxin 1A interacts with multiple exocytic proteins to regulate neurotransmitter release in vivo. *Neuron* **23:** 593–605.
25. Graham, M.E., A.W. Sudlow & R.D. Burgoyne. 1997. Evidence against an acute inhibitory role of nSec-1 (munc-18) in late steps of regulated exocytosis in chromaffin and PC12 cells. *J. Neurochem.* **69:** 2369–2377.
26. Dulubova, I. *et al.* 1999. A conformational switch in syntaxin during exocytosis: role of munc18. *EMBO J.* **18:** 4372–4382.
27. Munson, M. *et al.* 2000. Interactions within the yeast t-SNARE Sso1p that control SNARE complex assembly. *Nat. Struct. Biol.* **7:** 894–902.
28. Misura, K.M.S., R.H. Scheller & W.I. Weis. 2000. Three-dimensional structure of the neuronal-Sec1-syntaxin 1a complex. *Nature* **404:** 355–362.
29. Rowe, J. *et al.* 2001. Syntaxin 1A is delivered to the apical and basolateral domains of epithelial cells: the role of munc-18 proteins. *J. Cell Sci.* **114:** 3323–3332.
30. Medine, C.N. *et al.* 2007. Munc18-1 prevents the formation of ectopic SNARE complexes in living cells. *J. Cell Sci.* **120:** 4407–4415.
31. Graham, M.E., J.W. Barclay & R.D. Burgoyne. 2004. Syntaxin/Munc18 interactions in the late events during vesicle fusion and release in exocytosis. *J. Biol. Chem.* **279:** 32751–32760.
32. Toonen, R.F.G. *et al.* 2005. Munc18-1 stabilizes syntaxin 1, but is not essential for syntaxin 1 targeting and SNARE complex formation. *J. Neurochem.* **93:** 1393–1400.
33. Gulyas-Kovacs, A. *et al.* 2007. Munc18-1: sequential interactions with the fusion machinery stimulate vesicle docking and priming. *J. Neurosci.* **27:** 8676–8686.
34. Richmond, J.E., R.M. Weimer & E.M. Jorgensen. 2001. An open form of syntaxin bypasses the requirement for UNC-13 in vesicle priming. *Nature* **412:** 338–341.
35. Dulubova, I. *et al.* 2002. How Tlg2p/syntaxin 16 "snares" Vps45. *EMBO J.* **21:** 3620–3631.
36. Bracher, A. & W. Weissenhorn. 2002. Structural basis for the Golgi membrane recruitment of Sly1p by Sed5p. *EMBO J.* **21:** 6114–6124.
37. Carr, C.M. *et al.* 1999. Sec1p binds to SNARE complexes and concentrates at sites of secretion. *J. Cell Biol.* **146:** 333–344.
38. Scott, B.L. *et al.* 2004. Sec1p directly stimulates SNARE-mediated membrane fusion in vitro. *J. Cell Biol.* **167:** 75–85.
39. Carpp, L.N. *et al.* 2006. The Sec1p/Munc18 protein Vps45p binds its cognate SNARE proteins via two distinct modes. *J. Cell Biol.* **173:** 927–936.
40. Rickman, C. *et al.* 2007. Functionally and spatially distinct modes of MUNC18-syntaxin 1 interaction. *J. Biol. Chem.* **282:** 12097–12103.
41. Shen, J. *et al.* 2007. Selective activation of cognate SNAREpins by Sec1/Munc18 proteins. *Cell* **128:** 183–195.
42. Dulubova, I. *et al.* 2007. Munc18-1 binds directly to the neuronal SNARE complex. *Proc. Natl. Acad. Sci. USA* **104:** 2697–2702.
43. Burkhardt, P. *et al.* 2008. Munc18a controls SNARE assembly through its interaction with the syntaxin N-peptide. *EMBO J.* **27:** 923–933.
44. Rickman, C. & B. Davletov. 2005. Arachidonic acid allows SNARE complex formation in the presence of Munc18. *Chem. Biol.* **12:** 545–553.
45. Burgoyne, R.D. & A. Morgan. 2007. Membrane trafficking: three steps to fusion. *Current Biol.* **17:** R255–258.
46. Hu, S.H. *et al.* 2007. Structure of the Munc18c/Syntaxin4 N-peptide complex defines universal features of the N-peptide binding mode of Sec1/

Munc18 proteins. *Proc. Natl. Acad. Sci. USA* **104:** 8773–8778.
47. Fisher, R.J., J. Pevsner & R.D. Burgoyne. 2001. Control of fusion pore dynamics during exocytosis by Munc18. *Science* **291:** 875–878.
48. Barclay, J.W. *et al.* 2003. Phosphorylation of Munc18 by protein kinase C regulates the kinetics of exocytosis. *J. Biol. Chem.* **278:** 10538–10545.
49. Ciufo, L.F. *et al.* 2005. Munc18-1 regulates early and late stages of exocytosis via syntaxin independent protein interactions. *Mol. Biol. Cell.* **16:** 470–482.
50. Barclay, J.W. *et al.* 2004. Regulation of the fusion pore conductance during exocytosis by cyclin-dependent kinase 5. *J. Biol. Chem.* **279:** 41495–41503.
51. Graham, M.E. *et al.* 2008. A gain-of-function mutant of Munc18-1 stimulates secretory granule recruitment and exocytosis and reveals a direct interaction of Munc18-1 with Rab3. *Biochem. J.* **409:** 407–416.
52. Barclay, J.W. 2008. Munc-18-1 regulates the initial release rate of exocytosis. *Biophys. J.* **94:** 1084–1093.
53. Graham, M.E., R.J. Fisher & R.D. Burgoyne. 2000. Measurement of exocytosis by amperometry in adrenal chromaffin cells: effects of clostridial neurotoxins and activation of protein kinase C on fusion pore kinetics. *Biochimie* **82:** 469–479.
54. Burgoyne, R.D. & J.W. Barclay. 2002. Splitting the quantum: regulation of quantal release during vesicle fusion. *Trends Neurosci.* **26:** 176–178.
55. Neco, P. *et al.* 2008. Myosin II contributes to fusion pore expansion during exocytosis. *J. Biol. Chem.* **283**: 10949–57.
56. Peng, R. & D. Gallwitz. 2004. Multiple SNARE interactions of an SM protein: Sed5p/Sly1p binding is dispensible for transport. *EMBO J.* **23:** 3939–3949.
57. Verhage, M. *et al.* 1997. Doc2 proteins in rat brain: complementary distribution and proposed function as vesicular adapter proteins in early stages of secretion. *Neuron* **18:** 453–461.
58. Coppola, T. *et al.* 2002. Pancreatic b-cell protein granuphilin binds rab3 and munc18 and controls exocytosis. *Mol. Biol. Cell.* **13:** 1906–1915.
59. Okamoto, M. & T.C. Sudhof. 1997. Mints, munc18-interacting proteins in synaptic vesicle exocytosis. *J. Biol. Chem.* **272:** 31459–31464.
60. Schutz, D. *et al.* 2005. A dual function for Munc-18 in exocytosis of PC12 cells. *Eur. J. Neurosci.* **21:** 2419–2432.
61. Duncan, R.R. *et al.* 2000. Is double C2 protein (DOC2) expressed in bovine adrenal medulla? A commercial anti-DOC2 monoclonal antibody recognizes a major bovine mitochondrial antigen. *Biochem. J.* **351:** 33–37.
62. Borg, J.P. *et al.* 1998. The X11alpha protein slows cellular amyloid precursor protein processing and reduces Abeta40 and Abeta42 secretion. *J. Biol. Chem.* **273:** 14761–14766.
63. Ho, C.S. *et al.* 2002. Synergistic effects of munc18a and X11 proteins on amyloid precursor protein metabolism. *J. Biol. Chem.* **277:** 27021–27028.
64. Ikin, A.F. *et al.* 2007. Evidence against roles for phorbol binding protein Munc13-1, ADAM adaptor Eve-1, or vesicle trafficking phosphoproteins Munc18 or NSF as phospho-state-sensitive modulators of phorbol/PKC-activated Alzheimer APP ectodomain shedding. *Mol. Neurodegener.* **2:** 23.
65. Torii, S. *et al.* 2004. Rab27 effector granuphilin promotes the plasma membrane targeting of insulin granules via interaction with syntaxin 1a. *J .Biol. Chem.* **279:** 22532–22538.
66. Tomas, A. *et al.* 2008. Munc 18-1 and granuphilin collaborate during insulin granule exocytosis. *Traffic* **9:** 813–832.
67. Ossig, R. *et al.* 1991. The yeast SLY gene products, suppressors of defects in the essential GTP-binding YPT1 protein, may act in endoplasmic reticulum to Golgi transport. *Mol. Cell. Biol.* **11:** 2980–2993.
68. Ballew, N., Y. Liu & C. Barlowe. 2005. A Rab requirement is not bypassed in SLY1-20 suppression. *Mol. Biol. Cell.* **16:** 1839–1849.
69. Dresbach, T. *et al.* 1998. A neuronal Sec1 homolog regulates neurotransmitter release at the squid giant synapse. *J. Neurosci.* **18:** 2923–2932.
70. Thurmond, D.C. *et al.* 2000. Munc18c function is required for insulin-stimulated plasma membrane fusion of GLUT4 and insulin-responsive amino peptidase storage vesicles. *Mol. Cell. Biol.* **20:** 379–388.
71. van Weering, J.R., R.F. Toonen & M. Verhage. 2007. The role of Rab3a in secretory vesicle docking requires association/dissociation of guanidine phosphates and Munc18-1. *PLoS ONE* **2:** e616.
72. Fujita, Y. *et al.* 1996. Phosphorylation of Munc-18/n-Sec1/rbSec1 by protein kinase C. *J. Biol. Chem.* **271:** 7265–7268.
73. Liu, J. *et al.* 2004. Fluorescence resonance energy transfer reports properties of syntaxin 1A interaction with Munc18-1 in vivo. *J. Biol. Chem.* **279:** 55924–55936.
74. Craig, T.J., G.J.O. Evans & A. Morgan. 2003. Physiological regulation of Munc18/nSec1 phosphorylation on serine-313. *J. Neurochem.* **86:** 1450–1457.
75. Nili, U. *et al.* 2006. Munc18-1 phosphorylation by protein kinase C potentiates vesicle pool replenishment in bovine chromaffin cells. *Neuroscience* **143:** 487–500.
76. Wierda, K.D. *et al.* 2007. Interdependence of PKC-dependent and PKC-independent pathways for presynaptic plasticity. *Neuron* **54:** 275–290.

77. Palmer, Z.J. *et al*. 2008. S-nitrosylation of syntaxin1 at Cys145 is a regulatory switch controlling Munc18-1 binding. **413:** 479–491.
78. Burgoyne, R.D. *et al*. 2001. Control of membrane fusion dynamics during regulated exocytosis. *Biochem. Soc. Trans.* **29:** 467–472.
79. Tareste, D. *et al*. 2008. SNAREpin/Munc18 promotes adhesion and fusion of large vesicles to giant membranes. *Proc. Natl. Acad. Sci. USA* **105:** 2380–2385.
80. Toonen, R.F. *et al*. 2006. Dissecting docking and tethering of secretory vesicles at the target membrane. *EMBO J.* **25:** 3725–3737.
81. Toonen, R.F. *et al*. 2006. Munc18-1 expression levels control synapse recovery by regulating readily releasable pool size. *Proc. Natl. Acad. Sci. USA* **103:** 18332–18337.
82. Toonen, R.F. & M. Verhage. 2007. Munc18-1 in secretion: lonely Munc joins SNARE team and takes control. *Trends Neurosci.* **30:** 564–572.
83. D'Andrea-Merrins, M. *et al*. 2007. Munc18c interaction with syntaxin 4 monomers and SNARE complex intermediates in GLUT4 vesicle trafficking. *J. Biol. Chem.* **282:** 16553–16566.
84. Khvotchev, M. *et al*. 2007. Dual modes of Munc18-1/SNARE interactions are coupled by functionally critical binding to syntaxin-1 N terminus. *J. Neurosci.* **27:** 12147–12155.
85. Zilly, F.E. *et al*. 2006. Munc18-bound syntaxin readily forms SNARE complexes with synaptobrevin in native plasma membranes. *PLoS Biol.* **4:** e330.
86. Connell, E. *et al*. 2007. Mechanism of arachidonic acid action on syntaxin-Munc18. *EMBO Reports* **8:** 414–419.
87. Latham, C.F. *et al*. 2007. Arachidonic acid potentiates exocytosis and allows neuronal SNARE complex to interact with Munc18a. *J. Neurochem.* **100:** 1543–1554.

Voltage-gated Potassium Channel as a Facilitator of Exocytosis

Lori Feinshreiber,[a] Dafna Singer-Lahat,[a] Uri Ashery,[b] and Ilana Lotan[a]

[a]*Department of Physiology and Pharmacology, Sackler Faculty of Medicine, Tel Aviv University, Tel-Aviv, Israel*

[b]*Department of Neurobiochemistry, Life Science Institute, Tel Aviv University, Tel-Aviv, Israel*

Voltage-gated ion channels are well characterized for their function in excitability signals. Accumulating studies, however, have established an ion-independent function for the major classes of ion channels in cellular signaling. During the last few years we established a novel role for Kv2.1, a voltage-gated potassium (Kv) channel, classically known for its role of repolarizing the membrane potential, in facilitation of exocytosis. Kv2.1 induces facilitation of depolarization-induced release through its direct interaction with syntaxin, a protein component of the exocytotic machinery, independently of the potassium ion flow through the channel's pore. Here, we review our recent studies, further characterize the phenomena (using chromaffin cells and carbon fiber amperometry), and suggest plausible mechanisms that can underlie this facilitation of release.

Key words: **exocytosis; Kv2.1 channel; chromaffin cells; PC12 cells; syntaxin; amperometry; large dense-core vesicles release; pore-independent function**

Nonconducting Functions of Voltage-gated Ion Channel

The primary function of voltage-gated ion channels in the nervous system is to generate the wide variety of different firing patterns that are required for processing sensory information and generating motor outputs.[1]

However, recent works have demonstrated that many ion channels can themselves directly influence biochemical events in ways that do not directly depend on their function as ion channels. Such control over cellular signaling has been reported for each of the major classes of ion channels that influence excitability, including the sodium, calcium, and potassium channels (as reviewed in Ref. 1).

Voltage-gated potassium (Kv) channels have classically been viewed to participate in exocytosis solely as inhibitors of neurotransmitter, neuropeptide, and hormone secretion through their ion-conducting function to repolarize membrane potential[2] in presynaptic terminals[3,4] and endocrine cells.[5] Among the cellular behaviors that Kv channels regulate independently of their conducting functions are mitogen-activated protein (MAP) kinase signaling [EAG, (Kv10.1) channels], cell proliferations (EAG, Kv11.1, Kv1.3 channels), and tyrosine phosphorylation (Kv1.3) (as reviewed in Ref. 1). Recently, a novel role for a Kv channel, Kv2.1, as facilitator of dense-core vesicle (DCV)-mediated secretion through a pore-independent function has been demonstrated (see in the following and Refs. 6, 7).

Kv Channels and SNARE Proteins

In secretory cells, such as neurons and neuroendocrine cells, Ca^{2+} influx triggers secretory vesicle exocytosis, releasing neurotransmitters or hormones.[8] Part of the exocytocic

Address for correspondence: Professor Ilana Lotan, Department of Physiology and Pharmacology, Sackler Faculty of Medicine, Tel Aviv University, 69978, Tel-Aviv, Israel. Voice: +972-3-6409863; fax: +972-3-6409113. ilotan@post.tau.ac.il

Mechanisms of Exocytosis: Ann. N.Y. Acad. Sci. 1152: 87–92 (2009).
doi: 10.1111/j.1749-6632.2008.03997.x

machinery are the SNARE (soluble N-ethylmaleimide-sensitive factor attachment protein receptor) proteins, which share a cytosolic SNARE motif domain essential for their interaction. Assembly of vesicle SNARE, VAMP, with the plasma membrane t-SNAREs, SNAP25 (synaptosomal-associated protein of 25 kDa), and syntaxin-1A, forms the SNARE complex, which initiates the fusion of the vesicle into the plasma membrane.

Physical interaction between the SNARE proteins, mainly syntaxin, and Kv channels (Kv1.1, Kv2.1) have been shown to modulate the gating function of the channels,[9–12] suggesting that SNARE proteins do not only participate in exocytosis *per se* but also regulate excitability by modulating the ion channels involved in secretion. Accumulating data from different cell types, including rat brain synaptosome,[9] *Xenopus* oocytes,[10] human embryonic kidney (HEK) cells,[13] β-cells,[13] cardiac mycocytes,[14] and esophageal smooth muscles,[15] have established a new paradigm that SNARE proteins can directly modulate the Kv channel superfamily.[8]

Here, however, we will focus on the physical aspect of the syntaxin–Kv channel interaction with regard to exocytosis rather than on the implications of the gating modulations caused by this interaction.

Kv2.1 Facilitation of Exocytosis by Interaction with Syntaxin

The Kv2.1 channel is abundantly expressed in the soma and dendrites of neurons[16–21] where it underlies most of the delayed rectifier current and could influence the release of neuropeptides and neurotrophins. Additionally, Kv2.1 channel is also expressed in neuroendocrine cells[22–24] where it is well positioned to regulate hormone release.

Earlier works have shown that Kv2.1 physically and dynamically interacts with syntaxin and with SNAP25 in PC12 cells, oocytes, β-cells, and *in vitro*.[13,25,26] The Kv2.1 syntaxin-interacting site is located at the cytoplasmatic C terminus between aa 411–522 (termed C1a domain).

Our main interest was to examine the significance of this interaction on modulation of secretion. First, we tested the role of endogenous Kv2.1 in modulation of secretion in cracked PC12 cells, which allowed penetration of peptides and can undergo MgATP priming and Ca^{2+} triggering to release norepinephrine (NE) stored in DCVs. Impairment of the Kv2.1–syntaxin interaction by introducing Kv2.1 syntaxin-binding peptide, Kv2.1-C1, to the cells significantly reduced NE release. Since secretion was measured from cells with permeable plasma membranes, the K^+ ion efflux and membrane potential changes were irrelevant. Hence, it was suggested that Kv2.1 enhances secretion under physiological condition in neuroendocrine cells, independently of its conducting function, through physical interaction with syntaxin (Table 1).[6]

Next, we examined the effect of overexpressed Kv2.1 on secretion from live PC12 cells by measuring secretion of GFP-tagged atrial natriuretic factor (ANF) after Ca^{2+} elevation. Overexpression of Kv2.1 enhanced the release of ANF. A similar enhancement effect was shown for overexpressed mutant nonconducting channel ($Kv2.1^{W365C/Y380T}$),[27] while mutant channel that lacks the syntaxin-binding site (Kv2.1ΔC1a) not only failed to enhance release but also slightly inhibited it. These findings have strengthened the notion that Kv2.1 channels enhance secretion from PC12 by binding a key component of the fusion machinery, syntaxin, through cytoplasmic domain-mediated protein–protein interaction and have led to the suggestion of a novel nonconduction role for the channel (Table 1).[7]

Enhancement of Release by Kv2.1 in Bovine Chromaffin Cells

Here, in an attempt to further characterize this phenomenon and establish Kv2.1 as

TABLE 1. Kv2.1 Effect on Release

Cell type	PC12 (Kv2.1 down-regulation)[6]		PC12 (Kv2.1 overexpression)[7]						Chromaffin (Kv2.1 overexpression)			
assay	NE release (%)		ANF release (%)						CA release (%)			
	Control (GST)	GST-Kv2.1C1	Control (GFP)	Kv2.1	Control (GFP)	$Kv2.1^{W365C/Y380T}$	Control (GFP)	Kv2.1ΔC1a	Control (EGFP)	Kv2.1	Control (EGFP)	$Kv2.1^{W365C/Y380T}$
	100	41*	100	123*	100	134*	100	86*	100	186*	100	170*

Abbreviations: ANF, atrial natriuretic factor; EGFP, enhanced green fluorescent protein; GFP, green fluorescent protein; GST, glutathione-S-transferase; NE, norepinephrine.

$^*P < 0.05$.

a common protein in the exocytosis machinery, in addition to its role in plasma membrane hyperpolarization, we overexpressed the channel in bovine chromaffin cells and measured large dense-core vesicle release by amperometry, which enables a highly sensitive measurement and high temporal resolution of secretion.[28]

Bovine chromaffin cells were infected with pSFV1-Kv2.1-IRES-EGFP construct (Kv2.1-expressing cells) or with pSFV1-EGFP alone (control cells). Cells were stimulated by 10-s focal application of high-K^+ solution and charge released from those cells was recorded for 100 s. Five seconds of basal secretion was recorded before stimulation (Fig. 1). Release of catecholamines (CA) in Kv2.1-expressing cells was significantly increased compared to control cells. Notably, the initial secretion during the first 5 s of the stimulation was only slightly larger and became significantly larger only at 10 s (Fig. 1A). Further analysis of secretion data in response to four consecutive stimulations in each cell revealed that the presence of the channel accelerated secretion compared to control cells (Fig. 1B). Expression of mutant nonconducting channel $Kv2.1^{W365C/Y380T}$ revealed a similar extent of enhancement (Fig. 1C), suggesting that a functional pore was not required for the Kv2.1-mediated facilitation of depolarization-evoked release.

These results established the nonconducting role of Kv2.1 as a facilitator protein of exocytosis in chromaffin cells and have characterized

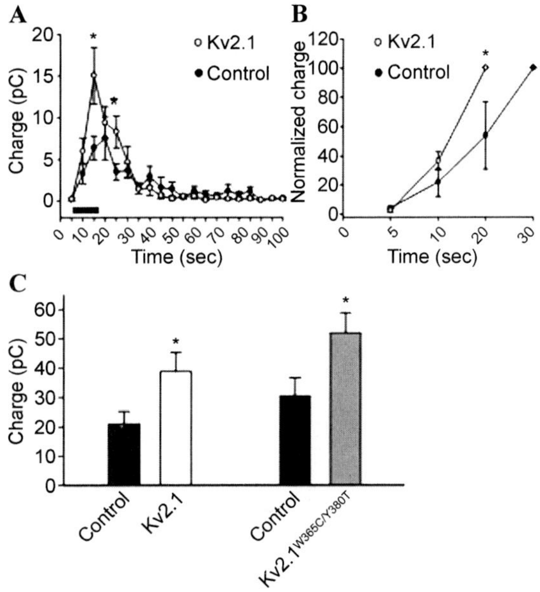

Figure 1. Kv2.1 enhances release evoked by K^+-induced depolarizations in chromaffin cells . (**A**) Average charge released from cells infected with Semliki Forest virus coding for IRES-EGFP (control cell; filled) or with virus coding for Kv2.1-IRES-EGFP (Kv2.1 cell; empty) in response to a 10-s focal application of high-K^+ solution (bar). Accumulated charge during 5 s was plotted. (**B**) Superimposed normalized and averaged charges released from control and Kv2.1-expressing cells. Cells were stimulated by four consecutive 10-s applications of high-K^+ solution with 2-min intervals (first application shown in A). Average values of charge released in control ($n > 9$) and Kv2.1-expressing ($n > 19$) cells were normalized to maximal value in each stimulation. (**C**) Averaged cumulative charge released during 20 s after onset of stimulation in control ($n = 22$; black bars) versus Kv2.1- ($n = 27$; empty bar) expressing cell and in control ($n = 11$) versus $Kv2.1^{W365C/Y380T}$- ($n = 21$; gray bar) expressing cell. Data are shown as mean ± SEM; $^*P < 0.05$.

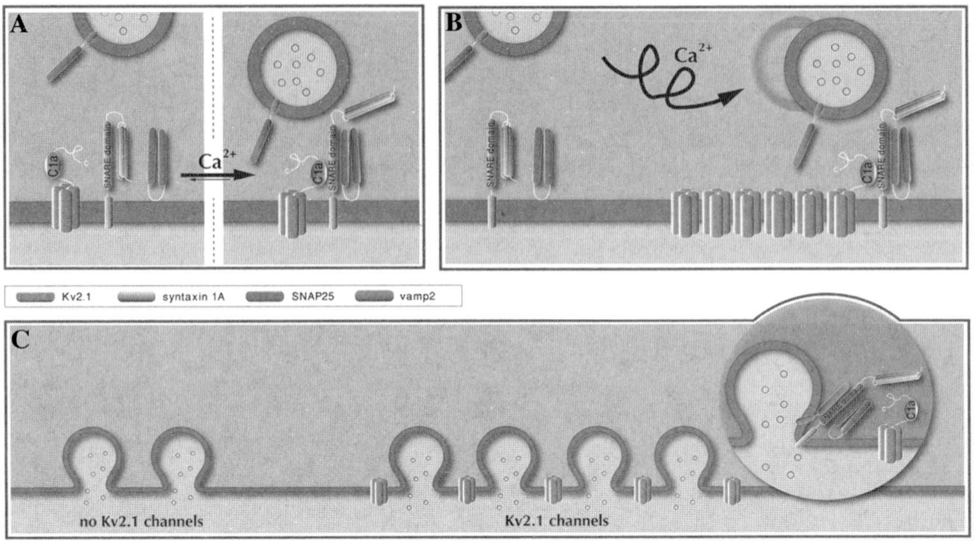

Figure 2. Plausible models for Kv2.1 enhancement of vesicle exocytosis in response to stimulation. (**A**) Ca^{2+}-dependent Kv2.1 binding to syntaxin stabilizes the t-SNARE (soluble *N*-ethylmaleimide-sensitive factor attachment protein receptor) complex. (**B**) Ca^{2+}-dependent relocation of vesicles to a high-density Kv2.1 cluster, which is a favorable site for fusion. (**C**) Both scenarios lead to favorable formation of ternary SNARE complexes at sites with Kv2.1 channels, resulting in an increased charge released. (Note that it has been recently shown that the ternary SNARE complex does not interact with Kv2.1.[45])

the time resolution of this enhancement effect during prolonged stimulation.

Plausible Models for Kv2.1 Facilitation of Release

As summarized in Table 1, Kv2.1 enhances secretion from neuroendocrine cells independently of its ion-conducting function via a physical interaction with syntaxin, a component of the exocytosis machinery. In order to put forward plausible models that can underlie this effect, several considerations were made based on recent data. Interaction of Kv2.1 C1a domain with syntaxin has been shown to underlie the Kv2.1-induced enhancement of neuropeptide release.[7] Kv2.1 has been shown to interact preferentially with the open conformation of syntaxin,[29] thus stabilizing this form of syntaxin, which has been implicated in vesicle docking[30] and priming (the latter via formation of binary t-SNARE complexes).[31,32]

We suggest two scenarios that might lead to increased charge released by Kv2.1 (Fig. 2). In the first scenario (Fig. 2A), the binding of Kv2.1 to open syntaxin is suggested to be Ca^{2+} dependent, based on the previous demonstration of a dynamic association between syntaxin and Kv2.1, which was significantly enhanced following Ca^{2+} triggering of primed cracked PC12 cells.[7] Interestingly, phosphorylation of sites within or near the C1a domain has been suggested to be linked to the Kv2.1-syntaxin interaction.[33] Moreover, dephosphorylation of these sites through a Ca^{2+}/calcineurin-dependent mechanism[21,34] has been shown to result in effects on activation and inactivation gating (hyperpolarizing shifts[35]) that resemble those of syntaxin binding.[26] Thus, Ca^{2+}-dependent dephosphorylation of Kv2.1 could be associated with syntaxin binding. Indeed, Ca^{2+}-dependent phosphorylation of some other proteins has been shown to modulate exocytosis. For example, protein kinase C (PKC) activation was shown to modulate exocytosis by phosphorylation of SNAP-25 at a specific amino acid residue in response to depolarizing stimuli in endocrine cells and neurons,[36–39] and phosphorylation of

Munc18-1 by PKC potentiates vesicle-pool replenishment following a depleting stimulation at a post-docking stage in chromaffin cells.[40] An alternative, but not mutually exclusive, scenario that could account for the Kv2.1 action relates to the recently demonstrated tendency of late-fusing (after 8 s of stimulation) chromaffin LDCVs. It can be suggested that these late-fusing vesicles are not in a favorable location for fusion at the time of stimulation and that they relocate just before fusion to a more favorable site.[41] Similar LDCV motions have been shown to be Ca^{2+} dependent in chromaffin cells[42] and in other tissues.[43,44] Notably, the time scale of relocation, on the order of 8 s, is compatible with the delayed development of the Kv2.1 facilitation of release (Fig. 1A). One possible fusion-favorable site might be an area with high-density Kv2.1 clusters[33] in which docking and/or t-SNARE complex formation are relatively intensified through the interaction of open syntaxin with Kv2.1 channels. Taken together, this scenario suggests relocation of LDCVs from a fusion-unfavorable location to a region of Kv2.1 clusters (Fig. 2B). Both scenarios lead to favorable formation of ternary SNARE complexes at sites with Kv2.1 channels and an increase in charge released (Fig. 2C).

Conflicts of Interest

The authors declare no conflicts of interest.

References

1. Kaczmarek, L.K. 2006. Non-conducting functions of voltage-gated ion channels. *Nat Rev. Neurosci.* **7:** 761–771.
2. Dodson, P.D. & I.D. Forsythe. 2004. Presynaptic K+ channels: electrifying regulators of synaptic terminal excitability. *Trends Neurosci.* **27:** 7–210.
3. Meir, A. *et al.* 1999. Ion channels in presynaptic nerve terminals and control of transmitter release. *Physiol. Rev.* **79:** 1019–1088.
4. Roeper, J. & O. Pongs. 1996. Presynaptic potassium channels. *Curr. Opin. Neurobiol.* **6:** 338–341.
5. MacDonald, P.E. *et al.* 2002. Inhibition of Kv2.1 voltage-dependent K+ channels in pancreatic beta-cells enhances glucose-dependent insulin secretion. *J. Biol. Chem.* **277:** 44938–44945.
6. Singer-Lahat, D., D. Chikvashvili & I. Lotan. 2008. Direct interaction of endogenous kv channels with syntaxin enhances exocytosis by neuroendocrine cells. *PLoS ONE* **3:** e1381.
7. Singer-Lahat, D. *et al.* 2007. K+ channel facilitation of exocytosis by dynamic interaction with syntaxin. *J. Neurosci.* **27:** 1651–1658.
8. Leung, Y.M. *et al.* 2007. SNAREing voltage-gated K +and ATP-sensitive K+ channels: tuning beta-cell excitability with syntaxin-1A and other exocytotic proteins. *Endocr. Rev.* **28:** 653–663.
9. Fili, O. *et al.* 2001. Direct interaction of a brain voltage-gated K+ channel with syntaxin 1A: functional impact on channel gating. *J. Neurosci.* **21:** 1964–1974.
10. Michaelevski, I. *et al.* 2002. Modulation of a brain voltage-gated K+ channel by syntaxin 1A requires the physical interaction of Gbetagamma with the channel. *J. Biol. Chem.* **277:** 34909–34917.
11. Cui, N. *et al.* 2004. H3 domain of syntaxin 1A inhibits KATP channels by its actions on the sulfonylurea receptor 1 nucleotide-binding folds-1 and -2. *J. Biol. Chem.* **279:** 53259–53265.
12. Pasyk, E.A. *et al.* 2004. Syntaxin-1A binds the nucleotide-binding folds of sulphonylurea receptor 1 to regulate the KATP channel. *J. Biol. Chem.* **279:** 4234–4240.
13. Leung, Y.M. *et al.* 2003. Syntaxin 1A binds to the cytoplasmic C terminus of Kv2.1 to regulate channel gating and trafficking. *J. Biol. Chem.* **278:** 17532–17538.
14. Yamakawa, T. *et al.* 2007. Interaction of syntaxin 1A with the N-terminus of Kv4.2 modulates channel surface expression and gating. *Biochemistry* **46:** 10942–10949.
15. Ji, J. *et al.* 2002. SNAP-25, a SNARE protein, inhibits two types of K channels in esophageal smooth muscle. *Gastroenterology* **122:** 994–1006.
16. Trimmer, J.S. 1993. Expression of Kv2.1 delayed rectifier K+ channel isoforms in the developing rat brain. *FEBS Lett.* **324:** 205–210.
17. Hwang, P.M. *et al.* 1993. Contrasting immunohistochemical localizations in rat brain of two novel K+ channels of the Shab subfamily. *J. Neurosci.* **13:** 1569–1576.
18. Rhodes, K.J. *et al.* 1995. Association and colocalization of K+ channel alpha- and beta-subunit polypeptides in rat brain. *J. Neurosci.* **15:** 5360–5371.
19. Du, J. *et al.* 1998. The K+ channel, Kv2.1, is apposed to astrocytic processes and is associated with inhibitory postsynaptic membranes in hippocampal and cortical principal neurons and inhibitory interneurons. *Neuroscience* **84:** 37–48.

20. Murakoshi, H. & J.S. Trimmer. 1999. Identification of the Kv2.1 K+ channel as a major component of the delayed rectifier K+ current in rat hippocampal neurons. *J. Neurosci.* **19:** 1728–1735.
21. Misonou, H. *et al.* 2004. Regulation of ion channel localization and phosphorylation by neuronal activity. *Nat. Neurosci.* **7:** 711–718.
22. MacDonald, P.E. *et al.* 2001. Members of the Kv1 and Kv2 voltage-dependent K(+) channel families regulate insulin secretion. *Mol. Endocrinol.* **15:** 1423–1435.
23. Yan, L. *et al.* 2004. Expression of voltage-gated potassium channels in human and rhesus pancreatic islets. *Diabetes* **53:** 597–607.
24. Wolf-Goldberg, T. *et al.* 2006. Target soluble N-ethylmaleimide-sensitive factor attachment protein receptors (t-SNAREs) differently regulate activation and inactivation gating of Kv2.2 and Kv2.1: Implications on pancreatic islet cell Kv channels. *Mol. Pharmacol.* **70:** 818–828.
25. MacDonald, P.E. *et al.* 2002. Synaptosome-associated protein of 25 kilodaltons modulates Kv2.1 voltage-dependent K(+) channels in neuroendocrine islet beta-cells through an interaction with the channel N terminus. *Mol. Endocrinol.* **16:** 2452–2461.
26. Michaelevski, I. *et al.* 2003. Direct interaction of t-SNAREs with the Kv2.1 channel: Modal regulation of channel activation and inactivation gating. *J. Biol. Chem.* **273:** 34320–34330.
27. Malin, S.A. & J.M. Nerbonne. 2002. Delayed rectifier K+ currents, IK, are encoded by Kv2 alpha -subunits and regulate tonic firing in mammalian sympathetic neurons. *J. Neurosci.* **22:** 10094–10105.
28. Yizhar, O. *et al.* 2004. Tomosyn inhibits priming of large dense-core vesicles in a calcium-dependent manner. *Proc. Natl. Acad. Sci. USA* **101:** 2578–2583.
29. Leung, Y.M. *et al.* 2005. Open form of syntaxin-1A is a more potent inhibitor than wild-type syntaxin-1A of Kv2.1 channels. *Biochem. J.* **387:** 195–202.
30. Hammarlund, M. *et al.* 2007. Open syntaxin docks synaptic vesicles. *PLoS Biol.* **5:** e198.
31. Fasshauer, D. *et al.* 1997. Structural changes are associated with soluble N-ethylmaleimide-sensitive fusion protein attachment protein receptor complex formation. *J. Biol. Chem.* **272:** 28036–28041.
32. An, S.J. & W. Almers. 2004. Tracking SNARE complex formation in live endocrine cells. *Science* **306:** 1042–1046.
33. Mohapatra, D.P., H. Vacher & J.S. Trimmer. 2007. The surprising catch of a voltage-gated potassium channel in a neuronal SNARE. *Sci. STKE* **2007:** pe37.
34. Misonou, H. *et al.* 2005. Calcium- and metabolic state-dependent modulation of the voltage-dependent Kv2.1 channel regulates neuronal excitability in response to ischemia. *J. Neurosci.* **25:** 11184–11193.
35. Park, K.S. *et al.* 2006. Graded regulation of the Kv2.1 potassium channel by variable phosphorylation. *Science* **313:** 976–979.
36. Shimazaki, Y. *et al.* 1996. Phosphorylation of 25-kDa synaptosome-associated protein. Possible involvement in protein kinase C-mediated regulation of neurotransmitter release. *J. Biol. Chem.* **271:** 14548–14553.
37. Gonelle-Gispert, C. *et al.* 2002. Phosphorylation of SNAP-25 on serine-187 is induced by secretagogues in insulin-secreting cells, but is not correlated with insulin secretion. *Biochem. J.* **368:** 223–232.
38. Nagy, G. *et al.* 2002. Protein kinase C-dependent phosphorylation of synaptosome-associated protein of 25 kDa at Ser187 potentiates vesicle recruitment. *J. Neurosci.* **22:** 9278–9286.
39. Pozzi, D. *et al.* 2008. Activity-dependent phosphorylation of Ser187 is required for SNAP-25-negative modulation of neuronal voltage-gated calcium channels. *Proc. Natl. Acad. Sci. USA* **105:** 323–328.
40. Nili, U. *et al.* 2006. Munc18-1 phosphorylation by protein kinase C potentiates vesicle pool replenishment in bovine chromaffin cells. *Neuroscience* **143:** 487–500.
41. Holz, R.W. & D. Axelrod. 2008. Secretory granule behaviour adjacent to the plasma membrane before and during exocytosis: total internal reflection fluorescence microscopy studies. *Acta Physiol. (Oxf.)* **192:** 303–307.
42. Allersma, M.W. *et al.* 2006. Motion matters: secretory granule motion adjacent to the plasma membrane and exocytosis. *Mol. Biol. Cell.* **17:** 2424–2438.
43. Shakiryanova, D. *et al.* 2007. Presynaptic ryanodine receptor-activated calmodulin kinase II increases vesicle mobility and potentiates neuropeptide release. *J. Neurosci.* **27:** 7799–7806.
44. Shakiryanova, D. *et al.* 2005. Activity-dependent liberation of synaptic neuropeptide vesicles. *Nat. Neurosci.* **8:** 173–178.
45. Tsuk, S. *et al.* 2008. Formation of the full SNARE complex eliminates interactions of its individual protein components with the Kv2.1 channel. *Biochemistry* **47:** 8342–8349.

SNAP-25 in Neuropsychiatric Disorders

Irene Corradini,[a,b] Claudia Verderio,[a,b] Mariaelvina Sala,[a] Michael C. Wilson,[c] and Michela Matteoli[a,b,d]

[a]*Department of Medical Pharmacology and* [b]*Consiglio Nazionale delle Richerche Institute of Neuroscience, University of Milan, Milano, Italy*

[c]*Department of Neurosciences, University of New Mexico Health Sciences Center, Albuquerque, New Mexico, USA*

[d]*Istituto di Ricovero e Cura a Carattere Scientifico, Fondazione Don C. Gnocchi, Milan, Italy*

SNAP-25 (synaptosomal-associated protein of 25 kDa) is a plasma membrane protein that, together with syntaxin and the synaptic vesicle protein VAMP/synaptobrevin, forms the SNARE (soluble *N*-ethylmaleimide-sensitive factor attachment protein receptor) docking complex for regulated exocytosis. SNAP-25 also modulates different voltage-gated calcium channels, representing therefore a multifunctional protein that plays essential roles in neurotransmitter release at different steps. Recent genetic studies of human populations and of some mouse models implicate alterations in SNAP-25 gene structure, expression, and/or function in contributing directly to these distinct neuropsychiatric and neurological disorders.

Key words: **SNAP-25; schizophrenia; epilepsy; ADHD; calcium channels**

SNAP-25: A Multifunctional Protein Regulating Synaptic Transmission

SNAP-25 (synaptosomal-associated protein of 25 kDa) is a soluble *N*-ethylmaleimide-sensitive factor attachment protein receptor (SNARE) protein that participates in the regulation of synaptic vesicle exocytosis. It is a membrane-bound protein anchored to the membranes of neurons via palmitoyl side chains located in the central region of the molecule and, together with syntaxin and the synaptic vesicle protein VAMP/synaptobrevin, constitutes the initial SNARE docking complex for regulated exocytosis. Clostridial toxins, which specifically cleave SNAP-25, have unequivocally demonstrated the requirement of the protein for vesicle exocytosis.[1–3] Furthermore, generation of SNAP-25 null mutant mice revealed that SNAP-25 is essential for evoked synaptic transmission, although it is not required for stimulus-independent neurotransmitter release.[4] Besides its well characterized role in regulating exocytosis, there is increasing evidence that SNAP-25 also modulates various voltage-gated ion channels.[5,6] In particular, SNAP-25 interacts with different types of voltage-gated calcium channels (VGCCs), including N type,[7] P/Q type,[8,9] and L type,[10] through a channel region known as the synaptic protein interaction (synprint) site. In line with these data, SNAP-25 negatively controls neuronal calcium responsiveness to depolarization[11] by specifically inhibiting, upon phosphorylation of Ser 187, neuronal VGCCs.[12] SNAP-25 therefore represents a multifunctional synaptic protein that plays an essential role in synaptic vesicle fusion and modulates calcium dynamics in response to depolarization (Fig. 1). Recently, evidence has accumulated that suggests that SNAP-25 is involved in different neuropsychiatric and neurological disorders. The present review

Address for correspondence: Michela Matteoli, PhD, Department of Medical Pharmacology, CNR Institute of Neuroscience, Via Vanvitelli 32 20129 Milan, Italy. Voice: +0039 02 50317097; fax: +0039 02 7490574. m.matteoli@in.cnr.it

Mechanisms of Exocytosis: Ann. N.Y. Acad. Sci. 1152: 93–99 (2009).
doi: 10.1111/j.1749-6632.2008.03995.x

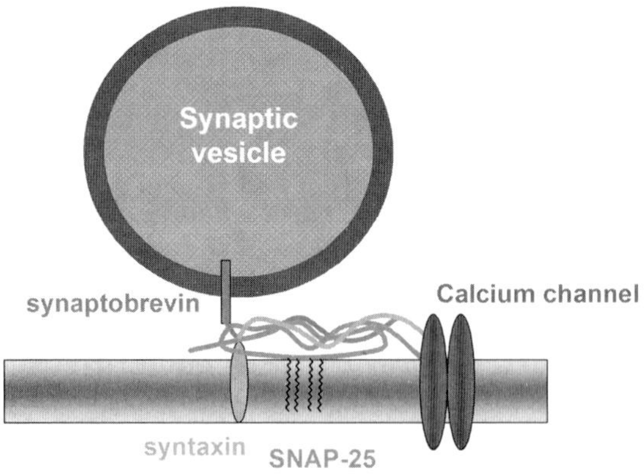

Figure 1. The scheme shows synaptosomal-associated protein of 25 kDa (SNAP-25) interactions with the partner SNAREs (soluble N-ethylmaleimide-sensitive factor attachment protein receptors) syntaxin I and synaptobrevin/VAMP 2 and with voltage-dependent calcium channels.

summarizes this recent literature, which indicates that the altered expression of the protein may produce abnormal behavioral phenotypes, including schizophrenia, attention-deficit/hyperactivity disorder (ADHD), and epilepsy.

Attention-deficit/Hyperactivity Disorder

ADHD [Mendelian Inheritance in Man (MIM) 143465] is one of the most prominent childhood neuropsychiatric disorders. It affects roughly one in 20 school-aged children in the USA,[13] although a recent analysis revealed an even higher worldwide prevalence of 8–12%,[14] with boys affected three or four times as frequently as girls.[15] The major symptoms are persistent age-inappropriate behaviors of inattention and/or impulsivity and hyperactivity. As many as two-thirds of children with ADHD have comorbid disorders, including language and communication deficits, disruption of learning and memory, anxiety and mood disorders, and Tourette's syndrome.[16] In about half of affected individuals, ADHD persists through adolescence into adulthood, with deleterious effects on educational, social, and occupational outcomes, and a higher risk of developing substance abuse.[14,17] One of the more striking aspects of ADHD is its clear inheritance, reflected by a twofold to eightfold increased risk for either parents or siblings of ADHD-diagnosed individuals and an estimate of 76% heritability based on twin studies.[18] Several candidate genes have been implicated and confirmed by linkage analysis. The multiplicity of these genes supports the idea that biological heterogeneity may be a central factor in the clinical variability of the disorder. A number of studies have also implicated the role of environmental factors, which have been estimated to account for the additional variability in penetrance reported in twin studies.[17–19] These environmental stressors have been shown to include low-birth-weight maternal exposure to alcohol and smoking during pregnancy, head injury, obstetric complications, and possibly lead exposure.[15]

The majority of genes evaluated for ADHD are involved in neurotransmission mediated by biogenic amines—dopamine, noradrenaline, and serotonin.[18] The rational for focusing on these genes is drawn largely from the well-documented therapeutic benefit of psychostimulant medication (e.g., methylphenidate and amphetamine) in ameliorating the behavioral impairments associated with ADHD.[20] Meta-analysis of linkage studies in case-controlled and family-based studies provide consistent evidence for the involvement of the dopamine D4 and D5 receptors, dopamine transporter, the synthetic enzyme dopamine β-hydroxylase, the serotonin transporter, and 5-hydroxytryptamine (HTR)1B receptor.[18] Among these genes, only those encoding for the D4 receptor[21,22] and the dopamine transporter[23] have been tested in mouse genetic models. These studies found significant neurochemical and behavioral effects only in homozygous null mutants.

SNAP-25 in ADHD

Besides the biogenic amine candidate genes, the gene encoding SNAP-25 has been identified as responsible for hyperkinetic behavior, based on analysis of the *coloboma* mutant mouse. This mutant mouse is heterozygous for a neutron irradiation-induced semidominant deletion mutation spanning 4.6 Mb on mouse chromosome 2 that encompasses 10–12 genes, including *Snap25*.[24] In *coloboma* mutant mice (*Cm*/+), deletion of the *Snap25* gene results in 50% lower amounts of the SNAP-25 mRNA and protein expression compared to wild-type mice. Interestingly, *coloboma* mutant mice possess several phenotypic characteristics that parallel ADHD symptoms. *Coloboma* mice exhibit normal circadian rhythm and, as children with ADHD, they are hyperactive during their active (nocturnal) phase, with locomotor activity averaging threefold the activity of control littermates.[25] Two key observations have led to the proposal that this mutant with reduced level of SNAP-25 expression represents a model for at least the hyperkinetic component of ADHD: 1) the hyperactivity was ameliorated by a low dose (4mg/kg, i.p.) of the psychostimulant amphetamine (AMPH) and 2) genetic rescue of the hyperactivity (but not the eye defect or head bobbing) was obtained by crossing the mutant with SNAP-25 overexpressing transgenic mice.[24] Interestingly, transgenic rescue of SNAP-25 function also restores normal dopaminergic transmission.[26] Based on this first observation and supported by further behavioral and neurochemical analyses in the *coloboma* mutant mouse,[27–32] polymorphisms at the *SNAP25* gene locus in humans have been examined and association of SNAP-25 with ADHD has been determined in a number of linkage studies,[33–39] a finding that has been further confirmed by meta-analysis.[18]

Schizophrenia

Schizophrenia (MIM 181500) is a common mental disorder with a prevalence of approximately 0.5–1%[13] and significant heritability estimated at about 85%.[40] The clinical symptoms can be classified into two main categories: psychotic or "positive" symptoms, including hallucinations, altered emotional activity, and disorganized behavior, and "negative" symptoms, such as delusions, reduced interest and motivation, and cognitive impairment. The disorder usually has its onset in early adulthood, although often cognitive and behavioral signs are present from childhood. Despite pharmacological treatments, outcomes are variable, and approximately two-thirds of affected individuals have persistent symptoms with only partial remission.[13] Schizophrenia is thought to be a neurodevelopmental disease characterized by defective connectivity in various brain regions during early life.[41–43] The neuropathophysiology of schizophrenia remains unclear, although alterations in dopaminergic and serotoninergic circuitry (as in the case of ADHD), as well as in glutamatergic transmission, have been strongly implicated. Numerous genetic linkage and association studies have identified regions of the genome that may harbor schizophrenia risk genes, but the complex genetics and symptomatology of the disease have limited progress in achieving a clear identification of the specific genetic determinants responsible for neurophysiological deficits that underpin the disorder.[44] Candidate genes that have been shown to be potentially associated with the disease include neuregulin, a protein belonging to the epidermal growth factor (EGF) family; dysbindin, a protein constituent of the dystrophin-associated protein complex; catechol-o-methyltransferase (COMT), one of several enzymes that degrade catecholamines; the DISC1 (disrupted-in-schizophrenia-1) encoding a brain protein likely to play a role in cytoskeletal scaffolding; RGS4 (regulator of G-protein signaling 4) that modulates the GTPase activity of heteromeric G-proteins; GRM3, a metabotropic glutamate receptor; and the gene *G72*, which encodes a protein in brain thought to modulate *N*-methyl-D-aspartate (NMDA) glutamate receptor function.[45] All these findings have led to the view of schizophrenia as a disorder of connectivity and synaptic signaling.[44]

SNAP-25 in Schizophrenia

A recent meta-analysis of 20 separate genome-wide linkage scans for schizophrenia susceptibility genes reported significant linkage to the chromosomal region 20p12.3-11, which contains *SNAP25*, and suggested 20p12.3-11 as a strong candidate region for the disease.[46] Evidence from several immunohistochemical and Western blot studies in postmortem brains implicated reductions in SNAP-25 expression in the etiology of schizophrenia. Decreased levels of SNAP-25 were found in the hippocampus[47,48] and in the frontal lobe Broadman's area 10 of patients with schizophrenia in comparison with controls.[49] In contrast, elevated levels of SNAP-25 were found in prefrontal lobe Broadman's area 9[49] and cingulate cortex.[50] It has been suggested that these discrepancies in altered expression between different brain regions may not only reflect the depressed functionality of certain neural circuits but also the hyperactivity of other pathways that may possibly result from compensatory mechanisms elicited through the neuropathophysiology of this disorder.[48]

The analysis of two genetic models has recently supported the involvement of SNAP-25 in the neuropsychopathology of schizophrenia. First, the mutant hDISC1 transgenic mouse, which has been proposed as an animal model of schizophrenia characterized by spontaneous locomotor activity, impaired spatial reference memory evaluated in the Morris water maze, and abnormal social interaction as shown by the decreased social nonaggressive activity, has been recently shown to lead to a significant reduction of SNAP-25 expression.[51] Second, a chemical mutagen induced dominant mutation (I67T missense) in a highly conserved domain of *Snap25* of the blind-drunk mutant mouse appears to give rise to a schizophrenic phenotype that includes impaired sensorimotor gating, an important component of the schizophrenia phenotype related to altered sensory processing, anxiety, and apathetic behaviors—phenotypic elements that appear to replicate aspects of the negative symptomatology of the disease.[52]

Epilepsy

Epilepsy is a group of heterogeneous neurologic disorders affecting almost 1% of the population. It is characterized by recurrent unprovoked episodes of seizures, from abnormal synchronous firing of groups of neurons, arising from periodic neuronal hyperexcitability. Clinical manifestations of epilepsy are varied, and, despite availability of a number of antiepileptic drugs, about one-third of epileptic patients are resistant to treatment. Several changes can occur in the brain undergoing pathological hyperactivity, at the level of both single cells and anatomically defined neuronal networks. Based on electroencephalographic recordings, seizures can be classified in focal (originating in a single area of the brain) or generalized (involving both brain hemispheres). Generalized epileptic syndromes can be classified as symptomatic (caused by identifiable factors) or idiopathic (without a clear ethiology and partially caused by genetic defects).[53] Generalized idiopathic epilepsies include childhood absence epilepsy and juvenile absence epilepsy. These are characterized by a short sudden impairment in consciousness and behavioral arrest, possibly accompanied by facial clonus. The electroencephalographic hallmark of absence epilepsy are spike-wave discharges (SWDs) that occur in the thalamocortical circuitry. Mutations in different genes, including genes codifying for GABA receptors, Na^+, K^+, and Ca^{2+} channels, have been associated with the familial forms of generalized idiopathic epilepsy with a Mendelian inheritance.[54–56] Analysis of absence epilepsy mouse models *tottering/tg*, *lethargic/lh*, and *stargazer/stg*, bearing mutations in $\alpha 1A$, $\beta 4$, and $\gamma 2$ Ca^{2+} channel subunits, respectively, revealed altered high voltage-activated (HVA) channels current density in *tg* and *stg* thalamic neurons and increased low voltage-activated (LVA) channels peak current and a depolarized shift of the steady-state inactivation curves in *tg*, *lh*, and *stg* thalamic neurons.[57] Furthermore, the association of mutations in human α1A (P/Q type) channels with absence epilepsy has been reported.[58]

SNAP-25 in Epilepsy

Interestingly, the *Cm/+* mouse, which has been implicated as a model of ADHD (see above), has also provided evidence for the involvement of SNAP-25 in epilepsy. In addition to their hyperkinetic activity, *Cm/+* mice display frequent spontaneous bursts of bilateral cortical SWDs that are accompanied by behavioral arrest, which is typical of absence epilepsy. These seizures could be completely blocked by intraperitoneal injection of the antiepileptic drug ethosuximide.[59] Analysis of calcium current amplitude in thalamocortical neurons of this mutant, moreover, revealed an increase in the peak current density of LVA currents in *Cm/+* cells compared to wild-type mice. It is important that this calcium increase is not itself seizure induced as it precedes the developmental onset of SWDs,[59] suggesting that the seizures may arise from abnormalities in calcium transients caused by a SNAP-25 deficiency in modulating presynaptic voltage-gated calcium channels.[12]

Conclusions

SNAP-25 is a multifunctional synaptic protein that plays an essential role in neurotransmitter release via the formation of the SNARE complex[4] and modulates calcium dynamics in response to depolarization.[11,12] While recent genetic studies of human populations and of some mouse models do implicate alterations in SNAP-25 gene structure, expression, and/or function in contributing directly to these distinct neuropsychiatric and neurological disorders, in other cases changes in SNAP-25 may represent a consequence of the neuropathologies of disease. Under this circumstance, effects on SNAP-25 expression could be downstream and secondary to maintaining synaptic homeostasis resulting from altered neural circuits or network function or from more general perturbations to brain development. Moreover, genetic studies have also revealed that candidate genes associated with neuropsychiatric disorders, including *SNAP25*, appear to contribute only a small effect towards the overall phenotype of disability (as reviewed in Ref. 18 for ADHD). Consequently, these genetic components are likely to reflect susceptibility genes that act together and in concert with environmental factors to affect the broad spectrum of deficits associated with mental health disorders. Further studies are required, therefore, to delineate the potential mechanisms through which alterations in SNAP-25 may play a direct role in the etiology as well as contribute to the pathology of ADHD, schizophrenia, epilepsy, and possibly other neuropsychiatric disorders.

Acknowledgments

Supported by the European Union EUSynapse Integrated Project, Telethon GP04196, Cariplo 2006.0779/10.9251, Compagnia di S.Paolo (Prog. 2005.1964) (MM), and by National Institutes of Health Grant MH48989 (MCW)

Conflicts of Interest

The authors declare no conflicts of interest.

References

1. Schiavo, G., M. Matteoli & C. Montecucco. 2000. Neurotoxins affecting neuroexocytosis. *Physiol. Rev.* **80:** 717–766.
2. Jahn, R., T. Lang & T.C. Südhof. 2003. Membrane fusion. *Cell* **112:** 519–533.
3. Jahn, R. & R.H. Scheller. 2006. SNAREs-engines for membrane fusion. *Nat. Rev. Mol. Cell Biol.* **7:** 631–643.
4. Washbourne, P., P.M. Thompson, M. Carta, *et al.* 2002. Genetic ablation of the t-SNARE SNAP-25 distinguishes mechanisms of neuroexocytosis. *Nat. Neurosci.* **5:** 19–26.

5. Catterall, W.A. 1999. Interactions of presynaptic Ca2+ channels and snare proteins in neurotransmitter release. *Ann. N. Y. Acad. Sci.* **868:** 144–159.
6. Zamponi, G.W. 2003. Regulation of presynaptic calcium channels by synaptic proteins. *J. Pharmacol. Sci.* **92:** 79–83.
7. Sheng, Z.H., J. Rettig, T. Cook, *et al.* 1996. Calcium-dependent interaction of N-type calcium channels with the synaptic core complex. *Nature* **379:** 451–454.
8. Rettig, J., Z.H. Sheng, D.K. Kim, *et al.* 1996. Isoform-specific interaction of the alpha1A subunits of brain Ca2+ channels with the presynaptic proteins syntaxin and SNAP-25. *Proc. Natl. Acad. Sci. USA* **9:** 7363–7368.
9. Martin-Moutot, N., N. Charvin, C. Leveque *et al.* 1996. Interaction of SNARE complexes with P/Q-type calcium channels in rat cerebellar synaptosomes. *J. Biol. Chem.* **22:** 6567–6570.
10. Wiser, O., M. Trus, A. Hernandez, *et al.* 1999. The voltage sensitive Lc-type Ca2+ channel is functionally coupled to the exocytotic machinery. *Proc. Natl. Acad. Sci. USA* **5:** 248–253.
11. Verderio, C., D. Pozzi, E. Pravettoni, *et al.* 2004. SNAP-25 modulation of calcium dynamics underlies differences in GABAergic and glutamatergic responsiveness to depolarization. *Neuron* **41:** 599–610.
12. Pozzi, D., S.B. Condliffe, Y. Bozzi, *et al.* 2008. Activity-dependent phosphorylation of Ser187 is required for SNAP-25-negative modulation of neuronal voltage-gated calcium channels. *Proc. Natl. Acad. Sci. USA* **105:** 323–328.
13. American Psychiatric Association. 1994. *Diagnostic and statistical manual of mental disorders*, 4th ed. American Psychiatric Association. Washington, DC.
14. Faraone, S.V., J. Sergeant, C. Gillberg, *et al.* 2003. The worldwide prevalence of ADHD: is it an American condition? *World Psychiatry* **2:** 104–113.
15. Biederman, J. & S.V. Faraone. 2005. Attention-deficit hyperactivity disorder. *Lancet* **366:** 237–248.
16. Cantwell, D.P. 1996. Attention deficit disorder: a review of the past 10 years. *J. Am. Acad. Child. Adolesc. Psychiatry* **35:** 978–987.
17. Spencer, T.J., J. Biederman & E. Mick. 2007. Attention-deficit/hyperactivity disorder: diagnosis, lifespan, comorbidities, and neurobiology. *J. Pediatr. Psychol.* **32:** 631–642.
18. Faraone, S.V., R.H. Perlis, A.E. Doyle, *et al.* 2005. Molecular genetics of attention deficit hyperactivity disorder. *Biol. Psychiatry* **57:** 1313–1323.
19. Thapar, A., M. O'Donovan & M.J. Owen. 2005. The genetics of attention deficit hyperactivity disorder. *Hum. Mol. Genet.* **14:** R275–282.
20. Castellanos, F.X. 1997. Toward a pathophysiology of attention-deficit/hyperactivity disorder. *Clin. Pediatr. (Phila.)* **36:** 381–393.
21. Falzone, T.L., D.M. Gelman, J.I. Young, *et al.* 2002. Absence of dopamine D4 receptors results in enhanced reactivity to unconditioned, but not conditioned, fear. *Eur. J. Neurosci.* **15:** 158–164.
22. Avale, M.E., T.L. Falzone, D.M. Gelman, *et al.* 2004. The dopamine D4 receptor is essential for hyperactivity and impaired behavioral inhibition in a mouse model of attention deficit/hyperactivity disorder. *Mol. Psychiatry* **9:** 718–726.
23. Gainetdinov, R.R., S.R. Jones & M.G. Caron. 1999. Functional hyperdopaminergia in dopamine transporter knock-out mice. *Biol. Psychiatry* **46:** 303–311.
24. Hess, E.J., K.A. Collins & M.C. Wilson. 1996. Mouse model of hyperkinesis implicates SNAP-25 in behavioral regulation. *J. Neurosci.* **16:** 3104–3111.
25. Hess, E.J. *et al.* 1992. Spontaneous locomotor hyperactivity in a mouse mutant with a deletion including the Snap gene on chromosome 2. *J. Neurosci.* **12:** 2865–2874.
26. Steffensen, S.C., S.J. Henriksen & M.C. Wilson. 1999. Transgenic rescue of SNAP-25 restores dopamine-modulated synaptic transmission in the coloboma mutant. *Brain Res.* **847:** 186–195.
27. Heyser, C.J., M.C. Wilson & L.H. Gold. 1995. Coloboma hyperactive mutant exhibits delayed neurobehavioral developmental milestones. *Brain Res. Dev. Brain Res.* **89:** 264–269.
28. Raber, J., P.P. Mehta, M. Kreifeldt, *et al.* 1997. Coloboma hyperactive mutant mice exhibit regional and transmitter-specific deficits in neurotransmission. *J. Neurochem.* **68:** 176–186.
29. Jones, M.D., M.E. Williams & E.J. Hess. 2001. Abnormal presynaptic catecholamine regulation in a hyperactive SNAP-25-deficient mouse mutant. *Pharmacol. Biochem. Behav.* **68:** 669–676.
30. Jones, M.D. & E.J. Hess. 2003. Norepinephrine regulates locomotor hyperactivity in the mouse mutant coloboma. *Pharmacol. Biochem. Behav.* **75:** 209–216.
31. Bruno, K.J., C.S. Freet, R.C. Twining, *et al.* 2007. Abnormal latent inhibition and impulsivity in coloboma mice, a model of ADHD. *Neurobiol. Dis.* **25:** 206–216.
32. Fan X. & E.J. Hess. 2007. D2-like dopamine receptors mediate the response to amphetamine in a mouse model of ADHD. *Neurobiol. Dis.* **26:** 201–211.
33. Barr, C.L., Y. Feng, K. Wigg, *et al.* 2000. Identification of DNA variants in the SNAP-25 gene and linkage study of these polymorphisms and attention-deficit hyperactivity disorder. *Mol. Psychiatry* **5:** 405–409.
34. Brophy, K., Z. Hawi, A. Kirley, *et al.* 2002. Synaptosomal associated protein 25 (SNAP-25) and attention deficit hyperactivity disorder (ADHD): evidence of

linkage and association in the Irish population. *Mol. Psychiatry* **7:** 913–917.

35. Mill, J., S. Curran, L. Kent, *et al.* 2002. Association study of a SNAP-25 microsatellite and attention deficit hyperactivity disorder. *Am. J. Med. Genet.* **114:** 269–271.
36. Kustanovich, V., B. Merriman, J. McGough, *et al.* 2003. Biased paternal transmission of SNAP-25 risk alleles in attention-deficit hyperactivity disorder. *Mol. Psychiatry* **8:** 309–315.
37. Mill, J., S. Richards, J. Knight, *et al.* 2004. Haplotype analysis of SNAP-25 suggests a role in the aetiology of ADHD. *Mol. Psychiatry* **9:** 801–810.
38. Feng, Y., J. Crosbie, K. Wigg, *et al.* 2005. The SNAP25 gene as a susceptibility gene contributing to attention-deficit hyperactivity disorder. *Mol. Psychiatry* **10:** 998–1005.
39. Choi, T.K., H.S. Lee, J.W. Kim, *et al.* 2007. Support for the MnlI polymorphism of SNAP25; a Korean ADHD case–control study. *Mol. Psychiatry* **12:** 224–226.
40. Cardno, A.G. & I.I. Gottesman. 2000. Twin studies of schizophrenia: from bow-and-arrow concordances to star war Mx and functional genomics. *Am. J. Med. Genet.* **97:** 12–17.
41. Rehn, A.E. & S.M. Rees. 2005. Investigating the neurodevelopmental hypothesis of schizophrenia. *Clin. Exp. Pharmacol. Physiol.* **32:** 687–696.
42. Lang, U.E., I. Puls, D.J. Muller, *et al.* 2007. Molecular mechanisms of schizophrenia. *Cell. Physiol. Biochem.* **20:** 687–702.
43. Walsh, T., J.M. McClellan, S.E. McCarthy, *et al.* 2008. Rare structural variants disrupt multiple genes in neurodevelopmental pathways in schizophrenia. *Science* **320:** 539–543.
44. Harrison, P.J. & D.R.Weinberger. 2005. Schizophrenia genes, gene expression, and neuropathology: on the matter of their convergence. *Mol. Psychiatry* **10:** 40–68.
45. Kirov, G., M.C. O'Donovan & M.J. Owen. 2005. Finding schizophrenia genes. *J. Clin. Invest.* **115:** 1440–1448.
46. Lewis, C.M., D.F. Levinson, L.H. Wise, *et al.* 2003. Genome scan meta-analysis of schizophrenia and bipolar disorder, part II: Schizophrenia. *Am. J. Hum. Genet.* **73:** 34–48.
47. Young, C.E., K. Arima, J. Xie, *et al.* 1998. SNAP-25 deficit and hippocampal connectivity in schizophrenia. *Cereb. Cortex.* **8:** 261–268.
48. Thompson, P.M., S. Egbufoama & M.P. Vawter. 2003. SNAP-25 reduction in the hippocampus of patients with schizophrenia. *Prog. Neuropsychopharmacol. Biol. Psychiatry* **27:** 411–417.
49. Thompson, P.M., A.C. Sower & N.I. Perrone-Bizzozero. 1998. Altered levels of the synaptosomal associated protein SNAP-25 in schizophrenia. *Biol. Psychiatry* **43:** 239–243.
50. Gabriel, S.M., V. Haroutunian, P. Powchik, *et al.* 1997. Increased concentrations of presynaptic proteins in the cingulate cortex of subjects with schizophrenia. *Arch. Gen. Psychiatry* **54:** 559–566.
51. Pletnikov, M.V., Y. Ayhan, O. Nikolskaia, *et al.* 2008. Inducible expression of mutant human DISC1 in mice is associated with brain and behavioral abnormalities reminiscent of schizophrenia. *Mol. Psychiatry* **13:** 173–186.
52. Jeans, A.F., P.L. Oliver, L. Johnson, *et al.* 2007. A dominant mutation in *Snap25* causes impaired vesicle trafficking, sensorimotor gating, and ataxia in the blind-drunk mouse. *Proc. Natl. Acad. Sci. USA* **104:** 2431–2436.
53. Fisher R., W. van Emde Boas, W. Blume, *et al.* 2005. Epileptic seizures and epilepsy: definitions proposed by the International League Against Epilepsy (ILAE) and the International Bureau for Epilepsy (IBE). *Epilepsia* **46:** 470–472.
54. Mulley, J.C., I.E. Scheffer, S. Petrou, *et al.* 2003. Channelopathies as a genetic cause of epilepsy. *Curr. Opin. Neurol.* **16:** 171–176.
55. Khosravani, H. & G.W. Zamponi. 2006. Voltage-gated calcium channels and idiopathic generalized epilepsies. *Physiol. Rev.* **86:** 941–966.
56. Helbig, I., I.E. Scheffer, J.C. Mulley, *et al.* 2008. Navigating the channels and beyond: unravelling the genetics of the epilepsies. *Lancet Neurol.* **7:** 231–245.
57. Zhang, Y., M. Mori, D.L. Burgess, *et al.* 2002. Mutations in high-voltage-activated calcium channel genes stimulate low-voltage-activated currents in mouse thalamic relay neurons. *J. Neurosci.* **22:** 6362–6371.
58. Jouvenceau, A., L.H. Eunson, A. Spauschus, *et al.* 2001. Human epilepsy associated with dysfunction of the brain P/Q-type calcium channel. *Lancet* **358:** 801–807.
59. Zhang, Y., A.P. Vilaythong, D. Yoshor, *et al.* 2004. Elevated thalamic low-voltage-activated currents precede the onset of absence epilepsy in the SNAP25-deficient mouse mutant coloboma. *J. Neurosci.* **24:** 5239–5248.

Exocytosis, Mediatophore, and Vesicular Ca^{2+}/H^+ Antiport in Rapid Neurotransmission

Yves Dunant,[a] J. Miguel Cordeiro,[a,b] and Paula P. Gonçalves[b]

[a]*Neurosciences Fondamentales, Université de Genève, Centre Médical Universitaire, Geneva, Switzerland*

[b]*Department de Biologia, Universidade de Aveiro, Aveiro, Portugal*

In rapid synapses, neurotransmitter quanta are emitted in less than 100 μs, often at a high frequency. Using fast cryofixation of synapses, we found a very brief (2–3 ms) change affecting intramembrane particles in presynaptic membrane. Vesicle openings also occurred but after a significant delay. The particle change is most probably linked to mediatophore, a proteolipid of 220 kDa. Mediatophore aggregates were demonstrated in active zones of the presynaptic membrane. Reconstituted in liposomes, *Xenopus* oocytes, and neuroblastoma cells, mediatophore releases acetylcholine in a Ca^{2+}-dependent and quantal manner, mimicking physiological release. In restricted presynaptic "nanodomains," Ca^{2+} concentration explosively reaches a high level and then vanishes with a time constant of 300–400 μs. Among the processes contributing to the fast phase of Ca^{2+} buffering, a vesicular Ca^{2+}/H^+ antiport plays a major role. Energized by the Vesicular-ATPase-dependent proton gradient, the antiport has a low affinity for Ca^{2+}. We inactivated the Ca^{2+}/H^+ antiport using bafilomycin A1, which annihilates the proton gradient. As a result, the postsynaptic potential was increased in duration for about 3 ms, an effect caused by persistence of transmitter release. A similar change was obtained by replacing extracellular Ca^{2+} by strontium, which inhibits the antiport. The antiport function, therefore, is to abbreviate the presynaptic Ca^{2+} signal, making transmitter release briefer. This allows transmission to operate at high frequency. Following a brief period of stimulation, calcium transiently accumulates in synaptic vesicles where it is exchanged against transmitter. Calcium is subsequently cleared from the terminal, most probably by exocytosis.

Key words: **synaptic transmission; transmitter release; mediatophore; freeze-fracture morphology; Ca^{2+}/H^+ antiport; synaptic vesicles; exocytosis; calcium; strontium**

Introduction

Cells use exocytosis as a major way to secrete or excrete a variety of substances, especially those that cannot be present at a high concentration in the cytosol, such as polypeptides, monoamines, and Ca^{2+} ions. In this symposium, it has been demonstrated that there are different mechanisms, or modes, of exocytosis, for example, full fusion, "kiss-and-run." Exocytosis also takes place in rapid synapses, that is, in synapses that transmit impulses within milliseconds and often repetitively at a high frequency. However, we shall present evidence that the role and time course of exocytosis in these synapses are different than classically believed. The electric organ of *Torpedo* contains myriads of cholinergic synapses (approximately 4–6 × 10^{11}), which are typical examples of rapid synapses. The duration of a single electric discharge is 2–4 ms even at 15°C,

Address for correspondence: Yves Dunant, Département des Neurosciences fondamentales CMU, CH-1211 Genève 4, Switzerland. Voice & fax: +41 21 946 30 41. a_y.dunant@bluewin.ch

Mechanisms of Exocytosis: Ann. N.Y. Acad. Sci. 1152: 100–112 (2009).
doi: 10.1111/j.1749-6632.2008.04000.x

and the fish can use a frequency over 100 Hz for attack or defense. The brevity of transmission is ensured by the rapid hydrolysis of acetylcholine (Ach) by acetylcholinesterase, which is highly concentrated in the cleft, and also by the very short open time of nicotinic receptors (0.6 ms only), shorter than those of neuromuscular junctions.[1] But, as noticed by Sir Bernard Katz,[2] "time is gained at the expense of sensitivity." To be so fast, cholinesterase and nicotinic receptors must have a low affinity for acetylcholine. Similarly, we shall see that the presynaptic actors directly involved in rapid transmitter release have to work at a high speed but with low affinity.

Morphological Description of Synaptic Transmission at a Time Resolution of 1 ms

Conventional optical techniques or chemical fixation are far too slow for capturing the changes affecting presynaptic and postsynaptic membranes during transmission of a single impulse. For this purpose, ultrafast cryofixation is appropriate because it allows vitrification of a thin tissue layer at the required velocity.[3–7] Figures 1 and 2 illustrate such experiments performed on the *Torpedo* nerve–electroplaque junction under the most physiological conditions (a single impulse, 16°C, no cryoprotection, no 4-aminopyridine). Fifteen to 20 small fragments of electric tissue were mounted on the plunger of a homemade freezing stimulator. The plunger was then released, and, during its fall, single electrical stimuli were applied at 1-ms time intervals to each specimen successively. All specimens were frozen at exactly the same moment when the plunger fell into contact with a viscous isopentane-propane mixture maintained at −190°C. The precise time of impact was monitored by a 30-μm thermocouple placed at the level of specimens. During the run, we recorded the postsynaptic potential generated by one of the tissue specimens. Cryofixed specimens were separated from the plunger un-

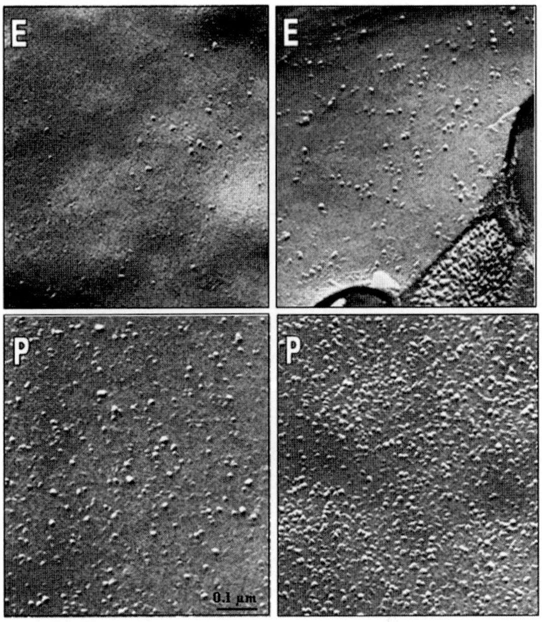

At rest 2 ms

Figure 1. Ultrastructure changes occurring at the *Torpedo* nerve-electroplaque synapse during transmission of a single nerve impulse. Ultrarapid freezing was performed in the absence of any drug or cryoprotection. The upper panels show replicas from the E-face (external leaflet), and the lower panels show replicas from the P-face (protoplasmic leaflet) of the presynaptic membrane. On the left-hand side, tissue was cryofixed under resting conditions; on the right-hand side, it was cryofixed 2 ms after the stimulus. Synaptic transmission was accompanied in the presynaptic membrane by an abrupt increase in the number of intramembrane particles (IMPs) larger than 10 nm. The replica of the stimulated presynaptic E-face shows in addition a small area of the postsynaptic membrane P-face with the characteristic array of nicotinic receptors. See further explanations in Figure 2 and in the text.

der liquid nitrogen and transferred one by one into the freeze-fracture apparatus. They were fractured very superficially; platinum-carbon replicas were made under low temperature and high vacuum conditions. They were then examined using transmission electron microscopy.

Transmission of the synaptic signal was accompanied by an abrupt increase in the number of intramembrane particles (IMPs) in both the P- and E leaflets of the presynaptic membrane (Figs. 1 and 2). The duration of the IMP

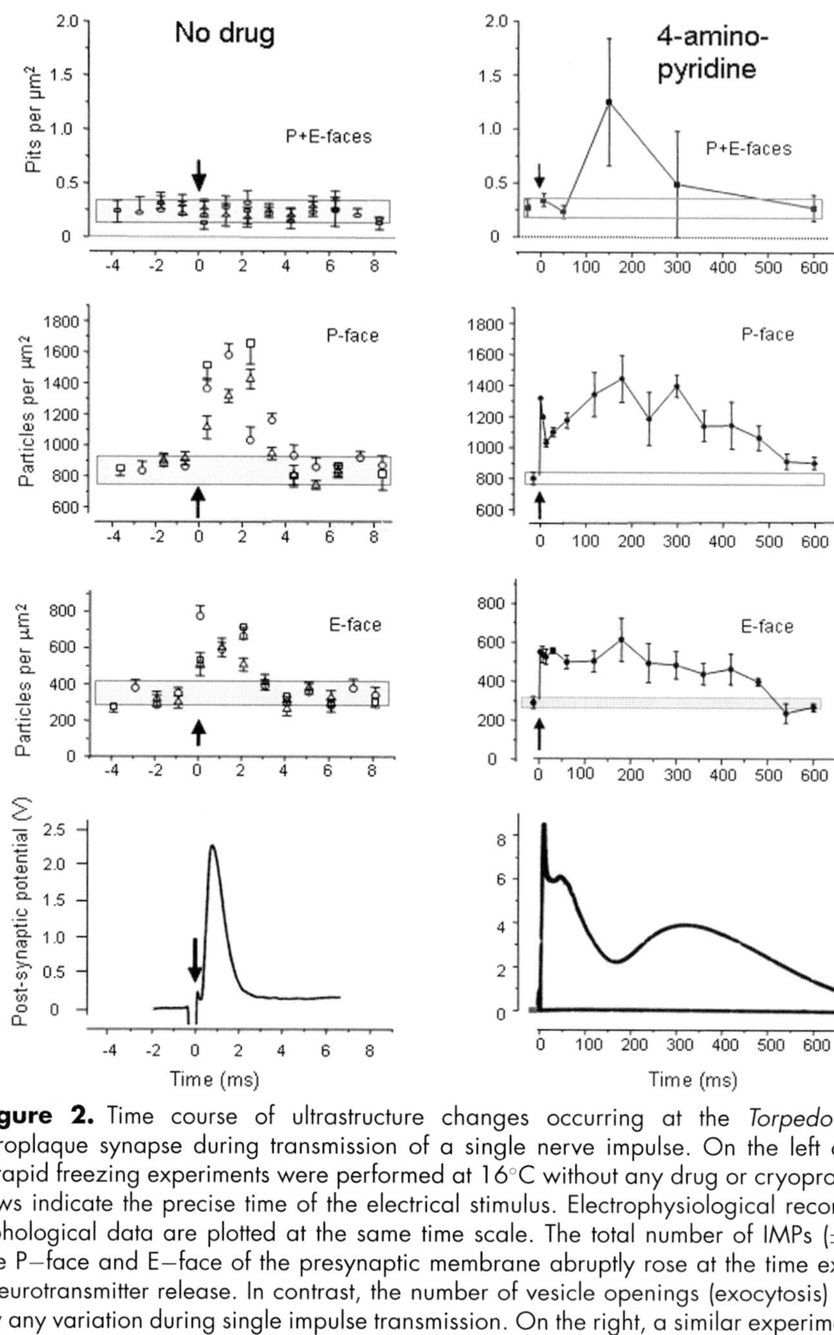

Figure 2. Time course of ultrastructure changes occurring at the *Torpedo* nerve-electroplaque synapse during transmission of a single nerve impulse. On the left column, ultrarapid freezing experiments were performed at 16°C without any drug or cryoprotection. Arrows indicate the precise time of the electrical stimulus. Electrophysiological records and morphological data are plotted at the same time scale. The total number of IMPs (± SEM) in the P–face and E–face of the presynaptic membrane abruptly rose at the time expected for neurotransmitter release. In contrast, the number of vesicle openings (exocytosis) did not show any variation during single impulse transmission. On the right, a similar experiment was carried out on tissue treated with 0.1 mmol/L 4-aminopyridine. The time course of both the electrical discharge and the IMPs changes was markedly prolonged, but it remained remarkably synchronous. An increase in the number of vesicle openings (pits) occurred transiently at approximately 200 ms after the stimulus.

change was very brief, about 3 ms, that is, in the range expected from the electrophysiological signal. The new IMP population which suddenly appeared and disappeared had a diameter larger than 10 nm. At 5°C, both the postsynaptic potential and the IMP change were slowed down and delayed to the same extent. When the experiment was carried out in a high-Mg^{2+} low-Ca^{2+} solution, there was no synaptic transmission (no transmitter release) and no IMP change.[6] Vesicle openings (pits) were rather scarce in the presynaptic membrane at rest (approximately 0.25/μm^2). Their density did not increase during signal transmission or during the following ms (it is recalled that transmission of a single impulse requires the release of approximately 1.3 quanta of ACh per μm^2). We shall see later in experiments involving longer stimulations that a substantial increase of the pit density takes place during the minute that follows activity.

Treatment of *Torpedo* synapses with 100 μmol/L 4-aminopyridine (4-AP) did not cause any noticeable morphological or functional changes as long as the nerves were not stimulated. In response to a single stimulus, however, the synapses generated a giant postsynaptic potential, which oscillated for several hundreds of milliseconds. This was a result of a huge persistence of transmitter release, probably consecutive to a combination of K^+ channels blockade and vesicular Ca^{2+}/H^+ antiport inhibition (see below). Rapid freezing was performed during the course of a 4-AP-provoked giant discharge. Again, the same population of IMPs $\geq$10 nm occurred in both leaflets of the presynaptic membrane. As expected, the duration of the phenomenon was longer in 4-AP treated synapses. Actually, the IMP change faithfully espoused the time course of the giant postsynaptic potential, from the onset to the end. Vesicle openings occurred during the giant potential, but 100 to 300 ms after the onset of the electrical discharge and the onset of IMP change (Fig. 2 right-hand side).[5] These results on *Torpedo* synapses should be compared to the well-known experiments of J. Heuser and colleagues on the frog neuromuscular junction. Under physiological conditions, they reported a transient increase in the number of large IMPs, thus a picture very similar to that found in the *Torpedo*. However in the presence of 4-AP, they obtained images of exocytosis, which occurred at active zones from the very beginning of the endplate potential.[8] Curiously, the occurrence of vesicle openings culminated after the end of the transmission signal and lasted much longer.[3,4] IMP changes were apparently not analyzed in 4-AP-treated neuromuscular junctions, but judging from careful examination of the published pictures, they most probably took place as well.

IMP changes accompanying transmitter release were also demonstrated using mammalian brain synaptosomes,[9] *Torpedo* synaptosomes,[10,11] and cultured cell lines.[12] Their significance in the process of rapid transmitter release will be examined in the next section.

As for the density of exocytosis images in the presynaptic membrane, it clearly increases in a Ca^{2+}-dependent manner as a function of stimulation in other synaptic preparations, such as in the *Torpedo*. But this increase was not found to coincide with the expected time of transmitter release.[7] We shall discuss the possible link of this late occurrence of exocytosis with processes of Ca^{2+} extrusion.

Mediatophore, the ACh-releasing Protein

Figure 3A shows a freeze-fracture replica displaying a *Torpedo* nerve terminal cryofixed during the course of nerve impulse transmission (mainly the E-face). Fracture-labeling was performed on the replica, using an antibody directed against the N-terminal region of a proteolipid called *mediatophore*. As seen in the picture, immunogold spots were regularly found in the vicinity of these large IMPs, which momentarily occur during release (see also Ref. 13).

Realizing that a protein fraction derived from nerve terminal membranes did perform Ca^{2+}-dependent ACh release, Maurice Israël and his colleagues searched for the component(s) supporting this property, using ACh-loaded proteoliposomes as a functional assay. In this way, they purified to homogeneity a proteolipid that rendered the liposome membrane competent for ACh release in a strictly

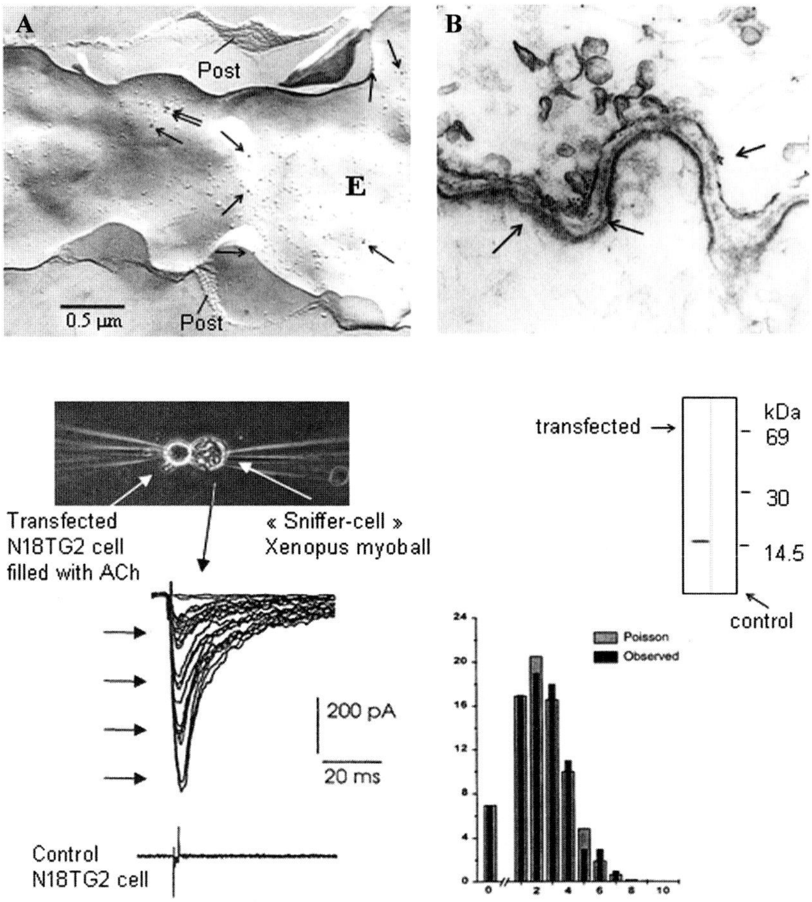

Figure 3. Localization and functional test of mediatophore. (**A**) A replica obtained from stimulated tissue was labeled using an antibody directed against the N-terminal of the mediatophore 15 kDa subunit. Immunogold spots decorate the large IMPs present at the external leaflet (E-face) of the active presynaptic membrane, as described in Figures 1 and 2. (**B**) Immunolabeling of mediatophore in a thin section of electric organ focused on a synaptic region. Aggregates of mediatophore are seen at the active zones of the presynaptic membrane. A few synaptic vesicles are also labeled. The lower part of the figure illustrates quantal acetylcholine (Ach) release induced by mediatophore transfection in a deficient neuroblastoma cell line (N18TG2). Control cells and cells transfected with the *Torpedo* 16-kDa mediatophore proteolipid were prefilled with ACh and then placed at the contact of a receptor-rich *Xenopus* myoball. On stimulation, the transfected—but not the control—cells generated curare-sensitive inward currents in the myoball. The number of responses containing 0, 1, 2, 3, or more ACh quanta were plotted in the histogram (filled symbols). Their distribution was compared to the theoretical distribution calculated from Poisson analysis (open symbols). For the immunoblot above, proteolipids were extracted from the membranes of control and mediatophore-transfected N18TG2 cells, and the 16 kDa subunit was revealed by a specific antibody.

Ca^{2+}-dependent manner. The proteolipid, named mediatophore, is a homo-oligomer of about 220 kDa, the subunit of which is 15–16 kDa. The subunit was cloned and sequenced in the *Torpedo*.[14,15] It turned out that it was a member of a proteolipid family, which includes the c-subunit of the membrane sector (V_0) of V—ATPase.[16] The localization of mediatophore was determined at the electron microscopic level in the synapses of the electric organ, using an antiserum (Fig. 3B). Mediatophore clusters were evidenced in the presynaptic membrane at spots facing the postsynaptic infoldings; that is, in the active zones where the neurotransmitter is expected to be released. Some labeling was also found in synaptic vesicles, as expected from the fact that the mediatophore 15–16 kDa subunit is very similar—if not identical—to the c component of V_0.[17]

Mediatophore molecules display a pentameric structure of a MW of 220 kDa. Each component of the pentamer should therefore be 44 kDa and composed of three subunits of 15 kDa. The proteolipid subunits have four intramembrane segments with the N-terminal probably exposed at the extracellular side of the membrane, explaining why it can be decorated with immunogold in fracture-labeling replicas on the E- but not the P-leaflet of the nerve ending membrane (Fig. 3A).

ACh-filled liposomes containing mediatophore in their membrane as unique protein release the transmitter in a Ca^{2+}-dependent manner, which in all aspects mimics the process observed in natural synapses. The protein exists under at least three configurations: open, closed, and desensitized. An abrupt rise of Ca^{2+} concentration at the inner face of the membrane transiently drives the protein from the closed to the open configuration. If Ca^{2+} rises only slowly in the terminal or remains in contact with the protein, mediatophores become desensitized, that is, they can no longer release Ach, even in response to a subsequent high Ca^{2+} challenge.[18] Such behavior (high-speed low-affinity activation, low-speed high-affinity desensitization) is typical for many channel proteins, including the nicotinic ACh receptors. It is precisely this type of kinetics that is predicted for processes directly involved in rapid neurotransmission.

Acetylcholine release and other cholinergic functions were successfully reconstituted in *Xenopus* oocytes upon injection of mRNAs extracted from cholinergic cells (*Torpedo* electromotoneurons). Like in the proteoliposome experiments, the precise dependency of ACh release upon Ca^{2+} concentration and the pharmacology of release closely matched the characteristics of native synapses. When antisense oligonucleotides directed against segments of the 15–16 kDa mRNA were injected together with the total poly(A)$^{+}$ mRNAs, both ACh release and 15–16 kDa expression in the oocyte membranes were selectively abolished. Interestingly, as much as 99% of the injected oocyte ACh content was hydrolyzed by disrupting the membranes in the presence of cholinesterase, demonstrating that almost all their ACh was cytosolic.[19,20] This confirms a large number of experiments carried out on intact synapses, which demonstrated that the cytosolic compartment of ACh is preferentially used and renewed upon activity.[21,22]

Proteoliposome and oocyte experiments clearly suggested direct involvement of mediatophore in transmitter release, but a crucial point remained to be demonstrated, that is, the pulsatile, or quantal, nature of ACh release found in rapid synapses.[23] This was achieved by expressing mediatophore in deficient cell lines.

Using any type of cells in culture, their ability to express Ca^{2+}-dependent ACh release can be tested as follows. The cells are loaded with ACh by simply adding the transmitter in the culture medium after cholinesterase inhibition. Extracellular ACh is subsequently washed and the cells can retain the loaded transmitter for hours. Filled in this way, some cell lines, such as N18TG-2, NIE115, or MMQ, were found unable to release ACh in response to a Ca^{2+} challenge. Others, such as C6BU-1, NG108-15, LTK,- or RBE4, were, on the contrary, fully competent for release. Importantly, the membrane of release-competent cell lines (C6BU-1) contained large amounts of the 15-kDa mediatophore proteolipid, while those of release-incompetent cells (N18TG-2) had only small amounts. In addition to biochemical assessment, transmitter release from cultured cells was measured electrophysiologically in real time as follows. A patch-clamped

Xenopus myoblast (myoball) was set in contact with the ACh-filled cell to be tested. Electrical stimulation was applied to the cell and quantal ACh release—if any—was recorded from the myoball. As a result, the release evoked from C6BU-1 cells was substantial and fluctuated as multiples of an elementary event. In contrast, N18TG-2 cells generated no quanta, or only had very low responses.[24]

The clearly positive demonstration was achieved as illustrated in Figure 3 (lower panel). We chose cells from the N18TG-2 line because they had a particularly low capacity for ACh release and were also deficient in choline acetyltransferase (ChAT), vesicular acetylcholine transporter (VAChT), synapsin, synaptophysin, as well as in synaptic vesicles. Using several different plasmids and transfection systems, we obtained a number of clones that expressed the *Torpedo* mediatophore. Transfected and control cells were tested for ACh release after being loaded in culture with the transmitter. As a result, the clones successfully transfected with the *Torpedo* mediatophore became release competent; they elicited inward currents in the myocyte in response to electrical stimuli. The currents peaked at preferential values, fluctuating as multiples of an elementary event of about 200 pA. This quantum size was evaluated to correspond to 800–1000 ACh molecules in this reconstituted system.[25,26] This quantal value corresponded to that of the smallest component of the quantum recorded in intact *Torpedo* synapses.[27] Hence, mediatophore transfection endowed incompetent cells with a quantal ACh release mechanism very similar to that found in the natural cholinergic terminals.

The Vesicular Ca^{2+}/H^+ Antiport Shapes the Time Course of Rapid Transmission

Rapid neurotransmission is like lightning: a spark of calcium in the nerve terminal, a spark of transmitter in the cleft, and the signal is over.

The rate of Ca^{2+} entry is the highest at the end of the presynaptic depolarization.[28] The spark of free Ca^{2+} is extremely brief and restricted to "nanodomains" situated at the inner mouth of Ca^{2+} channels.[29] At these spots, Ca^{2+} concentration explosively reaches high levels and subsequently decays according to two different kinetics. The first phase of decay has a time constant as fast as 300–400 μs.[30] Afterwards, slower mechanisms further reduce Ca^{2+} to a submicromolar range.[31]

Docked at the active zones of nerve terminals,[32] synaptic vesicles are in a strategic position for sequestrating calcium. It has long been demonstrated that cholinergic and other vesicles have calcium-binding capabilities and are able to accumulate Ca^{2+} by ATP-dependent mechanisms.[10,33] More recently, Gonçalves and colleagues[34] discovered that calcium is taken up by mammalian brain synaptic vesicles via two distinct processes: a Ca^{2+}-ATPase pump and Ca^{2+}/H^+ antiport (Fig. 4). The vesicular Ca^{2+}-pump has a high affinity for Ca^{2+} ($K_{0.5} = 0.6$ μmol/L) and is inhibited by vanadate. Its maximum velocity is observed at 25 μmol/L Ca^{2+} and pH 7.4, while larger Ca^{2+} concentrations inhibit the pump. The vesicular pump differs from the reticulum Ca^{2+} pump by having distinct kinetics; the reticulum pump has a still higher affinity for Ca^{2+} ($K_{0.5} = 0.017$ μmol/L) and is inhibited at 25 μmol/L Ca^{2+}, that is, at a concentration corresponding to the maximum velocity of the vesicular pump.

As for the rapid phase of presynaptic Ca^{2+} buffering, diffusion, binding to cytosolic proteins, Na^+/Ca^{2+} exchange at the plasma membrane, and other processes are certainly involved.[35,36] However, we present here evidence that the Ca^{2+}/H^+ antiport is crucial for rapid termination of the Ca^{2+} spark. In vesicles isolated from the mammalian brain, the Ca^{2+}/H^+ antiport displays a very low affinity for Ca^{2+} ($K_{0.5} = 217$ μmol/L; maximal activity at 500–600 μmol/L). As expected, the antiport velocity depends on the ΔpH through the vesicular membrane, thus it relies on the

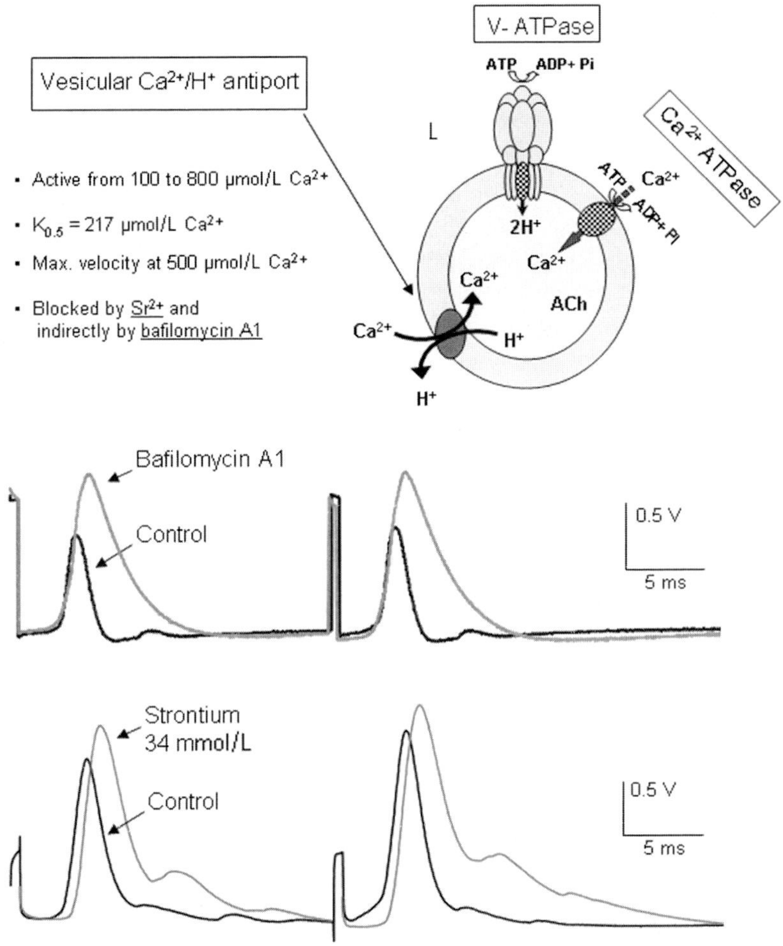

Figure 4. Description and physiological importance of the Ca^{2+}/H^+ antiport. The upper cartoon illustrates the properties of the antiport compared to properties of the vesicular Ca^{2+}-ATPase (see text). The electrophysiological traces were postsynaptic potentials recorded from small pieces of electric organ in response to paired-pulse stimulation. They show that indirect inhibition of the antiport (using bafilomycin A1) or direct inhibition (using strontium) markedly prolonged the time course of the electric response. Therefore, blockade of rapid Ca^{2+} buffering results in a prolongation of transmitter release and thereby in prolongation of the postsynaptic potential.

activity of V-ATPase (inhibition by bafilomycin; see Fig. 4, upper panel).[37,38]

The existence of a Ca^{2+}/H^+ antiport in the membrane of synaptic vesicles was demonstrated using a set of different biochemical techniques. We recently used electrophysiology for investigating whether it plays a significant role in the dynamics of synaptic transmission. We selected nerve-electroplaque transmission of *Torpedo*, a preparation especially appropriate for this purpose: the electric organ does not contract, transmission is very rapid, ACh release is quantal and fully characterized, and morphology and biochemistry can be carried out in parallel to electrophysiology on the same specimens. In a first series of experiments, we indirectly inhibited Ca^{2+}/H^+ antiport by using bafilomycin A1, a drug that specifically blocks V-ATPase. As a consequence, the vesicular proton gradient was annihilated and the antiport could no longer function. Figure 4 (middle panel) shows the electric discharges (postsynaptic potentials) generated by a small fragment of electrogenic tissue in response to paired stimuli applied at 20-ms intervals. Under control conditions the response is a relatively strong electric current, extremely well synchronized, which rose at about 2 ms after the stimulus. After incubation of the tissue with 2 μmol/L bafilomycin A1, the postsynaptic potentials were dramatically modified; bafilomycin A1 enhanced the response by increasing chiefly its duration. The potential area became approximately twice that of the control response previously measured on the same specimen (Fig. 4). The bafilomycin A1 effect was readily reversible when the drug was washed out. To make sure that the effect was presynaptic, we used a biochemical assay for measuring ACh release from the same tissue fragments. There was an excellent correspondence with electrophysiological data; the amount of ACh released in response to the paired pulse stimulation also increased significantly after bafilomycin treatment (in prep. and Ref. 39).

The second approach consisted of replacing calcium by strontium in the physiological solution. Strontium is not transported through the vesicular Ca^{2+}/H^+ antiport; it also prevents Ca^{2+} accumulation in the vesicles through this pathway.[40] It has also been known from the time of Locke[41] that Sr^{2+} can support neuromuscular transmission. There are two differences, however. Strontium is much less potent than calcium (about 10 times in our hands) and Sr^{2+} prolongs the time course of postsynaptic potentials, probably because of persistence of the ion within the terminals.[42] Figure 4 (lower panel) illustrates an experiment where Ca^{2+} (3.4 mmol/L) was replaced by Sr^{2+} (34 mmol/L). As with bafilomycin, the time course of the response was markedly prolonged. Unlike bafilomycin, however, the second response of the paired pulse stimulation was more enhanced that the first one. This was expected from the long persistence of those Sr^{2+} ions that entered during transmission of the first impulse.

These data shed light on the function of the vesicular Ca^{2+}/H^+ antiport in fast synaptic transmission. Rapidly reducing the high Ca^{2+} concentration in the active zone nanodomains, it abbreviates the duration of phasic transmitter release, thereby shortening the duration of the synaptic signal. This constitutes a real advantage for synapses operating at high frequency.

Exocytosis and Calcium Extrusion

Thus, synaptic vesicles provide a rapid and efficient sink in the clearance of Ca^{2+} from the cytosol. We present here evidence that following repetitive stimulation, calcium accumulates in vesicles, displaces ACh from the proteoglycan matrix, and is subsequently expelled from the terminal, probably by exocytosis. The latter processes are important but they do not need to operate at a submillisecond time scale.

Ancient observations suggested that the more synaptic vesicles *in situ* accumulate calcium, the more they lose vesicular acetylcholine. If nerve stimulation is prolonged over the point of transmission failure or low-frequency stimulation is pursued during a long period of time, calcium progressively accumulates while the level of vesicular ACh correlatively decreases.[43,44] Most probably ACh is displaced by Ca^{2+} from binding sites provided by the intravesicular proteoglycan matrix.[45] ACh will escape from the vesicle either by inversion of the vesicular ACh transporter or via the V_0 sector of V-ATPase (it is recalled that the c-subunit of V_0 is similar to the mediatophore proteolipid; it might pass ACh upon Ca^{2+} activation). In this view, vesicular ACh would be delivered from the vesicle into the cytosol at active zones, preventing local exhaustion of the rapidly turning-over free pool of ACh.

Calcium enters nerve terminals during a brief stimulation (for instance 12 s at 100 Hz) and transiently accumulates in synaptic vesicles. Then it returns to resting level during the minutes following the burst (Fig. 5). A substantial proportion of the accumulated calcium

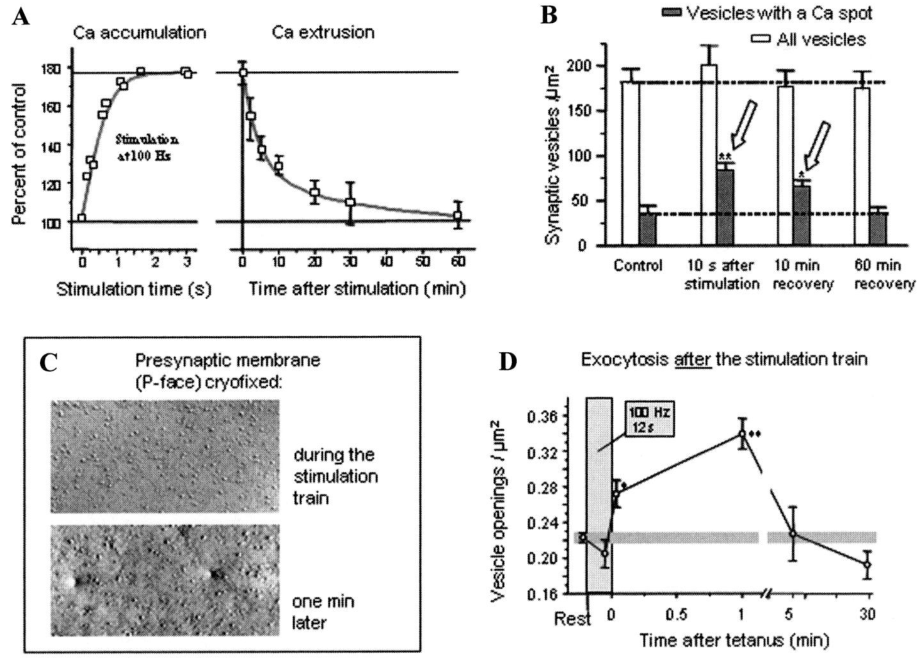

Figure 5. (**A**) Calcium accumulation and extrusion in *Torpedo* electric organ synapses during and after a brief period of stimulation. (**B**) Density of synaptic vesicles before and after the same protocol of stimulation. The total number of vesicles did not vary, but the number of vesicles containing an identifiable calcium precipitate significantly rose after the stimulation train. (**C and D**) Electrogenic tissue was cryofixed and replicas prepared (as in Fig. 1) before, during, and after the stimulation train. The number of vesicle openings (exocytosis or endocytosis) did not change during the stimulation period but it rose very significantly during the following period, in correspondence with the phase of maximum calcium clearance. $^*P < 0.05$; $^{**}P < 0.01$.

is stored in vesicles at the end of stimulation, as shown by the following observation. Under resting conditions a small proportion of synaptic vesicles were found to contain a "calcium spot" that could be identified by using electron microscopy imaging. During and after the stimulation train, the total number of vesicles did not change, but the proportion of vesicles containing a calcium spot significantly increased after the tetanus; it subsequently returns to control value within 30–60 min, in parallel to the decline of the surplus calcium accumulated in terminals (Fig. 5C).[43,46,47] Transient Ca^{2+} accumulation in synaptic vesicles after activity was also demonstrated in cholinergic and other synapses.[48,49] Vesicular Ca^{2+} accumulation was accompanied by a similar increase in the number of calcium spots in the reticulum of presynaptic and postsynaptic structures.

Calcium is subsequently cleared from vesicles, most probably by exocytosis. Indeed, while the density of vesicle openings was not found to increase at the very moment of synaptic transmission (as seen above), it clearly rose during minutes following the train of stimulation, that is, just at the period when the rate of calcium clearance was high (Fig. 5D).[47] The fact that calcium might be delivered to the synaptic cleft via vesicle exocytosis was suggested for a central nervous system synapse as a way to refurnish the ions in this limited space where they are exhausted after prolonged activity.[50]

Conclusion

Rapid transmission is supported by a chain of fast processes working at a high speed, therefore displaying a low affinity. Among them are: ion channel opening, Ca^{2+} activation of mediatophore, Ca^{2+}/H^{+} exchange in vesicles, nicotinic receptor opening, and transmitter hydrolysis in the cleft. In most of these processes, activation is followed by desensitization if the ligand or the stimulus (for instance a voltage change) persists. The above list is, of course, not complete since other synaptic proteins operate at a low affinity and might be included in the list (synaptotagmin could be a candidate). These rapid reactions are assisted by a large number of slower processes often working with higher affinities and which are vital for regulation, fine tuning, and maintenance of the synapse. The latter category encompasses metabotropic presynaptic and postsynaptic receptors, glial elements, Ca^{2+}- and V-ATPases, SNARE (soluble N-ethylmaleimide-sensitive factor attachment protein receptor) proteins, ion channels modulating membrane potential, plasmalemmal and vesicular transporters, and enzymes involved in neurotransmitter synthesis and storage as well as in the energy supply of the synapses. Alteration of any of these regulators is, of course, expected to alter the main chain of rapid reactions, thereby compromising synaptic transmission. For instance, inhibition of Ca^{2+} pumps leads, after some delay, to a steady rise in free Ca^{2+} concentration, a condition that will desensitize both Ca^{2+} channels and mediatophores. Transmission stops in this case (unpublished observations). This does not mean, however, that the transmitter is released through Ca^{2+} pumps.

At fast synapses, this complex network of rapid and slow mechanisms accomplishes the most fascinating task one can imagine—to use chemical reactions for the flash-like transmission of nerve impulses.

Conflicts of Interest

The authors declare no conflicts of interest.

References

1. Sakmann, B., C. Methfessel, M. Mishina, *et al.* 1985. Role of acetylcholine receptor subunits in gating of the channel. *Nature* **318:** 538–543.
2. Katz, B. 1989. Looking back at the neuromuscular junction. In *Neuromuscular Junction.* L.C. Sellin, R. Libelius & S. Thesleff, Eds.: 3–9. Elsevier Science Publisher, Amsterdam.
3. Heuser, J.E., T.S. Reese, M.J. Dennis, *et al.* 1979. Synaptic vesicle exocytosis captured by quick freezing and correlated with quantal transmitter release. *J. Cell. Biol.* **81:** 275–300.
4. Heuser, J.E. & T.S. Reese. 1981. Structural changes after transmitter release at the frog neuromuscular junction. *J. Cell. Biol.* **88:** 564–580.
5. Garcia-Segura, L.M., D. Muller & Y. Dunant. 1986. Increase in the number of presynaptic large intramembrane particles during synaptic transmission at the *Torpedo* nerve- electroplaque junction. *Neuroscience* **19:** 63–79.
6. Muller, D., L.M. Garcia-Segura, A. Parducz & Y. Dunant. 1987. Brief occurrence of a population of large intramembrane particles in the presynaptic membrane during transmission of a nerve impulse. *Proc. Natl. Acad. Sci. USA* **84:** 590–594.
7. Dunant, Y. 2000. Quantal acetylcholine release : Vesicle fusion or intramembrane particles? *Microscopy Research and Technique* **49:** 38–46.
8. Torri-Tarelli, F., F. Grohovaz, R. Fesce & B. Ceccarelli. 1985. Temporal coincidence between synaptic vesicle fusion and quantal secretion of acetylcholine. *J. Cell Biol.* **101:** 1386–1399.
9. Takano, Y. & H.O. Kamiya. 1979. Effect of ionic environment on densities of membrane-associated particles in presynaptic membranes observed in freeze-fractured synaptosomes. *Experientia* **35:** 1076–1078.
10. Israël, M., R. Manaranche, J. Marsal, *et al.* 1980. ATP-dependent calcium uptake by cholinergic synaptic vesicles isolated from *Torpedo* electric organ. *J. Membr. Biol.* **54:** 115–126.
11. Egea, G., J.E. Esquerda, R. Calvet, *et al.* 1987. Structural changes at pure cholinergic synaptosomes during transmitter release by A 23187 in *Torpedo marmorata. Cell Tissue Res.* **248:** 207–214.
12. Bugnard, E., N. Taulier, A. Bloc, *et al.* 2002. Quantal transmitter release by glioma cells: quantification of intramembrane particle changes. *Neuroscience* **113:** 125–135.
13. Morel, N., Y. Dunant & M. Israel. 2001. Neurotransmitter release through the V0 sector of V-ATPase. *J. Neurochem.* **79:** 485–488.
14. Birman, S., F.M. Meunier, B. Lesbats, *et al.* 1990. A 15 kD proteolipid found in mediatophore

preparations from *Torpedo* presents high sequence homology with the bovine chromaffin granule protonophore. *FEBS Lett.* **261:** 303–306.

15. Israël, M., B. Lesbats, N. Morel, *et al.* 1984. Reconstitution of a functional synaptosomal membrane possessing the protein constituents involved in acetylcholine translocation. *Proc. Natl. Acad. Sci. USA* **81:** 277–281.
16. Nelson, N. 1992. The vacuolar H(+)-ATPase–one of the most fundamental ion pumps in nature. *J. Exp. Biol.* **172:** 19–27.
17. Brochier, G., M. Israël & B. Lesbats. 1993. Immunolabelling of the presynaptic membrane of *Torpedo* electric organ nerve terminals with an antiserum towards the acetylcholine releasing protein mediatophore. *Biol. Cell* **78:** 145–154.
18. Israël, M., F.M. Meunier, N. Morel & B. Lesbats. 1987. Calcium-induced desensitization of acetylcholine release from synaptosomes or proteoliposomes equiped with mediatophore, a presynaptic membrane protein. *J. Neurochem.* **49:** 975–982.
19. Cavalli, A., L. Eder-Colli, Y. Dunant, *et al.* 1991. Release of acetylcholine from *Xenopus* oocytes injected with nRNAs from cholinergic neurons. *EMBO J.* **10:** 1671–1675.
20. Cavalli, A., Y. Dunant, C. Leroy, *et al.* 1993. Antisense probes against mediatophore block transmitter release in oocytes primed with neuronal mRNAs. *Eur. J. Neurosci.* **5:** 1539–1544.
21. Israël, M., Y. Dunant & R. Manaranche. 1979. The present status of the vesicular hypothesis. *Prog. Neurobiol.* **13:** 237–275.
22. Dunant, Y. & M. Israël. 2000. Neurotransmitter release in rapid synapses. *Biochimie* **82:** 289–302.
23. Katz, B. 1969. *The release of neural transmitter substances.* University Press, Liverpool, p. 60.
24. Falk-Vairant, J., F.M. Meunier, B. Lesbats, *et al.* 1996. Cell lines expressing an acetylcholine release mechanism, correction of a release-defficient cell by mediatophore transfection. *J. Neurosci. Res.* **45:** 195–201.
25. Falk-Vairant, J., P. Corrèges, L. Eder-Colli, *et al.* 1996. Quantal acetylcholine release induced by mediatophore transfection. *Proc. Natl. Acad. Sci. USA* **93:** 5203–5207.
26. Bloc, A., E. Bugnard, Y. Dunant, *et al.* 1999. Acetylcholine synthesis and quantal release reconstituted by transfection of mediatophore and choline acetyltransferase cDNAs. *Eur. J. Neurosci.* **11:** 1523–1534.
27. Girod, R., P. Corrèges, J. Jacquet & Y. Dunant. 1993. Space and time characteristics of transmitter release at the nerve-electroplaque junction of Torpedo. *J. Physiol. Lond.* **471:** 129–157.
28. Llinas, R., I.Z. Steinberg & K. Walton. 1981. Relationship between presynaptic calcium current and postsynaptic potential in squid giant synapse. *Biophys. J.* **33:** 323–352.
29. Llinas, R., M. Sugimori & R.B. Silver. 1992. Microdomains of high calcium concentration in a presynaptic terminal. *Science USA* **256:** 677–679.
30. Yazejian, B., X.P. Sun & A.D. Grinnell. 2000. Tracking presynaptic Ca^{2+} dynamics during neurotransmitter release with Ca^{2+}-activated K+ channels. *Nat. Neurosci.* **3:** 566–571.
31. Castonguay, A. & R. Robitaille. 2001. Differential regulation of transmitter release by presynaptic and glial Ca^{2+} internal stores at the neuromuscular synapse. *J. Neurosci.* **21:** 1911–1922.
32. Couteaux, R. & M. Pécot-Dechavassine. 1970. Vésicules synaptiques et poches au niveau des "zones actives" de la jonction neuromusculaire. *C. R. Acad. Sci. Paris* **271:** 2346–2349.
33. Michaelson, D.M., I. Ophir & I. Angel. 1980. ATP-stimulated Ca^{2+} transport into cholinergic *Torpedo* synaptic vesicles. *J. Neurochem.* **35:** 116–124.
34. Gonçalves, P.P., S.M. Meireles, P. Neves & M.G. Vale. 2000. Distinction between Ca(2+) pump and Ca(2+)/H(+) antiport activities in synaptic vesicles of sheep brain cortex. *Neurochem. Int.* **37:** 387–396.
35. Roberts, W.M. 1994. Localization of calcium signals by a mobile calcium buffer in frog saccular hair cells. *J. Neurosci.* **14:** 3246–3262.
36. Neher, E. 1998. Vesicle pools and Ca^{2+} microdomains: New tools for understanding their roles in neurotransmitter release. *Neuron* **20:** 389–399.
37. Gonçalves, P.P., S.M. Meireles, C. Gravato & M.G. Vale. 1998. Ca^{2+}-H^{+}-Antiport activity in synaptic vesicles isolated from sheep brain cortex. *Neurosci. Lett.* **247:** 87–90.
38. Gonçalves, P.P., S.M. Meireles, P. Neves & M.G. Vale. 1999. Synaptic vesicle Ca^{2+}/H+ antiport: dependence on the proton electrochemical gradient. *Brain Res. Mol. Brain Res.* **71:** 178–184.
39. Cordeiro, J.M., V. Bancila, A. Bloc, *et al.* 2004. A calcium-proton antiport in presynaptic nerve terminals. In *Cholinergic Mechanisms, Function and Dysfunction.* Israel Silman & et al. Eds.: 527–529. Martin Dunitz, London.
40. Gonçalves, P.P., S.M. Meireles, P. Neves & M.G. Vale. 1999. Ionic selectivity of the Ca^{2+}/H^{+} antiport in synaptic vesicles of sheep brain cortex Molec. *Brain Res.* **67:** 283–291.
41. Locke, F.S. 1894. Notiz über den Einfluss physiologischer Kochsalzlösung auf die elektrische. *Erregbarkeit von Muskel und Nerv Centralblatt für Physiologie* **5:** 166–167.
42. Xu-Friedman, M.A. & W.G. Regehr. 2000. Probing fundamental aspects of synaptic transamission with strontium. *J. Neurosci.* **20:** 4414–4422.

43. Babel-Guérin, E. 1974. Métabolisme du calcium et libération de l'acétylcholine dans l'organe électrique de la Torpille. *J. Neurochem.* **23:** 525–532.
44. Schmidt, R., H. Zimmermann & V.P. Whittaker. 1980. Metal ion content of cholinergic synaptic vesicles from the electric organ of *Torpedo*: effect of stimulation-induced transmitter release. *Neuroscience* **5:** 625–638.
45. Reigada, D., I. Diez-Perez, P. Gorostiza, *et al.* 2003. Control of neurotransmitter release by an internal gel matrix in synaptic vesicles. *Proc. Natl. Acad. Sci. USA* **100:** 3485–3490.
46. Parducz, A. & Y. Dunant. 1993. Transient increase in calcium in synaptic vesicles after stimulation. *Neuroscience* **52:** 27–33.
47. Parducz, A., F. Loctin, E. Babel-Guérin & Y. Dunant. 1994. Exo-endocycytotic images following tetanic stimulation at a cholinergic synapse. A role in calcium extrusion? *Neuroscience* **62:** 93–103.
48. Parducz, A., J. Toldi, F. Joo, *et al.* 1987. Transient increase of calcium in pre- and postsynaptic organelles of rat superior cervical ganglion after tetanizing stimulation. *Neuroscience* **23:** 1057–1061.
49. Buchs, P.A., L. Stoppini, A. Parducz, *et al.* 1994. A new cytochemical method for the ultrastructural localization of calcium in the central nervous system. *J. Neurosci. Meth.* **54:** 83–93.
50. Borst, J.G.G. & B. Sakmann. 1999. Depletion of calcium in the synaptic cleft of a calyx-type synapse in the rat brainstem. *J. Physiol. Lond.* **521:** 123–133.

Single Molecule Probing of SNARE Proteins by Atomic Force Microscopy

Wei Liu and Vladimir Parpura

Department of Neurobiology, Center for Glial Biology in Medicine, Atomic Force Microscopy and Nanotechnology Laboratories, Civitan International Research Center, Evelyn F. McKnight Brain Institute, University of Alabama, Birmingham, Alabama, USA

Atomic force microscopy (AFM) in force spectroscopy mode has recently emerged as a technique of choice for studying mechanical interactions between the proteins of the core soluble *N*-ethylmaleimide-sensitive fusion protein attachment protein attachment protein receptor (SNARE) complex. In these experiments, the rupture force, extension, spontaneous dissociation time, and interaction energy for SNARE protein–protein interactions can be obtained at the single molecule level. These measurements, which are complementary to results and conclusions drawn from other techniques, improve our understanding of the role of the SNARE complex in exocytosis.

Key words: **SNARE complex; exocytosis; atomic force microscope; single molecule; dynamic force spectroscopy; energy**

Introduction

Since its invention in 1986,[1] the atomic force microscope (AFM) has been widely used in biological research. This microscope utilizes an ultrasensitive microcantilever that can probe biological molecules, such as proteins and DNA, at the subnanometer vertical scale resolution.[2,3] Recently, AFM has emerged as the technique of choice for studying mechanical interactions between the proteins of the core soluble *N*-ethylmaleimide-sensitive fusion protein attachment protein receptor (SNARE) complex.[4–6] In contrast to traditional bulk chemical reaction experiments, AFM probing can report on events at the single molecule level. Here, pairs of interacting proteins are brought together in order to interact and then taken back apart. The magnitude of the stretching force of the intermolecular bonds, within piconewton (pN) range, and the stretching length, within nanometer (nm) range, can be measured. In this article we focus our attention on a subset of work that used AFM techniques to study intermolecular SNARE protein interactions.[4,6,7] We first discuss the AFM instrumentation in force spectroscopy mode, followed by a description of a directional protein deposition method. Various measurements achieved on SNARE proteins using these approaches are discussed.

Instrumentation

A basic AFM system consists of a base and a controller. The base includes a piezoelectric element with a sample plate located underneath a "head" which contains a built-in diode laser and a photo detector. The controller, directed by a computer, operates the equipment and performs analog–digital/digital–analog conversion. The AFM operates by dragging a probe tip across the sample. A laser beam is focused onto the back of the cantilever containing the integral tip (Fig. 1). The reflected laser signal is collected by the photo detector. This optical lever design effectively amplifies the minute deflection of cantilever/tip

Address for correspondence: Vladimir Parpura, Department of Neurobiology, 1719 6th Avenue South, CIRC 429, University of Alabama, Birmingham, AL 35294. Voice: +205-996-7369. vlad@uab.edu

Mechanisms of Exocytosis: Ann. N.Y. Acad. Sci. 1152: 113–120 (2009).
doi: 10.1111/j.1749-6632.2008.03991.x

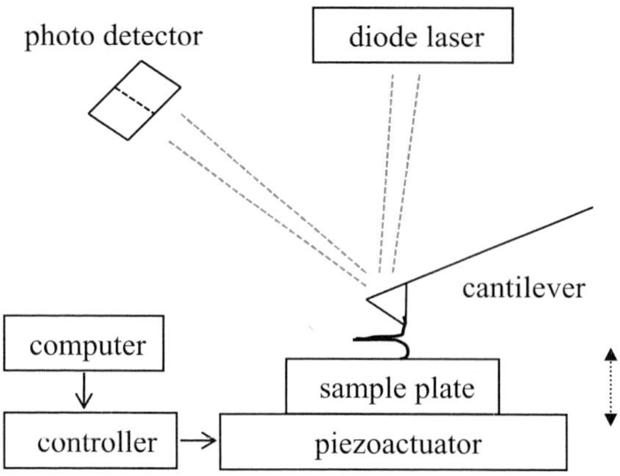

Figure 1. Schematic of experimental force spectroscopy setup. Recombinant proteins are attached to the cantilever tip and coverslip through a Ni^{2+}–six histidine residue coordination. The piezo is used to first move the coverslip, attached to the sample plate at the top of the piezo, up toward (approach, arrow pointing up) the cantilever tip to establish contact and then down (retraction, arrow pointing down) to exert force on the intermolecular bond. The detection of cantilever deflection is achieved using a laser beam reflected off the back of the cantilever, which is collected by a photo detector. This detector is interfaced with a computer via a controller, which also generates signals to control the movement of the piezoactuator.

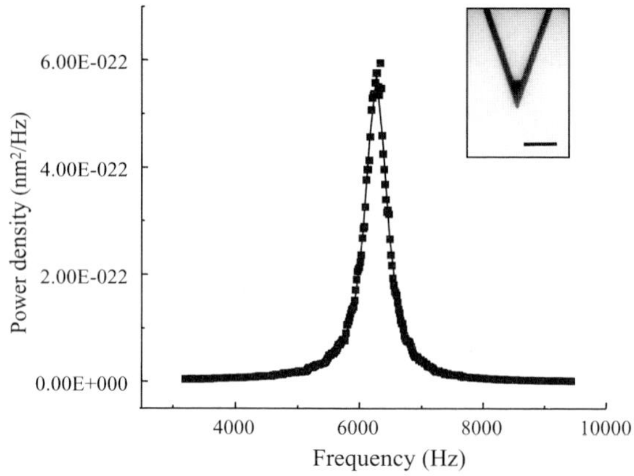

Figure 2. A typical power density spectrum of a cantilever. A bright field image of a triangular (320 μm-long) silicon nitride cantilever is shown in inset. Integration of power density over frequency gives the average square amplitude $\langle x^2 \rangle$ that can be used to calculate the spring constant k. Scale bar in inset, 100 μm.

resulting in a large excursion of the laser beam on the photo diode. The distance between the tip and the sample is controlled by the piezoactuator which expands or retracts depending on the voltage applied. AFM can operate in force spectroscopy mode. Here, by delivering a voltage ramp from the controller to the piezo element, the sample surface is brought into contact with the tip. If the contacting surfaces, the cantilever tip and the glass coverslip (or mica) attached to the sample plate, are functionalized with interacting proteins, then during the contact time, usually lasting approximately 0.5–3 s, a bond can be formed between a protein attached to the tip and an interacting protein attached to the substrate. When the tip is then retracted from the substrate, it bends from intermolecular bonding force. The deflection signal of this bending is recorded as a function of the distance the piezoactuator moved.

The interaction force F is proportional to the deflection magnitude Δx of the tip, $F = k * \Delta x$, where k is the spring constant of the cantilever. Thus, the calibration of k and Δx is the key element in adequately measuring the force and distance/extension. There are a few methods to obtain the spring constant of the cantilever.[8] Perhaps the most common method used is based on the detection of cantilever vibration from thermal noise. According to the equipartition theorem, thermal energy $1/2k_BT$ is equal to vibration energy $1/2k< x^2>$ of the cantilever at a given temperature, where k_B is the Boltzmann constant, T is absolute temperature, and x is vibration amplitude, so that the spring constant is:

$$k = k_B T/\langle x^2 \rangle. \tag{1}$$

One way to obtain the average square amplitude of vibration is by using the Fourier transform. One can measure the amplitude in the frequency domain and then convert it into the time domain by Fourier transform. A typical power density spectrum of cantilever vibration is shown in Figure 2. After integrating the peak, one can get $\langle x^2 \rangle$, which can then be used to calculate the spring constant k of the cantilever.

Most commonly used in biological applications cantilevers have spring constants of approximately 0.01 N/m, offering good force sensitivity, although stiffer cantilevers could be used. Calibration of the deflection magnitude Δx is actually linked to the calibration of the piezoactuator. This can be done by imaging a reference sample surface and then making appropriate adjustments to the equipment, usually via a computer-based interface. An alternate method is to use an interferometer to generate an interference pattern from the movement of the piezoactuator and then define the relationship between the wavelength of light used and the displacement of the piezoactuator.[9] Again, the adjustments to the equipment are performed by the click of a computer mouse. Generally, interferometric calibration is much more accurate. At this juncture, it is worth mentioning that the experiments investigating interactions between proteins require working in an aqueous environment, thus a fluid cell is a necessary accessory for the microscope.

Single Molecule Dynamic Force Spectroscopy

As alluded to earlier, the AFM makes it possible to study the interaction of proteins at the single molecule level when used in force spectroscopy mode. Here, an intermolecular bond between interacting proteins can be formed and broken repeatedly, while the force and extension required for rupturing this bond are recorded. Assuming that there is only one energy barrier to overcome during this process, then when an external force F is applied to the bond by the cantilever, the energy barrier E_B will be lowered by the applied force to $E_B - F * x_\beta$, where x_β is the barrier width (Fig. 3). This can be used to obtain spontaneous dissociation times of protein–protein interactions. According to thermal dynamics,[10,11] the off-rate of a bond k_{off} is related to the rupture force:

$$k_{off} = k_D * \exp(-(E_B - F * x_\beta)/k_B T), \quad (2)$$

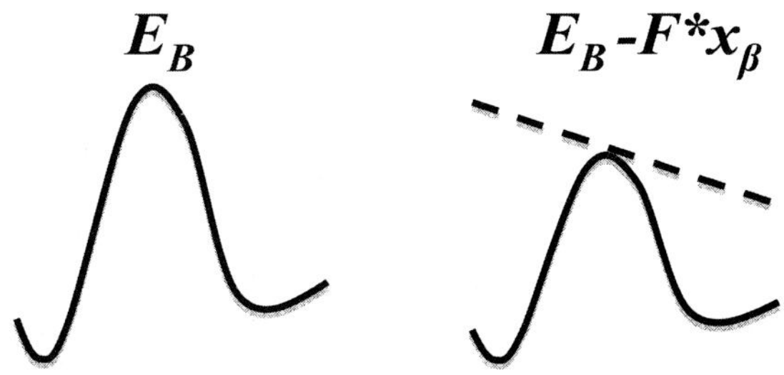

Figure 3. Energy landscape of a single activation barrier bond. When an external force F is applied, the energy barrier E_B is lowered (dashed line); x_β is the barrier width.

where k_D is a coefficient that is related to the thermal frequency of the bond. Consequently, the spontaneous dissociation rate k^0_{off} can be obtained by measuring rupture forces at different force loading rates, as given by chemical reaction rate theory. Here the applied force loading rate can be adjusted by varying the extension/retraction of the piezoelectric tube as $dF/dt = k * v$, where k is the spring constant of the cantilever and v is the movement speed of the sample plate. The relationship between the rupture force and the force loading rate can then be fitted to the equation:

$$F = k_B T/x_\beta * \ln((dF/dt) * x_\beta * \tau_0/k_B T). \quad (3)$$

When the applied force is set to zero, then the spontaneous lifetime τ_0 $(1/k^0_{off})$ of the bond can be obtained:

$$1/k^0_{off} = \tau_0 = k_B T/x_\beta/(dF/dt)_{F=0}. \quad (4)$$

Thus, when the forces needed to take apart interacting SNARE proteins are measured at various pulling speeds, the spontaneous lifetimes of the protein–protein interactions can be obtained. Indeed, this approach has been used for studying the interactions between SNARE proteins (see below).[4]

Besides being a powerful tool to study force and off-rate in single molecule interactions, dynamic force spectroscopy can also be used to obtain energetic information of a bond. Hence, from Eq. (2), the off-rate is a temperature-dependent parameter. If one measures off-rates at different temperatures, then one can obtain the interaction energy by fitting the relationship between the off-rate and inverse temperature, referred to as the Arrhenius plot (see below; Fig. 6). It is important to realize that here the interaction energy actually is the adhesion energy between interacting protein molecules. In dynamic force spectroscopy, the adhesion energy corresponds to the change in enthalpy (ΔH) of the intermolecular interactions according to $\Delta H = \Delta G + T\Delta S$, where ΔG is the change in free energy and ΔS is the change in entropy. Taking into account that ΔG is negligible over the relatively narrow temperature range used in biological experiments[12] studying exocytosis, i.e., from 277 K (+4°C) to approximately 307 K (34°C), one can calculate all three energy parameters from a fit to Eq. (2).

Sample Preparation

In AFM single-molecule probing experiments, the type of procedure used for depositing proteins on the tip and substrate is critical. A variety of methods have been used to attach proteins onto cantilevers and substrates, including covalent (e.g., Au-S) and noncovalent bonds (e.g., biotin–avidin-based attachment), as reviewed elsewhere.[13] For deposition of SNARE proteins two different approaches have been used thus far. Yersin *et al.* used glutaraldehyde cross-linking to attach SNARE proteins onto the AFM tip and mica surface.[5] The aminopropyltrietoxysilane (ATEPS)-treated mica/tip surface was exposed to protein solution containing glutaraldehyde so that amines in ATEPS and lysines in proteins were cross-linked, thus allowing proteins to covalently bind to the mica/tip. Although this approach can immobilize SNARE proteins on tips and mica, the amine cross-linker might also cross-link the lysines within a single protein molecule or among multiple molecules, resulting in large aggregates that can obscure the interpretation of data gathered by force spectroscopy. Preferably, one should employ a directional deposition of SNARE proteins via sterical coordination.[4,14] Here, both the cantilever and the glass coverslip were coated with a thin nickel film applied by thermal evaporation. After allowing the nickel to partly oxidize in air to form Ni^{2+}, recombinant SNARE proteins, containing tags of six consecutive histidines (His_6) at their C-termini, were applied in aqueous solution. As a result of the binding between Ni^{2+} and His_6, the proteins became attached. The measured strength of Ni^{2+}–His_6 bond was more than double the forces generated between proteins of the SNARE complexes. Thus, a Ni^{2+}–His_6 bond can be effectively used to attach SNARE proteins to the nickel-coated tips and coverslips in order to study them by force spectroscopy, as outlined below. Furthermore, this linking method helps SNARE proteins to interact freely in a more physiologically prevalent parallel fashion.[15–17] Once the attachment of proteins to the coverslips and tips is completed, the experiments can start right away, or alternatively, such functionalized tips and substrates can be stored in a refrigerator at 277 K (+4°C) overnight until used in experiments.

Measurements

As indicated earlier, proteins deposited on the tip and substrate (glass coverslip) will form a bond when they are brought into contact by the piezoelectric element. Then the piezo is used to retract the coverslip from the cantilever. The typical retraction part of a force curve is shown in Figure 4. Here, the tip, cofunctionalized with syntaxin 1A (Sx1A) and synaptosome-associated protein of 25 kDa isoform B (SNAP25B), was brought into contact with the glass coverslip, which was functionalized with synaptobrevin 2 (Sb2). Upon contact the ternary complex can form. The line *abc* is a result of direct contact between the coverslip and the cantilever tip. The section *def* was confirmed to be specific to this interaction as it was absent when either the cantilever tip or the coverslip or both remained unfunctionalized. As the tip and coverslip are

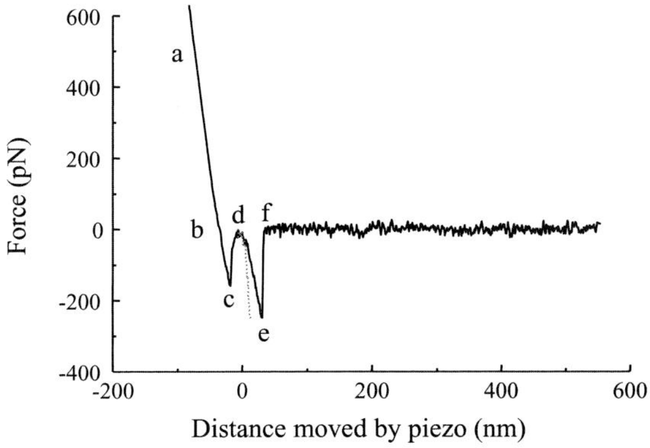

Figure 4. The retraction part of a typical force–distance (extension) curve using a Sx1A/SNAP25B-functionalized tip and a Sb2-functionalized coverslip. In the segments *ab* and *bc* the coverslip and the cantilever tip are still in contact. The *bcd* segment is from nonspecific interactions that are present even in the absence of any proteins. The Sx1A-SNAP25B-Sb2 intermolecular "bond" starts to be extended at point *d*. The increasing extension as the coverslip moves further away from the tip leads to an increased application of force on the intermolecular bond until it ruptures at point *e*. The segment *ef* is then the measure of the force (ordinate) necessary to remove Sx1A-SNAP25B-Sb2 interaction. The extension induced can be calculated from the *z*-axis distance moved by the piezo (abscissa) given by segment *de*. The dotted line indicates the extension after the distance has been corrected for the bending of the cantilever. Retraction velocity is 1.6 μm/s.

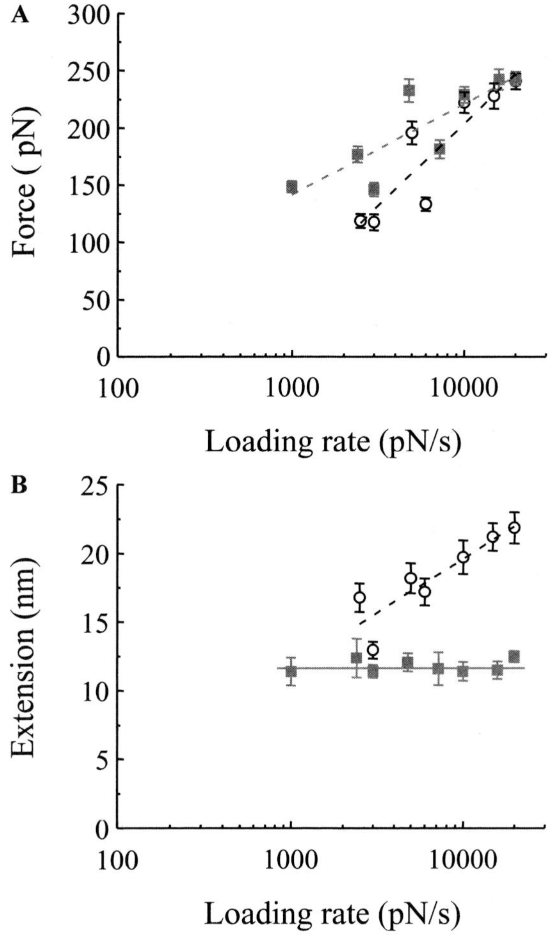

Figure 5. The force and extension values for the dissociation of SNARE proteins as a function of the force loading rate. (**A**) The force necessary to take apart Sx1A-Sb2 in the absence (open circles) or presence (squares) of SNAP25B increases exponentially with an increase in the loading rate. (**B**) The extension changes exponentially with loading rate when Sx1A-Sb2 interactions are ruptured, but it remains unchanged (constant at approximately12 nm) when measuring Sx1A-Sb2-SNAP25B interactions (adopted from Liu *et al.*[4]). The dashed lines indicate fits to the data, while the solid line indicates that the extension value is constant.

further separated, this leads to the increased application of force to the intermolecular bond until it ruptures at point *e* and the cantilever returns to its free equilibrium position *f*. The segment *ef* is the force necessary to rupture the Sx1A-SNAP25B-Sb2 bond, while the segment *de* reports on the extension of molecules within the individual ternary SNARE complex. It should be noted that the displacement of the piezoactuator in the segment *de* does not report on the true extension of the intermolecular interaction because of the bending of the cantilever from the application of force. Consequently, one needs to subtract such bending deflection from the displacement of the cantilever to obtain the correct value for the intermolecular extension (Fig. 5; dotted line). The outlined experiment is repeated, and, because of thermal fluctuation, one can obtain a distribution histogram of the rupture force of the bond with its mean and standard error rather than a single value. Using such an approach, Liu *et al.* obtained an interaction force of 237 ± 4 pN and 243 ± 5 pN for Sx1A-Sb2 and Sx1A-SNAP25B-Sb2 interactions, respectively, at a retraction velocity of 1.6 μm/s (corresponds to a force loading rate of 20 nN/s).[4] Additionally, the extension values of 23.0 ± 0.6 nm for Sx1A-Sb2 and 12.5 ± 0.4 nm for the ternary Sx1A-SNAP25B-Sb2 interactions were obtained.

To gain an insight into the nature of the interactions within the binary (Sx1A-Sb2) and ternary SNARE complexes, one needs to measure force and extension at the point of rupture of single intermolecular bonds as a function of the force loading rate (Fig. 5).[4] The measured rupture force increases exponentially with the loading rate. By extrapolating the loading force rate to zero, one can estimate dissociation rates which correspond to the spontaneous off-rates (k^0_{off}) when only a single barrier width to the transition state exists. In the case of Sx1A-Sb2 interaction, a spontaneous dissociation lifetime (τ_0) of 0.16 s can be obtained, while the ternary SNARE complex is substantially more stable, displaying a spontaneous dissociation lifetime of 2.1 s; both measurements were acquired at room temperature (24°C). The extension in the case of Sx1A-Sb2 exponentially increases as a function of the force loading rate. Thus, when combined with force measurements under various loads, this feature of Sx1A-Sb2 interaction points to a zipper-type nonlocalized interaction throughout the entire SNARE domains of these two proteins. In stark contrast, the extension measurements with the ternary SNARE complex are constant as the loading rate is varied. When this fact is combined with the observation that the rupture force increases exponentially with increasing loading rate, it implicates a cuffing role for SNAP25B when interacting with Sx1A and Sb2. This type of interaction refers to a strong and localized intermolecular bond localized at approximately12 nm extension, corresponding to the so-called 0 layer in X-ray crystallographic data.[18] Based on these measurements, Liu *et al.* further proposed that there could be two modes of intracellular vesicular positioning with respect to the plasma membrane.[4] For vesicle–membrane separation distances between 12–23 nm, a Sx1A-Sb2 bond could be formed. At distances less than 12 nm, the ternary SNARE complex oriented in parallel would play the major role in vesicular positioning and such interaction could last for an order of magnitude longer than the one mediated by Sx1A-Sb2 in the absence of SNAP25B.

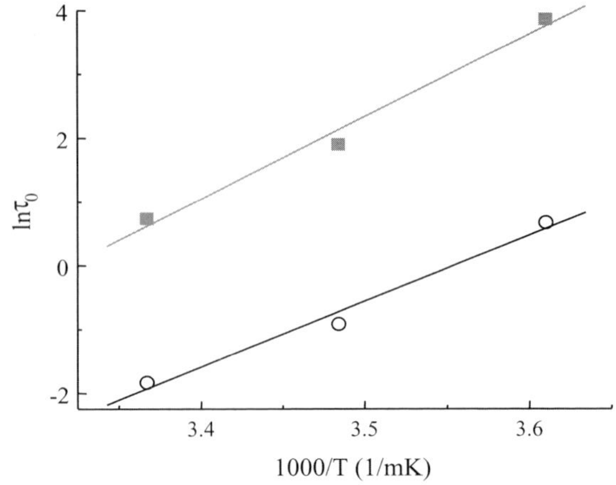

Figure 6. The natural logarithm of the mean lifetime τ_0 of the bound system plotted as a function of inverse temperature. Enthalpy (ΔH) and entropy (ΔS) can be obtained for the Sx1A-Sb2 in the absence (open circles) or presence (squares) of SNAP25B from the slope and intercept, respectively (modified form Liu *et al.*[6,7]).

Using the spontaneous dissociation lifetime of a bond obtained at various temperatures, information about the bond energy can be obtained. For example, Liu *et al.*[6] measured spontaneous lifetimes of binary Sx1A-Sb2 interactions at 277 K (+4°C), 287 K (+14°C) and 297 K (+24°C). A natural logarithm (Arrhenius) plot of spontaneous lifetime versus inverse temperature is shown in Figure 6. Since $k_D = k_a * \exp(\Delta S/k_B)$, Eq. (2) for the spontaneous dissociation rate can be rewritten as:

$$k^0_{off} = k_a * \exp(\Delta S/k_B) * \exp(-\Delta H/k_B T), \quad (5)$$

where k_a is $6.2 \times 10^{12}\ s^{-1}$ for the Eyring model or $3.3 \times 10^9\ s^{-1}$ for Kramer's model of chemical reactions for the interaction between proteins.[11,12] By using a linear fit to the data on an Arrhenius plot, one can obtain the slope and the intercept, corresponding to changes in enthalpy and entropy of the interaction, respectively. Hence, an enthalpy of the binary complex of 33 ± 6 $k_B T$ was obtained, while entropy values were $0.018 \pm 0.019\ k_BT \times K^{-1}$ or $0.042 \pm 0.019\ k_BT \times K^{-1}$ for the Eyring model or Kramer's model, respectively. Similar measurements for the ternary Sx1A-SNAP25B-Sb2 complex can be achieved with an enthalpy of $43 \pm 5\ k_BT$.[7] If the formation of binary and ternary complexes would lead to an energy release equivalent to that of their measured enthalpies, then, based on the 10 k_BT higher enthalpy measured for the ternary system along with the shorter extensions required to rupture the complex, this would implicate SNAP25B in profoundly increasing the probability for membrane fusion. Alternatively, if the energy release during the formation of complexes could not be applied toward vesicular fusions but rather is simply consumed in the tethering/docking of vesicles at the plasma membrane, this would indicate that there is a higher energy demand required to keep vesicles at half the distance and an order of magnitude longer duration in the presence of SNAP25B. Of course, these two possibilities accounting for enthalpy are not mutually exclusive.

In addition to using a temperature-dependent approach to measure energy parameters between interacting proteins as shown above, one can use the Jarzynski equality theorem,[19,20] where experimental data can be obtained at a single temperature point. Indeed, Liu *et al.*[6] obtained interaction energies of the Sx1A-Sb2 binary complex with values comparable to those obtained from the single molecule reaction rate theory.

The measurement of energy can also be used to cross-check for internal consistency in data on spontaneous lifetime. Here, from free energy ΔG, one can obtain the spontaneous lifetime according to:

$$k^0_{off} = k_a * \exp(-\Delta G/k_BT). \qquad (6)$$

Note that the free energy ΔG is used but not the enthalpy ΔH. For example, based on the ternary complex free energy of 23 k_BT from Liu *et al.*,[7] the spontaneous lifetime equates to approximately 3 s, a value that is in good agreement with the 2.1 s obtained from the force loading rate experiment.[4] Interestingly, if the activation energy ΔH of 43 k_BT is inappropriately applied instead of ΔG, then the "spontaneous lifetime" would be more than several days. Such a value would be consistent with reports of spontaneous lifetimes obtained using the surface force apparatus, as reported elsewhere.[21]

Concluding Remarks

The intent of this article was to provide a short overview of the use of the AFM in force spectroscopy mode to investigate exocytotic/SNARE proteins. Single molecule mechanical probing by AFM is an effective approach to study interactions between SNARE proteins. The interaction force, extension, spontaneous dissociation time, and energy of SNARE protein–protein interactions can be obtained. These results have improved our understanding of the role of the SNARE complex in exocytosis and are complementary to results and conclusions drawn from other techniques.

Acknowledgments

The authors' work is supported by the National Institute of Mental Health (MH 069791). We thank Dr. Erik B. Malarkey for comments on previous versions of this manuscript.

Conflicts of Interest

The authors declare no conflicts of interest.

References

1. Binnig, G., C.F. Quate & C. Gerber. 1986. Atomic force microscope. *Phys. Rev. Lett.* **56:** 930–933.
2. Florin, E.L., V.T. Moy & H.E. Gaub. 1994. Adhesion forces between individual ligand-receptor pairs. *Science* **264:** 415–417.
3. Lee, G.U., L.A. Chrisey & R.J. Colton. 1994. Direct measurement of the forces between complementary strands of DNA. *Science* **266:** 771–773.

4. Liu, W. *et al.* 2006. Single molecule mechanical probing of the SNARE protein interactions. *Biophys. J.* **91:** 744–758.
5. Yersin, A. *et al.* 2003. Interactions between synaptic vesicle fusion proteins explored by atomic force microscopy. *Proc. Natl. Acad. Sci. USA* **100:** 8736–8741.
6. Liu, W. *et al.* 2008. Comparative energy measurements in single molecule interactions. *Biophys. J.* **95:** 419–425.
7. Liu, W. *et al.* 2008. Single molecule measurements of interaction free energies between the proteins within binary and ternary SNARE complexes. *J. Nanoneurosci.* **1:** In Press.
8. Florin, E.L. *et al.* 1995. Sensing specific molecular-interactions with the atomic-force microscope. *Biosens. Bioelectron.* **10:** 895–901.
9. Chen, F. & U. Mohideen. 2001. Fiber optic interferometry for precision measurement of the voltage and frequency dependence of the displacement of piezoelectric tubes. *Rev. Sci. Instr.* **72:** 3100–3102.
10. Bell, G.I. 1978. Models for the specific adhesion of cells to cells. *Science* **200:** 618–627.
11. Evans, E. 2001. Probing the relation between force–lifetime–and chemistry in single molecular bonds. *Annu. Rev. Biophys. Biomol. Struct.* **30:** 105–128.
12. Atkins, P.W. 1990. *Physical chemistry*. W.H. Freeman and Company. New York.
13. Montana, V. *et al.* 2008. Single molecule probing of exocytotic protein interactions using force spectroscopy. *Croat. Chem. Acta* **81:** 31–40.
14. Liu, W. *et al.* 2003. Botulinum toxin type B micromechanosensor. *Proc. Natl. Acad. Sci. USA* **100:** 13621–13625.
15. Bowen, M.E. *et al.* 2004. Single molecule observation of liposome-bilayer fusion thermally induced by soluble N-ethyl maleimide sensitive-factor attachment protein receptors (SNAREs). *Biophys. J.* **87:** 3569–3584.
16. Lin, R.C. & R.H. Scheller. 1997. Structural organization of the synaptic exocytosis core complex. *Neuron* **19:** 1087–1094.
17. Hanson, P.I. *et al.* 1997. Structure and conformational changes in NSF and its membrane receptor complexes visualized by quick-freeze/deep-etch electron microscopy. *Cell* **90:** 523–535.
18. Sutton, R.B. *et al.* 1998. Crystal structure of a SNARE complex involved in synaptic exocytosis at 2.4 A resolution. *Nature* **395:** 347–353.
19. Jarzynski, C. 1997. Nonequilibrium equality for free energy differences. *Phys. Rev. Lett.* **78:** 2690–2693.
20. Jarzynski, C. 1997. Equilibrium free-energy differences from nonequilibrium measurements: A master-equation approach. *Phys. Rev. E.* **56:** 5018–5035.
21. Li, F. *et al.* 2007. Energetics and dynamics of SNAREpin folding across lipid bilayers. *Nat. Struct. Mol. Biol.* **14:** 890–896.

Identifying Critical Components of Native Ca^{2+}-triggered Membrane Fusion

Integrating Studies of Proteins and Lipids

Kendra L. Furber,[a,d] **Matthew A. Churchward,**[a,*] **Tatiana P. Rogasevskaia,**[a] **and Jens R. Coorssen**[a,b,c,d]

[a]*Department of Physiology and Biophysics, University of Calgary, Faculty of Medicine, Calgary, Canada*

[b]*Department of Biochemistry and Molecular Biology, University of Calgary, Faculty of Medicine, Calgary, Canada*

[c]*Department of Cell Biology and Anatomy, University of Calgary, Faculty of Medicine, Calgary, Canada*

[d]*Hotchkiss Brain Institute, University of Calgary, Faculty of Medicine, Calgary, Canada*

Ca^{2+}-triggered membrane fusion is the defining step of exocytosis. Despite realization that the fusion machinery must include lipids and proteins working in concert, only of late has work in the field focused more equally on both these components. Here we use isolated sea urchin egg cortical vesicles (CV), a stage-specific preparation of Ca^{2+}-sensitive release-ready vesicles that enables the tight coupling of molecular and functional analyses necessary to dissect molecular mechanisms. The stalk-pore hypothesis proposes that bilayer merger proceeds rapidly via transient, high-negative curvature, intermediate membrane structures. Consistent with this, cholesterol, a major component of the CV membrane, contributes to a critical local negative curvature that supports formation of lipidic fusion intermediates. Following cholesterol depletion, structurally dissimilar lipids having intrinsic negative curvature greater than or equal to cholesterol recover the ability of CV to fuse but do not recover fusion efficiency (Ca^{2+} sensitivity and kinetics). Conversely, cholesterol- and sphingomyelin-enriched microdomains regulate the efficiency of the fusion mechanism, presumably by contributing spatial and functional organization of other critical lipids and proteins at the fusion site. Critical proteins are thought to participate in Ca^{2+} sensing, initiating membrane deformations, and facilitating fusion pore expansion. Capitalizing on a novel effect of the thiol-reactive reagent iodoacetamide (IA), potentiation of the Ca^{2+} sensitivity and kinetics, a fluorescently tagged IA has been used to enhance fusion efficiency and simultaneously label the proteins involved. Isolation of cholesterol-enriched CV membrane fractions, using density gradient centrifugation, is being used to narrow the list of protein candidates potentially critical to the mechanism of fast Ca^{2+}-triggered membrane fusion.

Key words: **calcium; cholesterol; exocytosis; negative curvature; secretory vesicle; sphingomyelin; stalk-pore; microdomain; thiol-reactivity**

Introduction

Membrane fusion, the merger of two apposed native (e.g., proteo-lipid) bilayers, is a fundamental mechanism in the processes of membrane trafficking and exocytosis. As the defining step of regulated secretion,

Address for correspondence: Jens R. Coorssen, School of Medicine (Bldg 30), University of Western Sydney, Campbelltown, NSW 1797, Australia. j.coorssen@uws.edu.au or jcoorsse@ucalgary.ca

*Authors contributed equally to this study

Mechanisms of Exocytosis: Ann. N.Y. Acad. Sci. 1152: 121–134 (2009).
doi: 10.1111/j.1749-6632.2008.03993.x

Ca^{2+}-triggered membrane fusion has been under intensive investigation over the past two decades, yet the underlying molecular mechanisms still remain somewhat poorly understood. Despite realization that the fusion machinery must include lipids and proteins working in concert, only of late has work in the field focused more equally on both these components. The majority of evidence indicates that membrane merger proceeds via a lipidic rather than a proteinaceous fusion pore; as the lipidic components are much more abundant than proteins in the membrane, the energy requirements for bilayer merger are intimately linked with lipid rearrangements. Indeed, this is the hallmark of the stalk-pore hypothesis.[1,2] Proteins on the other hand are thought to be critical in targeting, priming, and vesicle docking and in the triggering and modulation of the fusion process. As the fusion steps require the involvement of both lipids and proteins, we have tailored our studies to analyze the roles of both of these molecular components. An integrated approach addressing both lipid and protein contributions promises to teach us substantially more about the mechanism by which Ca^{2+} triggers the merger of apposed native membranes.

Cortical Vesicle Model System

In regulated secretory cells, vesicles undergo several defined stages within the exocytotic pathway, including targeting, docking, priming, triggering, and fusion. This is followed by endocytosis and vesicle recycling for subsequent rounds of exocytosis. Thus, the major difficulty in studying Ca^{2+}-triggered membrane fusion is the fast, dynamic, and cyclic nature of the exocytotic pathway; analysis of the underlying molecular mechanisms at any specific step of the pathway is problematic, particularly in trying to tightly couple the functional and molecular analyses necessary to effectively dissect the mechanism. Given the critical role of compartmentalization in cellular functions and the high energy barriers involved in membrane merger, it is perhaps not unexpected that the underlying mechanism of triggered membrane fusion appears to be highly conserved.[3–7] Therefore, the use of *in vitro* model systems enables more direct analysis of the native fusion mechanism.

The urchin cortical vesicle (CV) preparation is a well-established stage-specific model for analyzing the molecular mechanisms underlying Ca^{2+}-triggered membrane fusion.[8–11] The cortical release reaction refers to the series of steps that give rise to the fertilization envelope.[12,13] Upon sperm binding, the release of Ca^{2+} from internal subplasmalemmal stores triggers thousands of fully primed and docked CV lining the inner surface of the plasma membrane (PM) to fuse, discharging CV content that then forms a physical/chemical barrier against polyspermy. Thus, the unfertilized oocyte is a rich source of fully docked, fusion-ready, secretory vesicles. Shearing the oocyte (in suspension or on a support surface, such as a microscope slide) yields the isolated cell surface complex (CSC) consisting of release-ready CV attached to fragments of PM; this stage-specific *in vitro* preparation requires only an increase in the free Ca^{2+} concentration ($[Ca^{2+}]_{free}$) to trigger fusion of the CV to the PM in the absence of cytosolic factors.[14–16] Taking a reductionist approach, Vogel and colleagues[17,18] were the first to show that isolated CV themselves undergo fast Ca^{2+}-triggered fusion with one another and with pure lipid membranes, indicating that CV contain the minimal essential components for intermembrane attachment, Ca^{2+} sensing, and membrane fusion. Thorough characterization indicates that homotypic CV fusion proceeds via the same mechanism as does CV–PM fusion.[19] This initially allows for the removal of the PM from the analytical equation, as it complicates quantitative molecular assessments with the presence of numerous excess membrane components not directly related to the fusion mechanism. The result is thus a high-yield, high-purity preparation extremely amenable to integrated functional and molecular analyses.

The packed protein content of the CV is optically dense, so fusion can be easily quantified using standard, high-throughput, light-scattering measurements.[18,20,21] Two routine assays are employed to assess fusion competence: percent fusion versus $[Ca^{2+}]_{free}$ (end point assay) and percent fusion versus time (kinetic assay). The data from end point assays are fit using a cumulative log-normal function in order to assess the extent and Ca^{2+} sensitivity of fusion.[22] The extent of fusion reflects the fundamental ability of the vesicles to fuse whereas Ca^{2+} sensitivity and kinetics are measures of triggering/fusion efficiency. Therefore, changes in fusion competence can quantitatively be correlated with changes in the molecular composition of CV membranes. This fully enables the tight coupling of functional and molecular analyses that is necessary for the effective dissection of the molecular mechanisms underlying Ca^{2+}-tiggered membrane fusion.

The Role of Lipids in Membrane Fusion

The stalk-pore model of membrane fusion describes the progression through a series of transient, high-curvature, intermediate membrane structures, resulting in the formation of a purely lipidic fusion pore.[1,2] Calculations of fundamental energetics, based on extensive work with model membranes[23–26] and viral fusion systems,[26–31] describe the structural requirements that minimize the energy needed for lipidic bilayer merger while preserving the topological relationship of the membranes and preventing the leakage of cellular or vesicular contents.[32–38] Briefly, close proximity between bilayers must be established—this involves overcoming strong repulsive hydration forces, giving rise to contact between very small localized areas of the apposing membranes. Localized membrane protrusions promote merger of the proximal monolayers and formation of the initial stalk intermediate (net negative curvature), which rapidly expands to form the hemifusion diaphragm. Subsequent pinching off of the distal monolayers yields the fusion pore (net positive curvature). Lipids that have a small head group and relatively large hydrophobic tail form concave structures and, by convention, are denoted as having intrinsic negative curvature, whereas lipids that have a large head group and relatively small hydrophobic tail form convex structures and are denoted as having intrinsic positive curvature. Thus, lipid compositions that facilitate the formation of these various stalk-pore intermediates decrease the energy requirement for lipid merger and enable membrane fusion to occur within the constraints of characterized biological systems. This requires molecular components having negative curvature in the contacting monolayers for the formation of the stalk and components having positive curvature in the distal monolayers to subsequently facilitate formation of the fusion pore.

Of particular interest is cholesterol (CHOL), an abundant component of native membranes that is substantially enriched in CV and other secretory vesicle membranes.[39,40] As a molecule of intrinsic negative curvature,[41] CHOL supports CV fusion by contributing to the formation of high curvature stalk-pore intermediates. Depletion of cholesterol from CV using the chelating reagent methyl-β-cyclodextrin (mβcd) results in a dose-dependent inhibition of the rate, Ca^{2+} sensitivity, and kinetics of fusion (Fig. 1).[39] Following CHOL depletion, the inherent ability of vesicles to fuse (fusion extent) is rescued by delivering structurally dissimilar membrane lipids having comparable negative curvature to CHOL (Fig. 2).[39,42] This recovery correlates linearly with both the quantity of each lipid delivered to the membrane and the specific intrinsic curvature of each of the lipids used (Fig. 2B).[42] Dioleoylphosphatidylethanolamine (DOPE), which has similar negative curvature as CHOL, is required at molar quantities equivalent to that of CHOL to elicit full recovery of fusion; in contrast, α-tocopherol (α-T) and a dioleoylglycerol (DOG), both

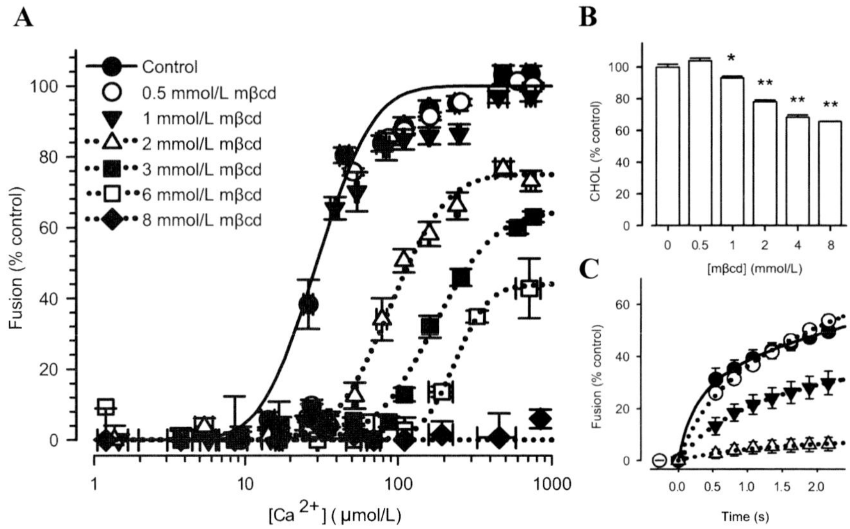

Figure 1. Cholesterol (CHOL) is a critical component of the fast, Ca^{2+}-triggered, native membrane fusion machinery. Effect of different concentrations of methyl-β-cyclodextrin (mβcd) on cortical vesicle (CV)–CV fusion. (**A**) Ca^{2+} activity curves following 30-min incubation with indicated concentrations of mβcd. (**B**) Quantification of total CV CHOL after mβcd treatments where *, ** indicate significant differences from control ($P < 0.05$) and other conditions ($P < 0.001$). (**C**) Kinetics of CV–CV fusion following treatments as indicated in (A), triggered with 71 ± 21 μM $[Ca^{2+}]_{free}$. (Reproduced from Ref. 39 with permission of the Company of Biologists.)

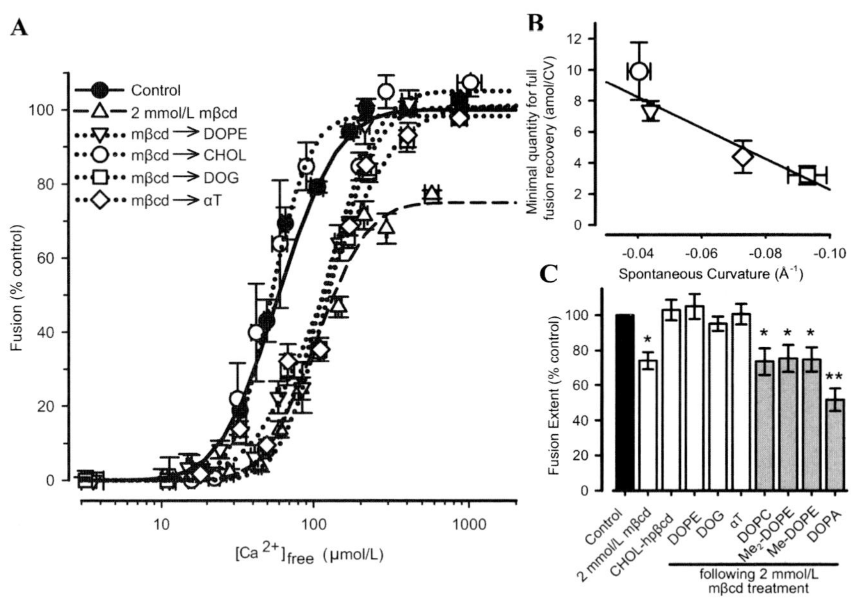

Figure 2. Lipids of intrinsic negative curvature rescue the ability of CHOL-depleted CV to fuse. (**A**) Ca^{2+} activity curves following 30-min incubation with 2 mM mβcd and subsequent 30-min incubations with structurally dissimilar negative curvature lipids as indicated. (**B**) The amount of each specific lipid species incorporated into the outer leaflet of CV membranes to fully recover the extent of fusion correlates linearly with the measured intrinsic curvature for each molecule. (**C**) Summary of CV–CV fusion recovery after 2 mM mβcd treatments where *, ** indicate significant differences from control ($P < 0.01$) and other conditions ($P < 0.001$). (Adapted from Ref. 42 with permission of the Biophysical Society.)

having an intrinsic negative curvature substantially higher than that of CHOL, are required at significantly lower levels (proportional to the amount of negative curvature each imparts to the outer membrane monolayer; Fig. 2B). Conversely, lipids having an intrinsic negative curvature less than that of CHOL do not support fusion (Fig. 2C). The exception is phosphatidic acid (here, DOPA) that not only does not support fusion but has been shown to consistently inhibit the merger process in CHOL-depleted CV.[42] In summary, these studies have established that there is a minimal specific negative curvature requirement for membrane merger—presumably facilitating initial formation of stalk-pore intermediates—in these native membranes.

In contrast to the recovery of fusion extent, the efficiency (Ca^{2+} sensitivity and kinetics) of fusion is not rescued by delivery of negative curvature molecules to the membranes of CHOL depleted CV. All parameters of fusion are, however, rescued by recovery of CHOL to native levels (Fig. 2A), indicating that CHOL plays a broader and more dynamic role in the mechanism of Ca^{2+}-triggered membrane merger. One well-studied role of CHOL in native membranes is that of a membrane organizer via the formation of CHOL- and sphingomyelin (SM)-enriched microdomains.[43–47] As certain membrane proteins are able to selectively associate with microdomains, another role of CHOL in membrane merger is suggested to involve the spatial organization of select protein and lipid components critical to the fusion mechanism at the merger site. Thus, CHOL may govern the interactions of these different critical molecular components and/or act as a cofactor of specific efficiency components, perhaps by promoting or regulating native functional conformations.[48–53]

Treating CV with various CHOL-altering reagents leads to differing effects on fusion ability and efficiency. Reagents that disrupt CHOL-rich microdomains, such as mβcd[54] (Fig. 1) and cholesterol oxidase[55,56] (Fig. 3A), potently and dose-dependently inhibit all three parameters of fusion: extent, Ca^{2+} sensitivity, and kinetics.[39] On the other hand, polyene antibiotics (amphotericin B and filipin; Fig. 3C, D) that sequester CHOL in the membrane, interfering with the functions of individual CHOL molecules without disrupting microdomains,[54] inhibit only the extent of fusion, leaving the Ca^{2+} sensitivity and kinetics generally unaltered.[39,42] These data have confirmed a separation of the fundamental ability to fuse from the Ca^{2+} sensitivity and rate of the fusion process. This suggests a model of a minimal fundamentally conserved fusion machine that is modulated by a variety of accessory proteins and lipids that act to promote the efficiency (e.g., Ca^{2+} sensitivity and kinetics) of the mechanism.[19,42] Thus, in contrast to a fundamental role for negative curvature, one possible interpretation is that microdomain integrity (and/or cofactor functions of CHOL) underlies the efficiency of fusion but not the ability of native vesicles to undergo Ca^{2+}-triggered membrane merger *per se*. Consistent with this is the observation that the selective hydrolysis of SM inhibits Ca^{2+} sensitivity and kinetics but not the extent of fusion (Fig. 3B).[57] Thus, the prefusion site may well be organized by CHOL- and SM-enriched microdomains, and these lipidic components may also serve as important cofactors for other molecular players.

Identifying Proteins Involved in Ca^{2+}-triggered Membrane Fusion

Although key features of the stalk-pore model are lipidic fusion intermediates, this model does not exclude (lipo)proteins from also playing a central role in the fusion mechanism.[58] Native Ca^{2+}-triggered membrane fusion is also known to be inhibited by thiol-reactive reagents[17–20,59–62] and protease treatments[63,64] that modify vesicular membrane proteins. Critical roles for proteins have been suggested in priming, docking, Ca^{2+} sensing, overcoming the hydration barrier, modulating membrane curvature (e.g., initiating

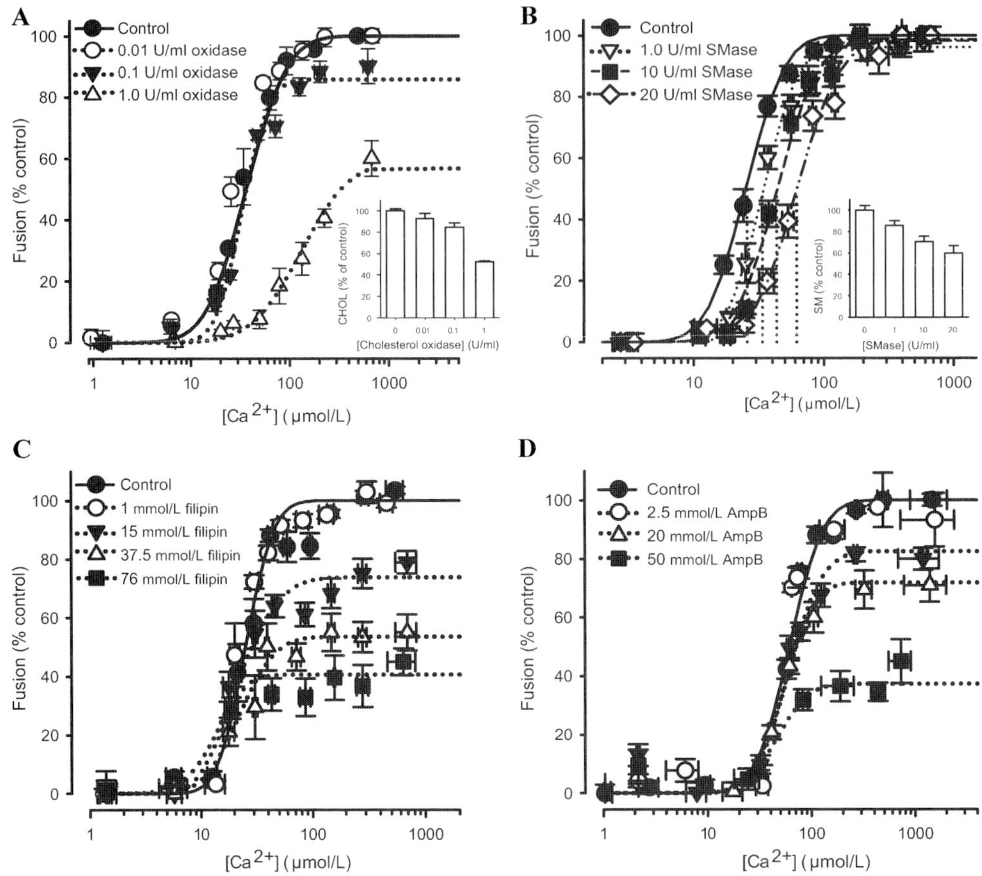

Figure 3. CHOL- and sphingomyelin (SM)-enriched microdomains regulate the efficiency of fusion. (**A**) Ca^{2+} activity curves following 30-min incubation with CHOL oxidase and quantification of CHOL after indicated treatments (inset). (**B**) Ca^{2+} activity curves following 30-min incubation with SMase (dotted vertical lines indicate progressive rightward shift in EC_{50} for Ca^{2+}) and quantification of SM after indicated treatments (inset). (**C**) Ca^{2+} activity curves following 30-min incubation with filipin as indicated. (**D**) Ca^{2+} activity curves following 30-min incubation with amphotericin B (AmpB) as indicated. (Adapted from Refs. 39 and 57 with permission of the Company of Biologists.)

deformations in bilayer structure), and promoting membrane line tension and thus rapid expansion of the fusion pore.[58,65–67] Although several proteins participating in the exocytotic pathway have been identified, there is still controversy regarding the precise function of these proteins and at which specific stage(s) they act.

The soluble *N*-ethylmaleimide-sensitive factor attachment protein receptor (SNARE) hypothesis suggests that Q-SNAREs on the target membrane and R-SNAREs on the vesicle membrane interact to help target and dock vesicles at specialized prefusion sites;[68,69] the rapid and transient formation of a coiled-coiled SNARE complex is postulated to then drive the merger of the two apposed lipid bilayers at the focal site of contact.[70] However, in the stage-specific CV preparation, SNARE complex assembly can precede the actual fusion step[19,71] and does not provide sufficient energy required for lipid bilayer rearrangement.[71] Furthermore, quantitative removal of SNARE cytosolic domains from CV by treatment with broad-spectrum proteases[64] does not obviate triggered fusion nor does blockade of SNARE interactions with excess recombinant binding proteins or antibodies.[63] This suggests that intermembrane SNARE interactions are not minimal essential components of the native membrane fusion reaction but act predominantly upstream in the docking and priming steps of exocytosis.[72] These conclusions are also supported by recent experiments in other model systems.[73–75] Nonetheless, the SNARE proteins may still contribute to the efficiency of native fusion,[64,74] possibly through interactions of their transmembrane domains.[76]

Our most current work employs thiol-reactive reagents, which bind to free sulfhydryl groups on cysteine residues, as a nonbiased approach to identifying proteins involved in fast Ca^{2+}-triggered membrane fusion. The most commonly used reagent, *N*-ethylmaleimide (NEM), has been shown to inhibit membrane fusion in various native model systems.[77–80] In the preparations derived from urchin oocytes, NEM and related reagents consistently inhibit the kinetics, Ca^{2+} sensitivity, and extent of both CV–PM[20,59–62] and CV–CV fusion[17–19,62] in a dose-dependent manner; treatments with $\leq$ 5 mmol/l leave CV irreversibly fusion incompetent. In this same dose range, Haggerty and Jackson[20] originally found that the common alkylating reagent, IA, had no effect on the extent of CV–PM fusion, and this reagent was never explored further. Thorough characterization of IA dose-response effects have now revealed that higher doses of this reagent actually enhance the Ca^{2+} sensitivity and rate of the triggered native membrane fusion mechanism.

IA was found to promote the efficiency of the fusion mechanism by acting at the Ca^{2+}-sensing step of the membrane merger process. Treatments with substantially higher doses of IA than previously used resulted in a leftward shift in Ca^{2+} sensitivity and a potentiation of fusion kinetics in both CV–PM and CV–CV (Fig. 4A) fusion paradigms.[81] There was no alteration in the quality of docking/intervesicle attachment in these assays, confirming that IA modifies the efficiency of the fusion mechanism itself rather than prefusion stages. Ca^{2+} mimetics, such as Sr^{2+}, are much less efficient than Ca^{2+} yet retain the ability to trigger fusion and are thus routinely used to probe the release mechanism.[19] Notably, there was a significantly larger leftward shift in Me^{2+} sensitivity when Sr^{2+} was used to trigger fusion (Fig. 4A) after IA treatment,[81] which is highly indicative that IA interacts with a thiol site that regulates fusion at the level of the Ca^{2+} sensor(s). Although the functional effects of thiol reagents have been well studied, the identities of the proteins they act on remain elusive. Taking advantage of the novel effect on the Ca^{2+}-sensing mechanism, we have identified a fluorescently labeled IA (Lucifer yellow IA; LYIA) that also effects both a leftward shift in Ca^{2+} sensitivity and potentiation of fusion kinetics (Fig. 4B). The analogous effect to that of IA alone and no effect of LY fluorophore itself on fusion, suggests that both IA and LYIA target the same thiol sites. Thus, LYIA will be used in future experiments to enhance fusion efficiency and simultaneously label the critical proteins involved.

Characterization of CHOL-rich Membrane Regions

Having characterized the specific roles of CHOL in the mechanism of Ca^{2+}-triggered fusion, we are now able to employ CHOL as a tool to identify additional protein and lipid components of the fusion machinery. CHOL can interact with proteins both through specific binding[49,82,83] and indirectly through CHOL- and SM-enriched microdomains;[43,45–47] given that microdomain integrity (e.g., ensured critical lipid–protein interactions, including cofactor functions) regulates the efficiency of fusion, one possibility is that proteins relevant to the fusion mechanism are organized within CHOL-enriched regions of the native membrane. There has, however, been an ongoing debate in the literature as to whether membrane isolates reflect the organization of the native membrane or if they are artifacts of the extraction procedure.[43,84–89] The majority of studies agree that microdomains (or even nanodomains) do exist as part of the native membrane structure,[90–94] yet the precise nature

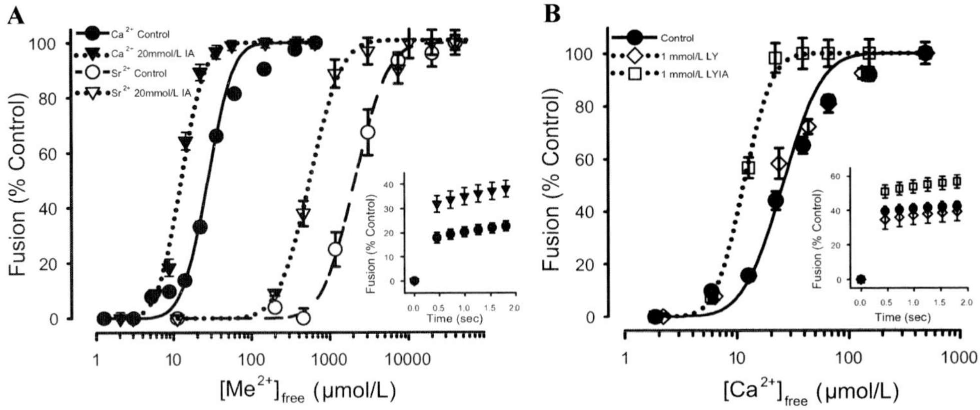

Figure 4. Iodoacetamide (IA) enhances the Ca^{2+}-triggered steps of membrane fusion. (**A**) Treatment with 20 mM IA for 20 min results in a leftward shift in both the Ca^{2+} and Sr^{2+} sensitivity of CV–CV fusion and potentiates the kinetics of triggered fusion in response to 35 ± 2 μM $[Ca^{2+}]_{free}$ (inset). (**B**) Treatments with 1 mM Lucifer yellow IA (LYIA) for 20 min yield a comparable leftward shift in Ca^{2+} sensitivity and potentiation of kinetics in response to 33 ± 3 μM $[Ca^{2+}]_{free}$ (inset), whereas treatment with the fluorophore (LY) alone has no effect on any fusion parameters. (**A** adapted from Ref. 81 with permission from Springer-Verlag.)

(i.e., number, size, composition) of these CHOL-enriched regions remains to be determined.[46,47] Nonetheless, isolation of such domains can be used as a prefractionation technique to enhance proteomic analyses. As mathematical modeling of Ca^{2+}-triggered CV fusion predicts an average of six to nine fusion complexes per CV,[39,59] the critical molecular components of Ca^{2+}-triggered membrane fusion are of low abundance, even when analyzing isolated organelles. Biochemical approaches to microdomain isolation can thus serve as a first step fractionation/concentration method to aid in enriching proteins and lipids that act at or near the fusion site.

The composition of CHOL-enriched microdomains is generally assessed by taking advantage of the differential detergent solubility of microdomains versus nonmicrodomain membranes. Nonionic detergents, such as Triton X-100,[88,95–100] lubrol,[96,98,100] and the Brij family,[88,90,97,100] have been used to analyze CHOL-rich microdomains from a wide variety of tissues and cell types. Much of the debate in the literature has focused on the relevance of detergent-resistant membrane microdomains (DRMs) to those observed *in vivo*. Isolation of Triton X-100 DRMs are typically carried out at 4°C, which may lead to a temperature-induced phase separation and aggregation of multiple CHOL-enriched membrane regions into a resulting detergent insoluble fraction.[43,89,99] Furthermore, Triton X-100 itself has distinct effects on the phase behavior of lipids in model membranes.[86,87] Another issue is that different detergent methods have been reported to isolate microdomains of varying composition,[90,96–98,100] leading to the hypothesis that there may be several distinct domains specialized for specific functions.[101] Here we recognize that these membrane isolates are unlikely to represent a single native species of microdomain but rather that the method does isolate membrane fractions of similar average lipid composition, resulting in an aggregate isolate of all CHOL-rich membrane regions. Consequently, isolation of CHOL-enriched membrane fractions remains a useful tool to isolate proteins that preferentially associate with CHOL and/or SM. As a result of, in part, the controversy surrounding the specific nature of detergent-resistant microdomains, recent nondetergent techniques have been developed that take advantage of the inherent differences in the density of CHOL-enriched versus nonenriched membranes by

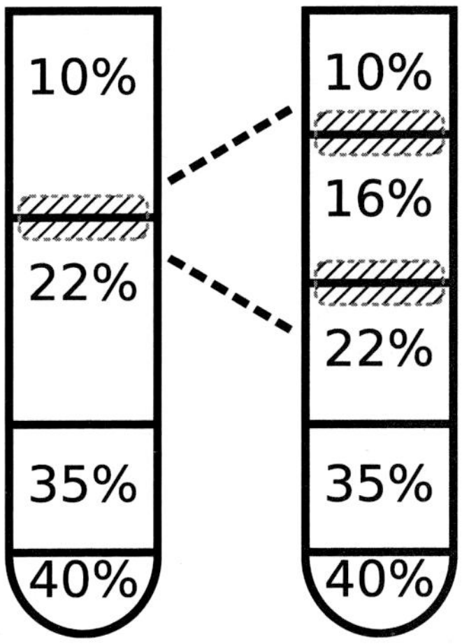

Figure 5. Schematic of the sucrose density gradients used to isolate CHOL-enriched CV membrane microdomain fractions. The region circled with dashed grey lines indicates the position of the buoyant CHOL-enriched fraction at the interface of 10% and 22% sucrose. Expanding the gradient by including a 16% sucrose step yields two distinct buoyant fractions collected at the 10–16% and 16–22% interfaces.

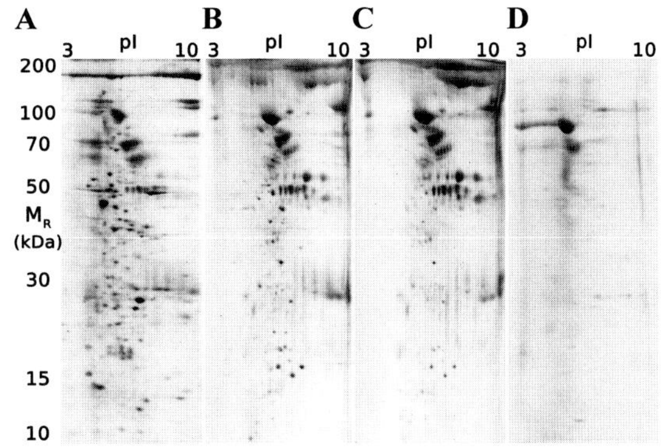

Figure 6. Resolution of proteins associated with CHOL-rich membrane fractions using 2-dimensional gel electrophoresis (2DE). Comparison of the **(A)** total CV membrane proteome to that of the **(B)** CHOL-rich membrane fraction isolated on sucrose density gradients shows an enrichment of proteins of lower abundance. **(C)** Treatment with the CHOL-binding antibiotic amphotericin B (105 μmol/L) results in no significant alterations in the complement of proteins that associate with the CHOL-rich membrane fraction. **(D)** Of particular interest are the proteins which become labeled upon treatment with 1 mmol/L LYIA; these are proteins potentially involved in Ca^{2+} sensing. Each gel image, except (A), is an average composite of $n = 2$ replicate gels, generated using the analysis program Delta2D (Decodon GmbH, Greifswald, Germany).

separating these fractions on sucrose gradients (Fig. 5).[102,103] Based on preliminary comparisons of multiple detergent- (including Triton X-100, lubrol, Brij 35, Brij 96, Brij 98) and nondetergent- (including Na_2CO_3 pH 11.0 and Tris pH 7.0) based protocols for isolating CHOL-enriched membrane fractions, we have chosen to analyze membrane microdomains using the nondetergent approach (Tris pH 7.0). Our ongoing work involving analyses of CV microdomain proteomes reveals that the nondetergent protocol yields a highly typical CHOL-enriched membrane fraction (e.g., fully comparable to that seen with multiple detergents) yet avoids any complicating factors that may arise from the presence of detergents (Churchward and Coorssen, unpublished observations).

Analysis of CV membrane proteomes, isolated as CHOL-enriched membrane fractions, yielded a distinct subset of proteins relative to the total CV membrane, including the selective enrichment of specific low-abundance proteins (Fig. 6A, B). We sought to determine if multiple CHOL-enriched fractions could be isolated from the CV membrane by increasing the number of steps in the resolving gradient from 10/22% sucrose to 10/16/22% sucrose. Two CHOL-enriched membrane fractions of differing densities were isolated at the two interfaces (10/16% and 16/22%) (Fig. 5); however, protein analysis of these revealed comparable protein spot patterns after resolution by 2-dimensional gel electrophoresis (2DE; data not shown). This indicates that the CV membrane may contain primarily (or perhaps only) one type of CHOL-enriched microdomain, which is perhaps not unexpected as it is a distinct organelle with a highly specific function. To further elucidate the composition of the CHOL-enriched regions of the CV membrane and confirm the effects of the different reagents used to probe the roles of CHOL in the intact CV,[39,42,57] we isolated CHOL-enriched membrane fractions from

CV treated with 2 mmol/L mβcd or with 105 μmol/L amphotericin B (ampB). Relative to control gels (Fig. 6B), ampB-treated samples yielded CHOL-enriched fractions of comparable protein composition (Fig. 6C); in contrast, treatment with mβcd resulted in the loss of any measurable CHOL-enriched fraction. This provides further evidence to support the idea that mβcd is capable of causing extensive disruption to native CHOL-enriched membrane regions, whereas ampB functions by binding CHOL within the membrane in a manner that does not cause disruption of such regions and the dispersal of associated protein components.[54]

As CHOL- and SM-enriched membrane regions are responsible for the spatial organization of the fusion site on CV, we sought to identify proteins involved in the fusion mechanism. Using the previously discussed LYIA treatments to simultaneously potentiate the Ca^{2+} sensitivity and kinetics of fusion and fluorescently label proteins involved in regulating fusion efficiency, we have isolated CV membrane proteins associated with CHOL-enriched microdomains as described above (Fig. 6D). In this manner we have reduced the complexity of the analysis to include only those proteins that associate with CHOL-enriched microdomains and are labeled with the LYIA fluorophore. This allows us to narrow the focus of future experiments to those proteins that are potentially involved in the Ca^{2+}-sensing steps of the fusion process.

Summary

As a process vital to multiple cellular functions, membrane fusion has been the subject of research in model systems from yeast to neurons. Decades of study are now culminating in a more focused research approach that promises substantial progress in understanding the molecular mechanisms underlying fast, Ca^{2+}-triggered, native membrane fusion as well as other critical steps in the exocytotic pathway. Integrated analytical paradigms will be the key. While not always receiving the same attention as proteins, it is abundantly clear that lipids play an active role in membrane fusion. In particular, CHOL has a dual role in fast Ca^{2+}-triggered fusion: (1) acts directly in bilayer merger by contributing localized negative curvature to the membrane; and (2) regulates the efficiency of fusion via microdomain organization and/or cofactor functions. Employing CHOL as "bait" in combination with thiol-reactive reagents, we are now targeting proteins that are involved in the triggering steps of the fusion mechanism. In parallel work we are also analyzing the roles of other lipidic components in the native process of regulated membrane merger. Using such a comprehensive approach, studying the contributions of both lipids and proteins to the fusion process will provide greater insight into the mechanism underlying fast Ca^{2+}-triggered membrane fusion and provide more specific identification of the critical molecular components that together constitute the native fusion machine.

Acknowledgments

The authors wish to thank D.M. Brandman, R.H. Butt, and Dr A.L. Yergey for helpful discussions and D. Bininda for assistance with aquatics. JRC acknowledges support from the Canadian Institutes of Health Research (CIHR), Alberta Heritage Foundation for Medical Research (AHFMR), and Natural Sciences and Engineering Research Council of Canada (NSERC). KLF is the recipient of postgraduate scholarship awards from AHFMR and CIHR, and MAC from AHFMR and NSERC.

Conflicts of Interest

The authors declare no conflicts of interest.

References

1. Kozlov, M.M. & V.S. Markin. 1983. [Possible mechanism of membrane fusion]. *Biofizika* **28:** 242–247.

2. Markin, V.S., M.M. Kozlov & V.L. Borovjagin. 1984. On the theory of membrane fusion. The stalk mechanism. *Gen. Physiol. Biophys.* **3:** 361–377.
3. Chavez, R.A., S.G. Miller & H.P. Moore. 1996. A biosynthetic regulated secretory pathway in constitutive secretory cells. *J. Cell Biol.* **133:** 1177–1191.
4. Coorssen, J.R., H. Schmitt & W. Almers. 1996. Ca2+ triggers massive exocytosis in Chinese hamster ovary cells. *EMBO J.* **15:** 3787–3791.
5. Dan, Y. & M.M. Poo. 1992. Quantal transmitter secretion from myocytes loaded with acetylcholine. *Nature* **359:** 733–736.
6. Knight, D.E. & M.C. Scrutton. 1986. Gaining access to the cytosol: the technique and some applications of electropermeabilization. *Biochem. J.* **234:** 497–506.
7. Knight, D.E., G.H. Von & C.M. Athayde. 1989. Calcium-dependent and calcium-independent exocytosis. *Trends Neurosci.* **12:** 451–458.
8. Vacquier, V.D. 1976. Isolated cortical granules: a model system for studying membrane fusion and calcium-mediated exocytosis. *J. Supramol. Struct.* **5:** 27–35.
9. Vogel, S.S., K. Delaney & J. Zimmerberg. 1991. The sea urchin cortical reaction. A model system for studying the final steps of calcium-triggered vesicle fusion. *Ann. N. Y. Acad. Sci.* **635:** 35–44.
10. Zimmerberg, J., J.R. Coorssen, S.S. Vogel & P.S. Blank. 1999. Sea urchin egg preparations as systems for the study of calcium-triggered exocytosis. *J. Physiol.* **520**(Pt 1): 15–21.
11. Zimmerberg, J., P.S. Blank, I. Kolosova, *et al.* 2000. A stage-specific preparation to study the Ca(2+)-triggered fusion steps of exocytosis: rationale and perspectives. *Biochimie* **82:** 303–314.
12. Anderson, E. 1968. Oocyte differentiation in the sea urchin, Arbacia punctulata, with particular reference to the origin of cortical granules and their participation in the cortical reaction. *J. Cell Biol.* **37:** 514–539.
13. Chandler, D.E. 1991. Multiple intracellular signals coordinate structural dynamics in the sea urchin egg cortex at fertilization. *J. Electron Microsc. Tech.* **17:** 266–293.
14. Baker, P.F. & M.J. Whitaker. 1978. Influence of ATP and calcium on the cortical reaction in sea urchin eggs. *Nature* **276:** 513–515.
15. Vacquier, V.D. 1975. The isolation of intact cortical granules from sea urchin eggs: calcium lons trigger granule discharge. *Dev. Biol.* **43:** 62–74.
16. Zimmerberg, J. & J. Liu. 1988. Ionic and permeability requirements for exocytosis in vitro in sea urchin eggs. *J. Membr. Biol.* **101:** 199–207.
17. Vogel, S.S., L.V. Chernomordik & J. Zimmerberg. 1992. Calcium-triggered fusion of exocytotic granules requires proteins in only one membrane. *J Biol. Chem.* **267:** 25640–25643.
18. Vogel, S.S. & J. Zimmerberg. 1992. Proteins on exocytic vesicles mediate calcium-triggered fusion. *Proc. Natl. Acad. Sci. USA* **89:** 4749–4753.
19. Coorssen, J.R., P.S. Blank, M. Tahara & J. Zimmerberg. 1998. Biochemical and functional studies of cortical vesicle fusion: the SNARE complex and Ca2+ sensitivity. *J. Cell Biol.* **143:** 1845–1857.
20. Haggerty, J.G. & R.C. Jackson. 1983. Release of granule contents from sea urchin egg cortices. New assay procedures and inhibition by sulfhydryl-modifying reagents. *J. Biol. Chem.* **258:** 1819–1825.
21. Zimmerberg, J., C. Sardet & D. Epel. 1985. Exocytosis of sea urchin egg cortical vesicles in vitro is retarded by hyperosmotic sucrose: kinetics of fusion monitored by quantitative light-scattering microscopy. *J. Cell Biol.* **101:** 2398–2410.
22. Blank, P.S., M.S. Cho, S.S. Vogel, *et al.* 1998. Submaximal responses in calcium-triggered exocytosis are explained by differences in the calcium sensitivity of individual secretory vesicles. *J. Gen. Physiol.* **112:** 559–567.
23. Chernomordik, L., A. Chanturiya, J. Green & J. Zimmerberg. 1995. The hemifusion intermediate and its conversion to complete fusion: regulation by membrane composition. *Biophys. J.* **69:** 922–929.
24. Chernomordik, L.V., G.B. Melikyan & Y.A. Chizmadzhev. 1987. Biomembrane fusion: a new concept derived from model studies using two interacting planar lipid bilayers. *Biochim. Biophys. Acta* **906:** 309–352.
25. Meers, P., K. Hong & D. Papahadjopoulos. 1988. Free fatty acid enhancement of cation-induced fusion of liposomes: synergism with synexin and other promoters of vesicle aggregation. *Biochemistry* **27:** 6784–6794.
26. Yeagle, P.L., F.T. Smith, J.E. Young & T.D. Flanagan. 1994. Inhibition of membrane fusion by lysophosphatidylcholine. *Biochemistry* **33:** 1820–1827.
27. Biswas, S., S.R. Yin, P.S. Blank & J. Zimmerberg. 2008. Cholesterol promotes hemifusion and pore widening in membrane fusion induced by influenza hemagglutinin. *J. Gen. Physiol* **131:** 503–513.
28. Chernomordik, L., E. Leikina, M.S. Cho & J. Zimmerberg. 1995. Control of baculovirus gp64-induced syncytium formation by membrane lipid composition. *J. Virol.* **69:** 3049–3058.
29. Chernomordik, L.V., S.S. Vogel, A. Sokoloff, *et al.* 1993. Lysolipids reversibly inhibit Ca(2+)-, GTP- and pH-dependent fusion of biological membranes. *FEBS Lett.* **318:** 71–76.
30. Chernomordik, L.V., E. Leikina, V. Frolov, P. Bronk & J. Zimmerberg. 1997. An early stage of membrane

fusion mediated by the low pH conformation of influenza hemagglutinin depends upon membrane lipids. *J. Cell Biol.* **136:** 81–93.

31. Vogel, S.S., E.A. Leikina & L.V. Chernomordik. 1993. Lysophosphatidylcholine reversibly arrests exocytosis and viral fusion at a stage between triggering and membrane merger. *J. Biol. Chem.* **268:** 25764–25768.
32. Kozlov, M.M., S.L. Leikin, L.V. Chernomordik, *et al.* 1989. Stalk mechanism of vesicle fusion. Intermixing of aqueous contents. *Eur. Biophys. J.* **17:** 121–129.
33. Kozlovsky, Y., L.V. Chernomordik & M.M. Kozlov. 2002. Lipid intermediates in membrane fusion: formation, structure, and decay of hemifusion diaphragm. *Biophys. J.* **83:** 2634–2651.
34. Kozlovsky, Y. & M.M. Kozlov. 2002. Stalk model of membrane fusion: solution of energy crisis. *Biophys. J.* **82:** 882–895.
35. Kozlovsky, Y., A. Efrat, D.A. Siegel & M.M. Kozlov. 2004. Stalk phase formation: effects of dehydration and saddle splay modulus. *Biophys. J.* **87:** 2508–2521.
36. Kuzmin, P.I., J. Zimmerberg, Y.A. Chizmadzhev & F.S. Cohen. 2001. A quantitative model for membrane fusion based on low-energy intermediates. *Proc. Natl. Acad. Sci. USA* **98:** 7235–7240.
37. Leikin, S.L., M.M. Kozlov, L.V. Chernomordik, *et al.* 1987. Membrane fusion: overcoming of the hydration barrier and local restructuring. *J. Theor. Biol.* **129:** 411–425.
38. Efrat, A., L.V. Chernomordik & M.M. Kozlov. 2007. Point-like protrusion as a prestalk intermediate in membrane fusion pathway. *Biophys. J.* **92:** L61–L63.
39. Churchward, M.A., T. Rogasevskaia, J. Hofgen, *et al.* 2005. Cholesterol facilitates the native mechanism of Ca2+-triggered membrane fusion. *J. Cell Sci.* **118:** 4833–4848.
40. Takamori, S., M. Holt, K. Stenius, *et al.* 2006. Molecular anatomy of a trafficking organelle. *Cell* **127:** 831–846.
41. Chen, Z. & R.P. Rand. 1997. The influence of cholesterol on phospholipid membrane curvature and bending elasticity. *Biophys. J.* **73:** 267–276.
42. Churchward, M.A., T. Rogasevskaia, D.M. Brandman, *et al.* 2008. Specific lipids supply critical negative spontaneous curvature–an essential component of native Ca2+-triggered membrane fusion. *Biophys. J.* **94:** 3976–3986.
43. Brown, D.A. & E. London. 2000. Structure and function of sphingolipid- and cholesterol-rich membrane rafts. *J. Biol. Chem.* **275:** 17221–17224.
44. Ramstedt, B. & J.P. Slotte. 2002. Membrane properties of sphingomyelins. *FEBS Lett.* **531:** 33–37.
45. Simons, K. & E. Ikonen. 1997. Functional rafts in cell membranes. *Nature* **387:** 569–572.
46. Simons, K. & W.L. Vaz. 2004. Model systems, lipid rafts, and cell membranes. *Annu. Rev. Biophys. Biomol. Struct.* **33:** 269–295.
47. Jacobson, K., O.G. Mouritsen & R.G. Anderson. 2007. Lipid rafts: at a crossroad between cell biology and physics. *Nat. Cell Biol.* **9:** 7–14.
48. Graziani, A., C. Rosker, S.D. Kohlwein, *et al.* 2006. Cellular cholesterol controls TRPC3 function: evidence from a novel dominant-negative knockdown strategy. *Biochem. J.* **396:** 147–155.
49. Kallen, J.A., J.M. Schlaeppi, F. Bitsch, *et al.* 2002. X-ray structure of the hRORalpha LBD at 1.63 A: structural and functional data that cholesterol or a cholesterol derivative is the natural ligand of RORalpha *Structure* **10:** 1697–1707.
50. Scanlon, S.M., D.C. Williams & P. Schloss. 2001. Membrane cholesterol modulates serotonin transporter activity. *Biochemistry* **40:** 10507–10513.
51. Taverna, E., E. Saba, J. Rowe, *et al.* 2004. Role of lipid microdomains in P/Q-type calcium channel (Cav2.1) clustering and function in presynaptic membranes. *J. Biol. Chem.* **279:** 5127–5134.
52. Criado, M., H. Eibl & F.J. Barrantes. 1982. Effects of lipids on acetylcholine receptor. Essential need of cholesterol for maintenance of agonist-induced state transitions in lipid vesicles. *Biochemistry* **21:** 3622–3629.
53. Fong, T.M. & M.G. Mcnamee. 1987. Stabilization of acetylcholine receptor secondary structure by cholesterol and negatively charged phospholipids in membranes. *Biochemistry* **26:** 3871–3880.
54. Awasthi-Kalia, M., P.P. Schnetkamp & J.P. Deans. 2001. Differential effects of filipin and methyl-beta-cyclodextrin on B cell receptor signaling. *Biochem. Biophys. Res. Commun.* **287:** 77–82.
55. Samsonov, A.V., I. Mihalyov & F.S. Cohen. 2001. Characterization of cholesterol-sphingomyelin domains and their dynamics in bilayer membranes. *Biophys. J.* **81:** 1486–1500.
56. Xu, X. & E. London. 2000. The effect of sterol structure on membrane lipid domains reveals how cholesterol can induce lipid domain formation. *Biochemistry* **39:** 843–849.
57. Rogasevskaia, T. & J.R. Coorssen. 2006. Sphingomyelin-enriched microdomains define the efficiency of native Ca2+-triggered membrane fusion. *J. Cell Sci.* **119:** 2688–2694.
58. Chernomordik, L.V. & M.M. Kozlov. 2003. Protein-lipid interplay in fusion and fission of biological membranes. *Annu. Rev. Biochem.* **72:** 175–207.
59. Vogel, S.S., P.S. Blank & J. Zimmerberg. 1996. Poisson-distributed active fusion complexes underlie

the control of the rate and extent of exocytosis by calcium *J. Cell Biol.* **134:** 329–338.

60. Jackson, R.C. & P.A. Modern. 1990. N-ethylmaleimide-sensitive protein(s) involved in cortical exocytosis in the sea urchin egg: localization to both cortical vesicles and plasma membrane. *J. Cell Sci.* **96**(Pt 2): 313–321.
61. Whalley, T. & A. Sokoloff. 1994. The N-ethylmaleimide-sensitive protein thiol groups necessary for sea-urchin egg cortical-granule exocytosis are highly exposed to the medium and are required for triggering by Ca2+. *Biochem. J.* **302**(Pt 2): 391–396.
62. Whalley, T., K. Timmers, J. Coorssen, *et al.* 2004. Membrane fusion of secretory vesicles of the sea urchin egg in the absence of NSF. *J. Cell Sci.* **117:** 2345–2356.
63. Szule, J.A., S.E. Jarvis, J.E. Hibbert, *et al.* 2003. Calcium-triggered membrane fusion proceeds independently of specific presynaptic proteins. *J Biol. Chem.* **278:** 24251–24254.
64. Coorssen, J.R., P.S. Blank, F. Albertorio, *et al.* 2003. Regulated secretion: SNARE density, vesicle fusion and calcium dependence. *J. Cell Sci.* **116:** 2087–2097.
65. Epand, R.M. & R.F. Epand. 2000. Modulation of membrane curvature by peptides *Biopolymers* **55:** 358–363.
66. Farsad, K. & C.P. De. 2003. Mechanisms of membrane deformation. *Curr. Opin. Cell Biol.* **15:** 372–381.
67. Lentz, B.R., V. Malinin, M.E. Haque & K. Evans. 2000. Protein machines and lipid assemblies: current views of cell membrane fusion. *Curr. Opin. Struct. Biol.* **10:** 607–615.
68. Sollner, T., S.W. Whiteheart, M. Brunner, *et al.* 1993. SNAP receptors implicated in vesicle targeting and fusion. *Nature* **362:** 318–324.
69. Sollner, T., M.K. Bennett, S.W. Whiteheart, *et al.* 1993. A protein assembly-disassembly pathway in vitro that may correspond to sequential steps of synaptic vesicle docking, activation, and fusion. *Cell* **75:** 409–418.
70. Weber, T., B.V. Zemelman, J.A. Mcnew, *et al.* 1998. SNAREpins: minimal machinery for membrane fusion. *Cell* **92:** 759–772.
71. Tahara, M., J.R. Coorssen, K. Timmers, *et al.* 1998. Calcium can disrupt the SNARE protein complex on sea urchin egg secretory vesicles without irreversibly blocking fusion. *J. Biol. Chem.* **273:** 33667–33673.
72. Szule, J.A. & J.R. Coorssen. 2003. Revisiting the role of SNAREs in exocytosis and membrane fusion. *Biochim. Biophys. Acta* **1641:** 121–135.
73. Chen, X., D. Arac, T.M. Wang, *et al.* 2006. SNARE-mediated lipid mixing depends on the physical state of the vesicles. *Biophys. J.* **90:** 2062–2074.
74. Dennison, S.M., M.E. Bowen, A.T. Brunger & B.R. Lentz. 2006. Neuronal SNAREs do not trigger fusion between synthetic membranes but do promote PEG-mediated membrane fusion. *Biophys. J.* **90:** 1661–1675.
75. Ungermann, C., K. Sato & W. Wickner. 1998. Defining the functions of trans-SNARE pairs. *Nature* **396:** 543–548.
76. Langosch, D., J.M. Crane, B. Brosig, *et al.* 2001. Peptide mimics of SNARE transmembrane segments drive membrane fusion depending on their conformational plasticity *J. Mol. Biol.* **311:** 709–721.
77. Block, M.R., B.S. Glick, C.A. Wilcox, *et al.* 1988. Purification of an N-ethylmaleimide-sensitive protein catalyzing vesicular transport. *Proc. Natl. Acad. Sci. USA* **85:** 7852–7856.
78. Frye, R.A. & R.W. Holz. 1985. Arachidonic acid release and catecholamine secretion from digitonin-treated chromaffin cells: effects of micromolar calcium, phorbol ester, and protein alkylating agents. *J. Neurochem.* **44:** 265–273.
79. Malhotra, V., L. Orci, B.S. Glick, *et al.* 1988. Role of an N-ethylmaleimide-sensitive transport component in promoting fusion of transport vesicles with cisternae of the Golgi stack. *Cell* **54:** 221–227.
80. Zheng, X. & J.A. Bobich. 1998. MgATP-dependent and MgATP-independent [3H]noradrenaline release from perforated synaptosomes both use N-ethylmaleimide-sensitive fusion protein. *Biochemistry* **37:** 12569–12575.
81. Furber, K.L., D.M. Brandman & J.R. Coorssen. 2008. Enhancement of the Ca2+-triggering steps of native membrane fusion via thiol-reactivity. *J. Chem. Biol.* [Epub: Oct. 2008].
82. Hamouda, A.K., D.C. Chiara, D. Sauls, *et al.* 2006. Cholesterol interacts with transmembrane alpha-helices M1, M3, and M4 of the Torpedo nicotinic acetylcholine receptor: photolabeling studies using [3H]Azicholesterol. *Biochemistry* **45:** 976–986.
83. Corbin, J., H.H. Wang & M.P. Blanton. 1998. Identifying the cholesterol binding domain in the nicotinic acetylcholine receptor with [125I]azidocholesterol. *Biochim. Biophys. Acta* **1414:** 65–74.
84. Babiychuk, E.B. & A. Draeger. 2006. Biochemical characterization of detergent-resistant membranes: a systematic approach. *Biochem. J.* **397:** 407–416.
85. Garner, A.E., D.A. Smith & N.M. Hooper. 2008. Visualization of detergent solubilization of membranes: implications for the isolation of rafts. *Biophys. J.* **94:** 1326–1340.
86. Heerklotz, H. 2002. Triton promotes domain formation in lipid raft mixtures. *Biophys. J.* **83:** 2693–2701.

87. Heerklotz, H., H. Szadkowska, T. Anderson & J. Seelig. 2003. The sensitivity of lipid domains to small perturbations demonstrated by the effect of Triton. *J. Mol. Biol.* **329:** 793–799.
88. Kim, K.B., S.I. Kim, H.J. Choo, *et al.* 2004. Two-dimensional electrophoretic analysis reveals that lipid rafts are intact at physiological temperature. *Proteomics* **4:** 3527–3535.
89. London, E. & D.A. Brown. 2000. Insolubility of lipids in triton X-100: physical origin and relationship to sphingolipid/cholesterol membrane domains (rafts). *Biochim. Biophys. Acta* **1508:** 182–195.
90. Brugger, B., C. Graham, I. Leibrecht, *et al.* 2004. The membrane domains occupied by glycosylphosphatidylinositol-anchored prion protein and Thy-1 differ in lipid composition. *J. Biol. Chem.* **279:** 7530–7536.
91. Gaus, K., E. Gratton, E.P. Kable, *et al.* 2003. Visualizing lipid structure and raft domains in living cells with two-photon microscopy. *Proc. Natl. Acad. Sci. USA* **100:** 15554–15559.
92. Mcintosh, A.L., A.M. Gallegos, B.P. Atshaves, *et al.* 2003. Fluorescence and multiphoton imaging resolve unique structural forms of sterol in membranes of living cells. *J. Biol. Chem.* **278:** 6384–6403.
93. Pralle, A., P. Keller, E.L. Florin, *et al.* 2000. Sphingolipid-cholesterol rafts diffuse as small entities in the plasma membrane of mammalian cells. *J. Cell Biol.* **148:** 997–1008.
94. Zhang, W., A.L. Mcintosh, H. Xu, *et al.* 2005. Structural analysis of sterol distributions in the plasma membrane of living cells. *Biochemistry* **44:** 2864–2884.
95. Sargiacomo, M., M. Sudol, Z. Tang & M.P. Lisanti. 1993. Signal transducing molecules and glycosyl-phosphatidylinositol-linked proteins form a caveolin-rich insoluble complex in MDCK cells. *J. Cell Biol.* **122:** 789–807.
96. Slimane, T.A., G. Trugnan, S.C. Van Ijzendoorn & D. Hoekstra. 2003. Raft-mediated trafficking of apical resident proteins occurs in both direct and transcytotic pathways in polarized hepatic cells: role of distinct lipid microdomains. *Mol. Biol. Cell* **14:** 611–624.
97. Radeva, G. & F.J. Sharom. 2004. Isolation and characterization of lipid rafts with different properties from RBL-2H3 (rat basophilic leukaemia) cells. *Biochem. J.* **380:** 219–230.
98. Roper, K., D. Corbeil & W.B. Huttner. 2000. Retention of prominin in microvilli reveals distinct cholesterol-based lipid micro-domains in the apical plasma membrane. *Nat. Cell Biol.* **2:** 582–592.
99. Brown, D.A. & J.K. Rose. 1992. Sorting of GPI-anchored proteins to glycolipid-enriched membrane subdomains during transport to the apical cell surface. *Cell* **68:** 533–544.
100. Gil, C., R. Cubi & J. Aguilera. 2007. Shedding of the p75NTR neurotrophin receptor is modulated by lipid rafts. *FEBS Lett.* **581:** 1851–1858.
101. Pike, L.J. 2004. Lipid rafts: heterogeneity on the high seas. *Biochem. J.* **378:** 281–292.
102. Luria, A., V. Vegelyte-Avery, B. Stith, *et al.* 2002. Detergent-free domain isolated from Xenopus egg plasma membrane with properties similar to those of detergent-resistant membranes. *Biochemistry* **41:** 13189–13197.
103. Song, K.S., S. LI, T. Okamoto, *et al.* 1996. Copurification and direct interaction of Ras with caveolin, an integral membrane protein of caveolae microdomains. Detergent-free purification of caveolae microdomains. *J. Biol. Chem.* **271:** 9690–9697.

The Fusion Pore and Vesicle Cargo Discharge Modulation

Nina Vardjan,[a,b] Matjaž Stenovec,[a,b] Jernej Jorgačevski,[b] Marko Kreft,[b,a] Sonja Grilc,[b,a] and Robert Zorec[b,a]

[a] *Celica Biomedical Center, Ljubljana, Slovenia*

[b] *Laboratory of Neuroendocrinology-Molecular Cell Physiology, Institute of Pathophysiology, Faculty of Medicine, University of Ljubljana, Ljubljana, Slovenia*

Exocytosis, the merger of the vesicle membrane with the plasma membrane, is thought to mediate the release of hormones and neurotransmitters from secretory vesicles. The work of Bernard Katz and colleagues decades ago considered that vesicle cargo discharge initially requires the delivery of secretory vesicles to the plasma membrane where vesicles dock and are primed for fusion with the plasma membrane. Then, upon stimulation, the vesicle and the plasma membranes fuse to form a transient fusion pore through which cargo molecules diffuse out of the vesicle lumen into the extracellular space. Katz and colleagues considered this process to occur in an all-or-none fashion. However, recent studies show that this may not be so simple. The aim of this overview is to highlight the novel findings that indicate that fusion pores are subject to regulations, which affect the release competence of a single vesicle. Here we discuss the elementary properties of spontaneous and stimulated peptidergic vesicle discharge, which appears to be modulated, at least in pituitary lactotrophs, by fusion pore conductance (pore diameter) and fusion pore gating (kinetics).

Key words: **exocytosis; hormone release; peptidergic vesicles; fusion pore**

Introduction

Although remarkable progress has been made in our understanding of the molecular mechanisms underlying regulated vesicle release of hormones and neurotransmitters in the recent past, the nature of exocytosis continues to amaze us. The secretory process consists of vesicle delivery to the plasma membrane where they dock and are primed for fusion with the plasma membrane (exocytosis) in response to a physiological stimulus, such as an increase in free intracellular calcium concentration. The central role of cytoplasmic calcium in triggering secretion, together with questions addressing the nature of exocytosis and vesicular contents discharge, was previously considered by Bernard Katz and coworkers.

Bernard Katz: Calcium Hypothesis and Transient Fusion

Although Katz and colleagues mainly used the neuromuscular junction and the giant synapse of the squid as model systems, they provided interesting conclusions:

> R. Miledi and I now put forward a more specific suggestion: namely that depolarization opens a gate to calcium ions (or-what amounts to the same thing-that depolarization makes available specific carriers for calcium ions in the membrane). As a consequence of this increased "calcium conductance", Ca ions can move-down a very high concentration gradient-towards the inside of the axon

Address for correspondence: Robert Zorec, PhD, Professor, University of Ljubljana, Faculty of Medicine, Institute of Pathophysiology, LN-MCP, Zaloška 4, 1000 Ljubljana, Slovenia. Voice: +386 1 543 70 20; fax: +386 1 543 70 36. robert.zorec@mf.uni-lj.si

Mechanisms of Exocytosis: Ann. N.Y. Acad. Sci. 1152: 135–144 (2009).
doi: 10.1111/j.1749-6632.2008.04007.x

membrane and thus reach the critical sites of the release reaction. We are suggesting that, at these sites, calcium is essential for the process which causes a transient fusion of axon and vesicular membranes and which leads to the release of a quantal packet of transmitter.

In experiments where the role of Ca^{2+} in triggering secretion was tested, Katz and Miledi were able to block synaptic transmission in the giant synapse of the squid by applying a pulse of strong depolarization to the presynaptic membrane.[1] The depolarization was of such high values that it made the membrane terminal at least 130 mV positive with respect to the outside, acting as an electric potential barrier. As soon as the barrier was dropped, a large surge of secretion was detected, as is seen in panels 5 to 8 of Figure 1 (adapted from Ref. 2, Fig. 23). This experiment provided compelling evidence to support the view that the entry of Ca^{2+} from the extracellular space is a prerequisite for the reaction that leads to the release of transmitter. Furthermore, the abrupt surge of secretion that followed the presynaptic depolarization also suggested that the vesicles had to be in a physiological state to respond swiftly to the rapid dissipation of the potential barrier and the ensuing change in axonal calcium. This question is still relevant today, especially in view of the molecular knowledge of the presynaptic release machinery. With the present knowledge of molecular structure of voltage-gated calcium channels[3] and the much more complex molecular anatomy of the secretory vesicle,[4] a question emerges as to how the supramolecular structure of a vesicle can attain full release competence, a transition from a state of no release to a state enabling the release surge of transmitter. In particular, one needs to question whether the vesicles that respond swiftly with release to a stimulus enter the high energy barrier of membrane merger; or their membrane is already fused to the plasma membrane but is in a release-incompetent state? Studies that provide experimental evidence about the transition between the states of vesicle release competence are most interesting, in particular those that can address properties of the fusion pore directly.

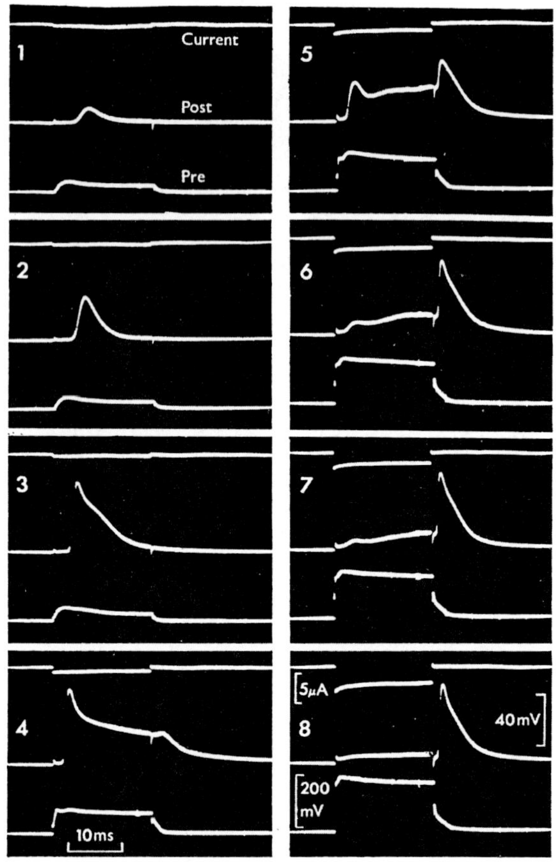

Figure 1. Suppression of transmitter release during a large "positive voltage step" of presynaptic membrane potential. Taken from Ref. 1,2, squid giant synapse treated with tetrodotoxin. Presynaptic terminal loaded with tetraethylammonium ions. Panels 1 to 8 display records obtained by increasing pulse intensity. In each panel, the bottom trace shows the presynaptic voltage step, the middle record shows postsynaptic response, while the current pulse is monitored in the top trace. Note progressive suppression of "on" response and replacement by "off" response as presynaptic voltage is increased from 100 to about 200 mV (records 5 to 8) (adapted from Katz[2] with the permission of The Liverpool University Press).

Fusion Pore Today

Classical view holds that upon stimulation of the fusion pore, an aqueous channel is formed between the vesicle and the plasma membrane through which cargo molecules diffuse out of the vesicle lumen into the cell exterior. Once formed, the fusion pore either closes (transient fusion), as already considered by Katz,[2] to allow the vesicle to be reused in the next round of exocytosis (kiss-and-run exocytosis)[5] or it fully widens leading to the complete merger of vesicle membrane with the plasma membrane (full fusion exocytosis).[6] The patch-clamp membrane capacitance measurements, introduced by Neher and Marty,[7] revealed that fusion pore can also fluctuate between an open and a closed state in the subsecond time domain (fusion pore flickering) before full fusion[8] or can retain the transient nature of opening and closing for several tens of minutes or longer, sometimes periodically ("the pulsing pore") as was observed in the anterior pituitary cells.[9–12] However, it is important to note that electrophysiological membrane capacitance measurements, where the properties of the fusion pore can be studied directly,[8,9,11,12] have some limitations. For example, it is difficult to conclude that the term *fusion pore closure* actually reflects the fission of a vesicle. Because of noise limitations, fusion pore closure may merely reflect narrowing of the fusion pore opening to a diameter that cannot be easily detected by the recording system (Ref. 13; J. Jorgacevski, personal observation). This consideration is important if one aims to understand the nature of transient repetitive fusion pore events, which can be observed in some cellular systems.

The aim of this overview is to highlight the novel findings that indicate that fusion pores are subject to regulations, which affect the release competence of a single vesicle. First, we will discuss spontaneous and stimulated release from a single peptidergic vesicle. Second, we will question whether the rate of cargo release from a single vesicle during spontaneous exocytosis is regulated by the fusion pore kinetics and the size of the fusion pore. Finally, we will discuss whether the stimulus can modulate the mode of peptidergic vesicle exocytosis.

Exocytotic Apparatus in Spontaneous and Stimulated Conditions

Basal secretion of vesicle cargo can be detected in the absence of cell stimulation in almost all secretory cell types, including neurons.[2] It is assumed to occur with low probability.[14] Spontaneous exocytotic events were in the past largely unstudied because of their infrequent occurrence and the belief that they exhibit similar properties to the stimulated events.[2] However, new methodological approaches revealed that spontaneous exocytosis differs from stimulated ones in many respects.

SNARE (soluble *N*-ethylmaleimide-sensitive factor attachment protein receptor) fusion proteins were shown to mediate both spontaneous and stimulated exocytosis (reviewed in Ref. 15); however, the deletions of SNARE proteins affect spontaneous and stimulated synaptic vesicle fusion to different extents.[16–18] Some results indicate that vesicles use different SNARE components when undergoing spontaneous and stimulated fusion.[19,20] Furthermore, SNARE proteins synaptotagmin-1/-2 are thought to trigger the release of vesicle cargo by binding of Ca^{2+} upon stimulation.[21–23] However, synaptotagmin-1/-2 were shown to restrict spontaneous release.[23] Recently, synaptotagmin-12, a Ca^{2+}-independent synaptic vesicle protein controlled by cAMP-dependent processes, has been identified as a possible selective modulator of spontaneous synaptic vesicle exocytosis.[24] Exocytotic apparatus at rest differs from that under stimulation in the distinct protein requirements for vesicle trafficking,[25] vesicle priming,[26] and possibly vesicle recycling.[27,28] Recent studies have pointed out the role of lipids, that is, cholesterol, in balancing spontaneous and stimulated neurotransmitter release by hindering spontaneous and sustaining evoked exocytosis or endocytosis.[29] Spontaneous and stimulated exocytosis also differ in the kinetics of the vesicular content discharge[9] and in the size of the fusion

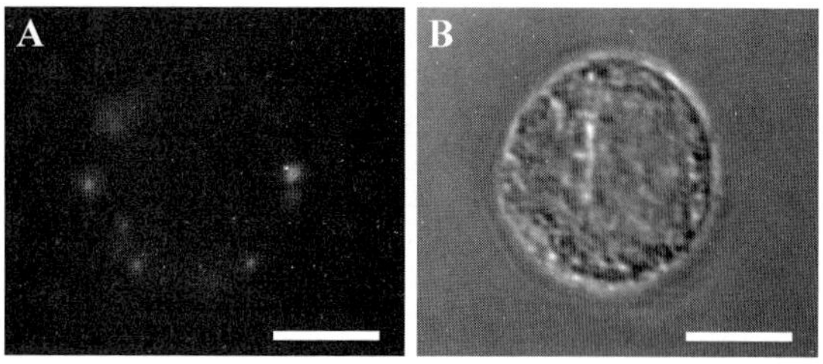

Figure 2. Spot-like fluorescence appearance of immunolabeled prolactin cores in isolated living lactotroph. (**A**) Confocal and (**B**) Differential interference contrast (DIC) image of secreted prolactin (PRL) visualized by anti-PRL antibody, biotinylated secondary antibodies, and quantum dots Qdot 655 (red) in a cell stimulated by bath application of K^+-enriched solution. Note several, bright, fluorescent spots (A) at cell perimeter, each likely reflecting immunolabeled prolactin core. The size of the quantum dot is estimated to 10–20 nm, roughly the same as in fluorescent proteins, indicating that the whole antibody assembly is taken into vesicles with fusion pore diameter of at least 20 nm. In control experiments, where adherence of Qdot 655 to the anti-PRL antibody was omitted or the cells were exposed only to Qdot 655, the fluorescent spots were not observed in lactotrophs. Scale bars = 5 μm. Extracellular solution contained (in mmol/L): 130 NaCl, 5 KCl, 8 $CaCl_2$, 1 $MgCl_2$, 10 D-glucose, and 10 HEPES/NaOH, at pH 7.2. K^+-enriched solution contained (in mmol/L): 35 NaCl, 100 KCl, 8 $CaCl_2$, 1 $MgCl_2$, 10 D-glucose, and 10 HEPES/NaOH, at pH 7.2. Chemicals were obtained from Sigma (Darmstadt, Germany). Pituitaries were obtained after decapitation from male Wistar rats and enriched for lactotrophs.[54] Cell-loaded coverslips were placed in a recording chamber (400 μL) on an inverted confocal microscope (Zeiss LSM 510; Jena, Germany) (objective x63, 1.4 NA, 488-nm Ar-Ion laser excitation and LP 585-nm emission filter). For immunocytochemistry cells were washed in extracellular solution and stimulated in 1 mL of K^+-enriched solution for 4 min at room temperature. Then, a wash was followed with 10% bovine serum albumin (BSA)/phosphate-buffered saline (PBS), and cells were incubated for 15 min at 4°C in 10% BSA/PBS containing a rabbit anti-PRL polyclonal antibody (1:40, 1:80; Chemicon, Temecula, CA). Subsequently, the cells were washed three times with PBS and exposed to 3 μg/mL biotinylated secondary antibodies [Biotin-xx goat anti-rabbit IgG(H+L); Invitrogen, Eugene, OR], 3 μg/mL in PBS with 10% BSA for 20 min at 4°C. Cells were then washed three times in PBS and incubated in 30 nmol/L Qdot 655ITK streptavidin conjugate in Qdot incubation buffer (Quantum Dot Corporation, Hayward, CA), for 30 min at 4°C. Cell-loaded coverslips were washed three times in PBS and finally transferred into extracellular solution for microscopic observation.

pore,[11] which will be discussed in the following sections.

Kinetics of Cargo Discharge from Spontaneous and Stimulated Peptidergic Vesicles

The rate[9,30] and the amount of vesicle cargo release[31] is regulated at the level of an individual secretory vesicle after the fusion pore opening (postfusion regulation).[32] To study the kinetic properties of spontaneous and stimulated peptidergic release,[9] we used lactotrophs, which are prolactin secreting cells of the anterior pituitary (Fig. 2; Ref. 33). We showed that stimulated hormone discharge from a single vesicle can be 10–20 times faster than spontaneous hormone discharge. Optical studies revealed that a fluorescently tagged peptide, atrial natriuretic peptide fused with the emerald green fluorescent protein (ANP-emd), which is similar in size to the hormone prolactin, is released from prolactin vesicles within seconds after stimulation. Simultaneously, vesicle loading with the styryl dye FM 4–64 was observed. In contrast, hormone release and FM 4–64 loading is slow (approximately 3 min) in spontaneously releasing vesicles, indicating that the fusion pore properties in resting and stimulated conditions differ (Fig. 3; Ref. 9).

Measurements of membrane capacitance in resting lactotrophs revealed regular, repetitive, transient fusion pore openings (the phenomenon termed *the pulsing pore*) appearing in bursts with a duration of up to 760 s.[9,11] The duration of a single transient event at rest was approximately 50 ms (Fig. 4; Refs. 9,11), similar to resting synapses of calyx of Held.[34] It has been proposed that the slow exchange of fluorescence probes through the fusion pore in resting lactotrophs (Fig. 3) may be constrained kinetically by a long-lasting, regular,

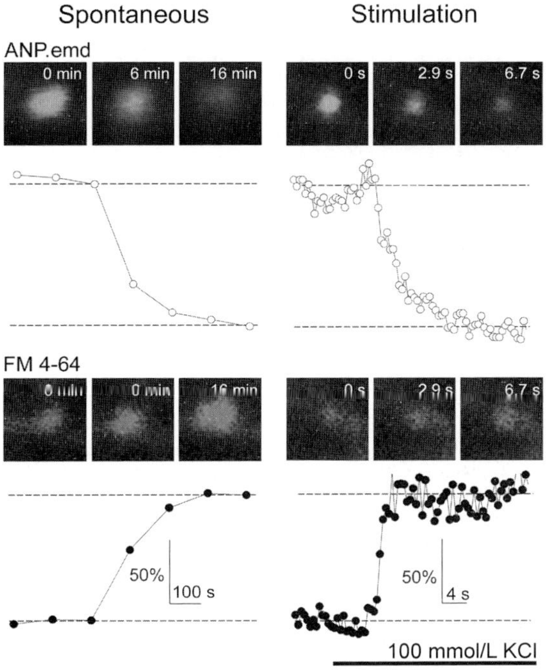

Figure 3. Slower peptidergic vesicle cargo release in spontaneous compared to stimulated conditions. Fluorescence peptide discharge [atrial natriuretic peptide fused with the emerald green fluorescent protein (ANP-emd); upper panels] and FM 4–64 dye loading (lower panels) into prolactin vesicles in spontaneous and in stimulated conditions. Top images show individual vesicle recorded in different time frames separately for the ANP.emd (green) and FM 4–64 (red) fluorescence. Traces below show the time course of relative fluorescence intensity changes of the two fluorescent probes. Note the differences in time scaling for spontaneous- (minutes) and stimulated- (seconds) releasing vesicles. Note that the time course of loading and unloading is similar (synchronous) in spontaneously active vesicles whereas it is different in stimulated vesicles (scale bar = 0.5 μm; modified from Stenovec and coworkers[9] with the permission of The Federation of American Societies for Experimental Biology).

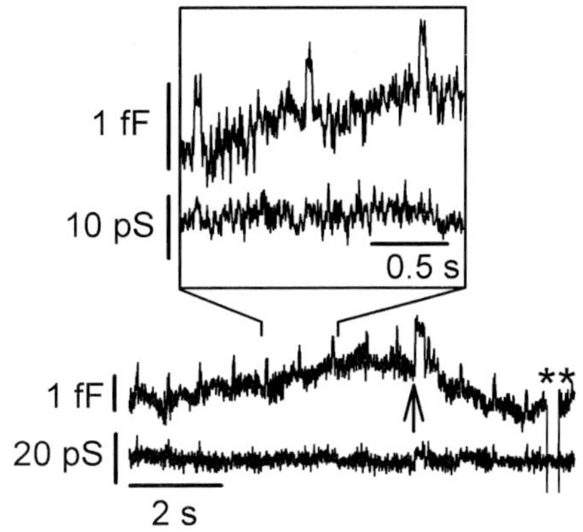

Figure 4. Repetitive transient fusion pore openings in resting lactotophs. Fusion pore openings of a single vesicle in resting pituitary cells were recorded by the patch-clamp capacitance measurements. Inset shows expanded segments of the trace, showing individual transient fusion events. Note that changes in membrane capacitance (C_m) (top trace, imaginary part of admittance) in the inset are not correlated with changes in the real part of admittance (lower trace), indicating the correct phase angle setting of the lock-in amplifier by the calibration pulse (**) at the end of the trace. Arrow indicates a fusion pore event with higher amplitude and a lower frequency of occurrence (appearing every ca. 12 s), suggesting the presence of two vesicles in this cell-attached membrane. Note the pulsing behavior of the fusion pore with fusion pore dwell time of around 50 ms (adapted from Stenovec and coworkers[9] with the permission of The Federation of American Societies for Experimental Biology).

fusion pore gating.[9] Transient vesicle fusion can limit or prevent peptide hormone release because of the large molecular size of peptides and consequent low diffusional mobility of peptide molecules.[35–37] Similarly, fusion pore flickering can limit dopamine release from synaptic terminals in neurons.[38] This indicates that, despite the relatively high diffusional mobility of small chemical transmitters, fusion pore gating may also be involved in the regulation of small chemical transmitter release in synapses.[39]

Subnanometer Fusion Pore Diameters in Resting Peptidergic Vesicles Expand upon Stimulation

The slower vesicle cargo release observed in resting lactotrophs (Fig. 3) might also reflect a narrow fusion pore of spontaneously releasing vesicles. Vesicle fusion was monitored by confocal microscopy in resting and stimulated lactotrophs expressing synaptopHluorin (spH), a pH-sensitive GFP that enables optical discrimination between unfused and fused vesicles.[40] Simultaneously, the permeation of

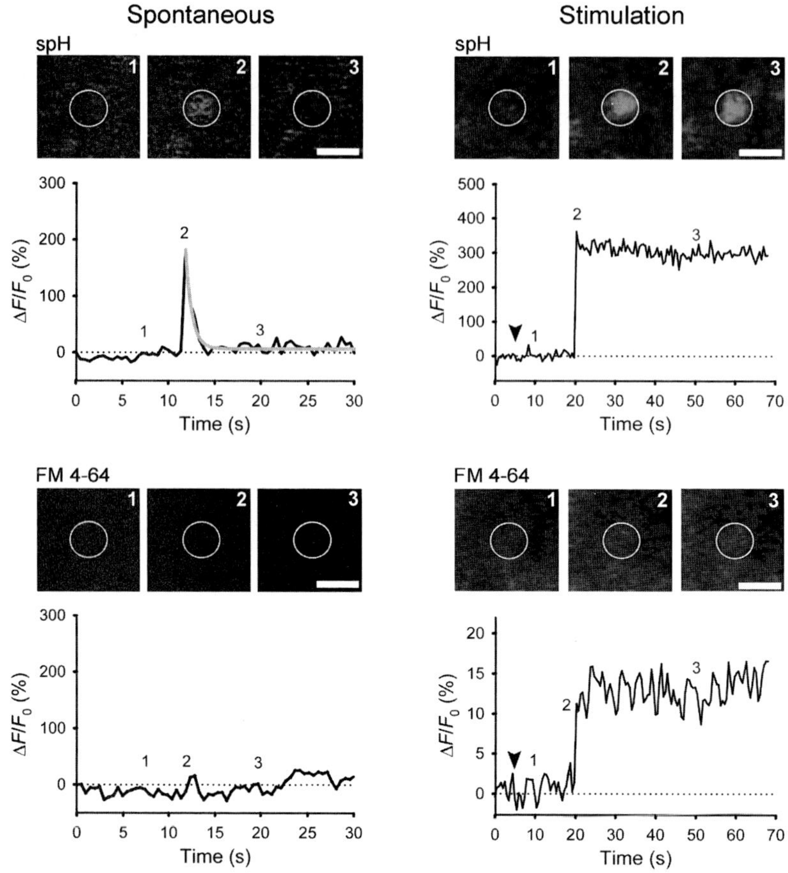

Figure 5. Narrow fusion pores in spontaneous exocytosis of peptidergic vesicles. Sequential images of a single vesicle and changes in synaptopHluorin (spH) and FM 4–64 fluorescence at the vesicle site at rest (left) and after stimulation (right). Exocytosis resulted in a rapid increase in spH fluorescence, followed by a rapid exponential decline (gray line; spontaneous) or a persistent elevation (after stimulation). Note that stimulation triggered loading of FM 4–64 dye into vesicles likely because of larger pore diameter. Numbers on plots correspond to times when images were recorded. White circles indicate sites of exocytic events; scale bar = 1 μm; arrowheads indicate the onset of stimulation. Modified from Vardjan and coworkers[11] with the permission of The Society for Neuroscience.

FM 4–64 dye and HEPES molecules through the fusion pores was monitored.[11] Fusion pore openings associated with a change in spH fluorescence were observed in more than 50% of the spontaneous exocytotic events (Fig. 5; Ref. 11). In these events the fusion pore was permeable to protons, as indicated by spH fluorescence changes, but impermeable to FM 4–64 (Fig. 5, left; molecular diameter = ca. 0.9 nm) and HEPES (molecular diameter = ca. 0.5 nm) molecules. Thus, fusion pore diameter in resting peptidergic vesicles is less than 0.5 nm, which is smaller than the size of peptide hormone prolactin stored in these vesicles (prolactin molecular diameter = ca. 5.2 nm; Ref. 11). Membrane capacitance measurements confirmed subnanometer fusion pores in resting lactotrophs,[11,12,41] suggesting that exocytosis without release of vesicle cargo may occur before delivery of the stimulus ("unproductive exocytosis").

Upon stimulation, greater than 70% of exocytotic events exhibited a larger FM 4–64 permeable pore (> 0.9 nm; Fig. 5, right) consistent with previous fusion pore permeation studies.[35,36,42,43] In capacitance measurements, stimulation decreased the fraction of events with the lowest measurable fusion pore diameter of 3.2 nm from 25% to 2% reflecting much wider pore diameters upon stimulation and likely a massive vesicle cargo discharge in the majority of these events.[11] Similarly, fusion pore expansion was observed in chromaffin cells upon stimulation, which likely improves the efficiency of release of small classical transmitters.[43,44] Moreover, in stimulated lactotrophs the pore size can reversibly open to several nanometers in diameter because quantum dots coupled to prolactin antibodies are introduced into retrieving prolactin vesicles (Fig. 2).

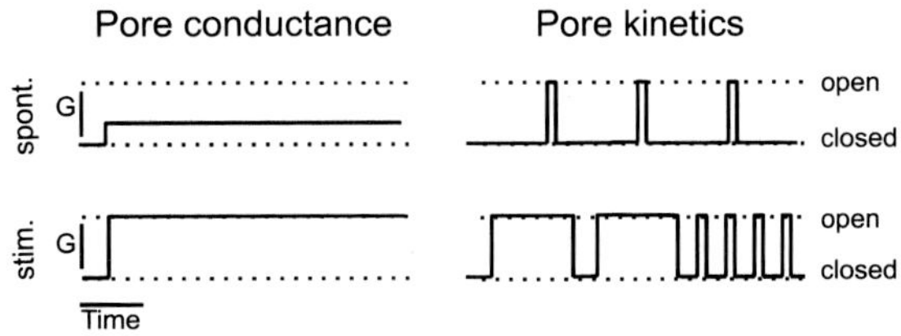

Figure 6. Fusion pore properties before and after stimulation. The permeation of molecules through the fusion pore depends on fusion pore conductance (G; diameter) and/or fusion pore kinetics. A wider fusion pore (higher G) and/or faster frequency of fusion pore openings with longer fusion pore dwell times (faster kinetic of fusion pore openings) leads to a faster more effective release of peptides from vesicles in stimulated events (modified from Stenovec and coworkers[9] with the permission of The Federation of American Societies for Experimental Biology).

Stimulus Modulates Kinetic Properties of a Fusion Pore

Transient fusion is accepted as a mode of vesicle cargo release both in central neurons and neuroendocrine cells, but the prevalence of this mechanism in comparison to full fusion is still under discussion.[45]

In lactotrophs transient exocytosis is the predominant mode of exocytosis in spontaneous and stimulated vesicles.[11] Optical studies on lactotrophs expressing spH revealed that 65% of spontaneous events were transient (Fig. 5, left), some of them were repetitive.[11] In 35% of all events, spH fluorescence persisted after fusion for more than 100 s. After stimulation the majority of exocytotic events (> 93%) were persistent (Fig. 5, right; Ref. 11). In persistent events the lateral diffusion of spH signal from the site of exocytosis into the plasma membrane was not detected,[11] suggesting that vesicle exocytosis occurred in a non-full-collapse mechanism.[46,47] Since FM 4–64 loading was fourfold slower in persistent events than in transient events, spH signal likely persisted because of a rapidly flickering fusion pore[9,38] and not because of a long fusion pore lifetime.[48] This hypothesis was confirmed by capacitance measurements that revealed that transient fusion pore openings with the burst duration of more than 100 s are the predominant mode of spontaneous and stimulated exocytosis in lactotrophs.[9,11] Optical monitoring of fusion pore dynamics is limited by the characteristics of the spH construct. Thus the effect of fusion pore closure is poorly reflected in these studies because of the reacidification, which is in comparison to the fusion pore closure observed as an off-step in capacitance measurements rather slow ($\tau = 4$–5 s) for synaptic vesicles.[49] These results are in contrast to previous studies on chromaffin cells where increased cell stimulation shifted the mode of exocytosis from transient to full fusion, resulting in the modulation of the amount of cargo release from a single vesicle.[43,50,51]

Capacitance measurements also revealed that, in lactotrophs, transient vesicle fusion occurred fourfold more frequently after stimulation with a twofold longer fusion pore dwell time and a wider pore diameter.[11] Therefore a stimulus prolongs the effective opening of a fusion pore and expands its subnanometer diameter to enable hormone secretion without full fusion (Fig. 5; Ref. 11).

Future Directions

Recent findings indicate that fusion pores are subject to regulations that affect the release competence of a single vesicle. Current knowledge of regulatory mechanisms on the postfusion stage through the regulation of fusion pore diameter and/or fusion pore open duration is still very limited.[32] In lactotrophs the release of vesicle cargo may be restrained kinetically and/or a result of a narrow fusion pore. Under stimulation, the preformed fusion pore may retain the transient nature but with a longer dwell time, increased frequency of reopenings, and an increased fusion pore diameter, which facilitate the vesicle cargo release (Fig. 6). Fusion pore dynamics are important, as any modulation of release processes could have impacts on specific physiological functions of resting and stimulated cells.

A key question to be answered in the future is related to the nature of the repetitive transient fusion events. Do they represent cycles of merger/fission of a single vesicle? Given that membrane merger involves a high energy barrier, such a repetitive cycle is unlikely. Perhaps the transient fusion events represent fluctuations of an open fusion pore between states where the pore is extremely narrow, with a neck formed by the fused vesicle and plasma membrane. A narrow fusion pore is structurally made of membrane domains with high local curvature, which may be attained by highly anisotropic molecules with negative curvatures.[52] In support of such distinct membrane structure adjacent to fused vesicles versus plasma membrane, it was reported that plasma membrane sites where prolactin vesicles become engaged in exocytosis are distinct from the plasma membrane areas devoid of docked prolactin vesicles.[53] Finally, the basis of energetic stability of such special curved structures is to be addressed in the future as well.

Acknowledgments

This work was supported by grants P3 310 381, Z3 7476 from the Ministry of Higher Education, Sciences and Technology of the Republic of Slovenia, NS36665 from the National Institutes of Health, European Commission (EC) grant #QLG3 2001-2004, EC support DECG, CLG3-CT-2001-02004, and EC GROWBETA QLG1-CT-2001-02233.

Conflicts of Interest

The authors declare no conflicts of interest.

References

1. Katz, B. & R. Miledi. 1967. A study of synaptic transmission in the absence of nerve impulses. *J. Physiol.* **192:** 407–436.
2. Katz, B. 1969. *The Release of Neural Transmitter Substances*. Liverpool University Press, Liverpool.
3. Jones, S. 1998. Overview of voltage-dependent calcium channels. *J. Bioenerg. Biomembr.* **30:** 299–312.
4. Takamori, S. *et al.* 2006. Molecular anatomy of a trafficking organelle. *Cell* **127:** 831–846.
5. Ceccarelli, B., W. Hurlbut & A. Mauro. 1973. Turnover of transmitter and synaptic vesicles at the frog neuromuscular junction. *J. Cell Biol.* **57:** 499–524.
6. Heuser, J. & T. Reese. 1973. Evidence for recycling of synaptic vesicle membrane during transmitter release at the frog neuromuscular junction. *J. Cell Biol.* **57:** 315–344.
7. Neher, E. & A. Marty. 1982. Discrete changes of cell membrane capacitance observed under conditions of enhanced secretion in bovine adrenal chromaffin cells. *Proc. Natl. Acad. Sci. USA* **79:** 6712–6716.
8. Fernandez, J., E. Neher & B. Gomperts. 1984. Capacitance measurements reveal stepwise fusion events in degranulating mast cells. *Nature* **312:** 453–455.
9. Stenovec, M. *et al.* 2004. Slow spontaneous secretion from single large dense-core vesicles monitored in neuroendocrine cells. *FASEB J.* **18:** 1270–1272.
10. Vardjan, N. *et al.* 2007. Elementary properties of spontaneous fusion of peptidergic vesicles: fusion pore gating. *J. Physiol.* **585:** 655–661.
11. Vardjan, N. *et al.* 2007. Subnanometer fusion pores in spontaneous exocytosis of peptidergic vesicles. *J. Neurosci.* **27:** 4737–4746.
12. Jorgacevski, J. *et al.* 2008. Hypotonicity and peptide discharge from a single vesicle. *Am. J. Physiol. Cell Physiol.* **295:** C624-C631.
13. Debus, K. & M. Lindau. 2000. Resolution of patch capacitance recordings and of fusion pore conductances in small vesicles. *Biophys. J.* **78:** 2983–2997.
14. Murthy, V. & C. Stevens. 1999. Reversal of synaptic vesicle docking at central synapses. *Nat. Neurosci.* **2:** 503–507.

15. Jahn, R. & R. Scheller. 2006. SNAREs–engines for membrane fusion. *Nat. Rev. Mol. Cell Biol.* **7:** 631–643.
16. Deitcher, D. *et al.* 1998. Distinct requirements for evoked and spontaneous release of neurotransmitter are revealed by mutations in the Drosophila gene neuronal-synaptobrevin. *J. Neurosci.* **18:** 2028–2039.
17. Schoch, S. *et al.* 2001. SNARE function analyzed in synaptobrevin/VAMP knockout mice. *Science* **294:** 1117–1122.
18. Washbourne, P. *et al.* 2002. Genetic ablation of the t-SNARE SNAP-25 distinguishes mechanisms of neuroexocytosis. *Nat. Neurosci.* **5:** 19–26.
19. Deák, F. *et al.* 2004. Synaptobrevin is essential for fast synaptic-vesicle endocytosis. *Nat. Cell Biol.* **6:** 1102–1108.
20. Hua, S. *et al.* 1998. Different VAMP/synaptobrevin complexes for spontaneous and evoked transmitter release at the crayfish neuromuscular junction. *J. Neurophysiol.* **80:** 3233–3246.
21. Geppert, M. *et al.* 1994. Synaptotagmin I: a major Ca2+ sensor for transmitter release at a central synapse. *Cell* **79:** 717–727.
22. Kreft, M. *et al.* 2003. Synaptotagmin I increases the probability of vesicle fusion at low [Ca2+] in pituitary cells. *Am. J. Physiol. Cell Physiol.* **284:** C547–554.
23. Pang, Z. *et al.* 2006. Genetic analysis of synaptotagmin 2 in spontaneous and Ca2+-triggered neurotransmitter release. *EMBO J.* **25:** 2039–2050.
24. Maximov, A. *et al.* 2007. Synaptotagmin-12, a synaptic vesicle phosphoprotein that modulates spontaneous neurotransmitter release. *J. Cell Biol.* **176:** 113–124.
25. Wucherpfennig, T., M. Wilsch-Bräuninger & M. González-Gaitán. 2003. Role of Drosophila Rab5 during endosomal trafficking at the synapse and evoked neurotransmitter release. *J. Cell Biol.* **161:** 609–624.
26. Virmani, T. *et al.* 2005. Phorbol esters target the activity-dependent recycling pool and spare spontaneous vesicle recycling. *J. Neurosci.* **25:** 10922–10929.
27. Sara, Y. *et al.* 2005. An isolated pool of vesicles recycles at rest and drives spontaneous neurotransmission. *Neuron* **45:** 563–573.
28. Groemer, T. & J. Klingauf. 2007. Synaptic vesicles recycling spontaneously and during activity belong to the same vesicle pool. *Nat. Neurosci.* **10:** 145–147.
29. Wasser, C. *et al.* 2007. Cholesterol-dependent balance between evoked and spontaneous synaptic vesicle recycling. *J. Physiol.* **579:** 413–429.
30. Albillos, A. *et al.* 1997. The exocytotic event in chromaffin cells revealed by patch amperometry. *Nature* **389:** 509–512.
31. Angleson, J. *et al.* 1999. Regulation of dense core release from neuroendocrine cells revealed by imaging single exocytic events. *Nat. Neurosci.* **2:** 440–446.
32. Rahamimoff, R. & J. Fernandez. 1997. Pre- and postfusion regulation of transmitter release. *Neuron* **18:** 17–27.
33. Walker, A. & M. Farquhar. 1980. Preferential release of newly synthesized prolactin granules is the result of functional heterogeneity among mammotrophs. *Endocrinology* **107:** 1095–1104.
34. Sun, J., X. Wu & L. Wu. 2002. Single and multiple vesicle fusion induce different rates of endocytosis at a central synapse. *Nature* **417:** 555–559.
35. Barg, S. *et al.* 2002. Delay between fusion pore opening and peptide release from large dense-core vesicles in neuroendocrine cells. *Neuron* **33:** 287–299.
36. Tsuboi, T. & G. Rutter. 2003. Multiple forms of "kiss-and-run" exocytosis revealed by evanescent wave microscopy. *Curr. Biol.* **13:** 563–567.
37. Obermüller, S. *et al.* 2005. Selective nucleotide-release from dense-core granules in insulin-secreting cells. *J. Cell Sci.* **118:** 4271–4282.
38. Staal, R., E. Mosharov & D. Sulzer. 2004. Dopamine neurons release transmitter via a flickering fusion pore. *Nat. Neurosci.* **7:** 341–346.
39. Alvarez de Toledo, G., R. Fernández-Chacón & J. Fernández. 1993. Release of secretory products during transient vesicle fusion. *Nature* **363:** 554–558.
40. Miesenböck, G., D. De Angelis & J. Rothman. 1998. Visualizing secretion and synaptic transmission with pH-sensitive green fluorescent proteins. *Nature* **394:** 192–195.
41. Spruce, A. *et al.* 1990. Properties of the fusion pore that forms during exocytosis of a mast cell secretory vesicle. *Neuron.* **4:** 643–654.
42. Takahashi, N. *et al.* 2002. Fusion pore dynamics and insulin granule exocytosis in the pancreatic islet. *Science* **297:** 1349–1352.
43. Fulop, T., S. Radabaugh & C. Smith. 2005. Activity-dependent differential transmitter release in mouse adrenal chromaffin cells. *J. Neurosci.* **25:** 7324–7332.
44. Elhamdani, A., H. Palfrey & C. Artalejo. 2001. Quantal size is dependent on stimulation frequency and calcium entry in calf chromaffin cells. *Neuron* **31:** 819–830.
45. LoGiudice, L. & G. Matthews. 2006. The synaptic vesicle cycle: is kissing overrated? *Neuron* **51:** 676–677.
46. Taraska, J. *et al.* 2003. Secretory granules are recaptured largely intact after stimulated exocytosis in cultured endocrine cells. *Proc. Natl. Acad. Sci. USA* **100:** 2070–2075.
47. Thorn, P. & I. Parker. 2005. Two phases of zymogen granule lifetime in mouse pancreas: ghost granules linger after exocytosis of contents. *J. Physiol.* **563:** 433–442.

48. Ohara-Imaizumi, M. *et al*. 2002. Monitoring of exocytosis and endocytosis of insulin secretory granules in the pancreatic beta-cell line MIN6 using pH-sensitive green fluorescent protein (pHluorin) and confocal laser microscopy. *Biochem. J.* **363:** 73–80.
49. Atluri, P. & T. Ryan. 2006. The kinetics of synaptic vesicle reacidification at hippocampal nerve terminals. *J. Neurosci.* **26:** 2313–2320.
50. Elhamdani, A., F. Azizi & C. Artalejo. 2006. Double patch clamp reveals that transient fusion (kiss-and-run) is a major mechanism of secretion in calf adrenal chromaffin cells: high calcium shifts the mechanism from kiss-and-run to complete fusion. *J. Neurosci.* **26:** 3030–3036.
51. Harata, N., A. Aravanis & R. Tsien. 2006. Kiss-and-run and full-collapse fusion as modes of exo-endocytosis in neurosecretion. *J. Neurochem.* **97:** 1546–1570.
52. Churchward, M. *et al*. 2008. Specific lipids supply critical negative spontaneous curvature–an essential component of native Ca2+-triggered membrane fusion. *Biophys. J.* **94:** 3976–3986.
53. Gonçalves, P. *et al*. 2008. Prolactin Secretion Sites Contain Syntaxin-1 and Differ from GM1 Rafts in Rat Lactotrophs. *Endocrinology* **149:** 4948–4957.
54. Ben-Tabou, S., E. Keller & I. Nussinovitch. 1994. Mechanosensitivity of voltage-gated calcium currents in rat anterior pituitary cells. *J. Physiol.* **476:** 29–39.

SNAP-25 and Gene-targeted Mouse Mutants

Christina Bark

The Rolf Luft Center for Diabetes and Endocrinology, Department of Molecular Medicine and Surgery, Karolinska Institutet, Karolinska University Hospital, Stockholm, Sweden

The evolutionary conserved soluble *N*-ethylmaleimide-sensitive factor attachment protein receptor (SNARE) fusion machinery is the operational unit in the release of neurotransmitters and hormones from excitable cells. The SNARE core complex consists of three proteins named SNAP-25 (synaptosomal-associated protein of 25kD), syntaxin 1, and VAMP (vesicle-associated membrane protein)/synaptobrevin. Syntaxin 1 is, together with SNAP-25, localized to the plasma membrane, whereas VAMP/synaptobrevin is a component of secretory vesicles. In concert with the SNAREs, accessory factors govern the docking and priming of secretory vesicles prior to trans-SNARE complex formation and ultimately Ca^{2+}-triggered fusion pore opening at the plasma membrane. The synaptic SNAP-25 protein exists as two closely related protein variants, named SNAP-25a and SNAP-25b. SNAP-25a and SNAP-25b are both encoded from a single copy gene and generated by obligate alternative splicing between two similar exon 5 sequences. Exon 5 spans a region of SNAP-25 that is subject to posttranslational palmitoylation and implicated in membrane anchoring of this cytosolic protein. The alternative splicing is strictly developmentally and neuroanatomically regulated, but the biological relevance of the distinct expression of these two similar protein variants is still a question of debate. However, recent findings in gene-targeted mouse mutants have started to unravel the importance that physiological levels of total SNAP-25 protein are present and, importantly, that this is accompanied by a balanced expression of SNAP-25a and SNAP-25b.

Key words: **SNARE; membrane fusion; regulated exocytosis**

Regulated membrane fusion and the discharge of signal substances to the extracellular space is the terminal step of a specialized membrane trafficking pathway in all secretory cells. In neuronal networks, electrical activity is processed into a chemical signal when an action potential reaches the nerve ending of the sending cell. The depolarization of the nerve terminal opens voltage-gated Ca^{2+} channels and the local cytoplasmic rise in Ca^{2+} mediates, via intracellular Ca^{2+}-sensing proteins, the signal for fusion of transmitter-containing vesicles with the plasma membrane. The release of neurotransmitters predominantly takes place at junctions between presynaptic and postsynaptic cells, termed *synapses*, and the secreted transmitter is converted to an electrochemical signal when reaching the target receptor on the postsynaptic cell. Ca^{2+}-triggered membrane fusion regulates both secretion in a quantitative and qualitative manner. One of the last steps prior to the secretion event requires the formation of a trans-membrane SNARE (soluble N-sensitive factor attachment protein receptor) complex.[1,2] The 7S core SNARE complex consists of the vesicle component vesicular-associated membrane protein (VAMP)/synaptobrevin as well as the plasma membrane proteins SNAP-25 (synaptosomal-associated protein of 25 kD) and syntaxin 1.[1–3] The SNARE complex is believed to mediate the fusion of several types of vesicles. In nerve terminals, two classes of

Address for correspondence: Christina Bark, The Rolf Luft Center for Diabetes Research and Endocrinology, L1:03, Department of Molecular Medicine and Surgery, Karolinska Institutet, Karolinska University Hospital, SE-171 76 Stockholm, Sweden. Voice: +46-8-51779452; fax: +46-8-51779450. Christina.Bark@ki.se

Mechanisms of Exocytosis: Ann. N.Y. Acad. Sci. 1152: 145–153 (2009).
doi: 10.1111/j.1749-6632.2008.04009.x

Ca^{2+}-regulated secretory vesicles can be found: small synaptic vesicles (SVs) that usually are 35–50 nm in diameter and store classical neurotransmitters and larger membrane vesicles 70–200 nm in diameter, called large dense-core vesicles (LDCVs). LDCVs predominantly carry neuropeptides and biogenic amines, but sometimes classical neurotransmitters co-exist with the other signaling molecules in LDCVs.[4,5] The small SVs are found in clusters of between 50 and 1000 vesicles and are located in the proximity of the presynaptic release sites in axon terminals.[6] The electron-dense LDCV are less spatially organized and fuse with the plasma membrane outside the active zone.[7] Despite that different releasable vesicles require different magnitudes and frequency of stimulation to fuse with the plasma membrane and discharge their contents, both types of vesicle fusion are mediated by SNARE proteins.

The precise, and apparently multiple, roles the different SNARE proteins play prior to, during, and possible also after vesicle and plasma membrane fusion are still not fully elucidated, but the amount of information is increasing quickly.[8,12] The current favored hypothesis is that SNARE proteins operate as engines for fusion as they evidently have intrinsic capabilities to perform this process.[13,14]

SNAP-25

SNAP-25 is a 206-amino acid 25-kD protein and was originally identified as a cDNA corresponding to an mRNA abundantly expressed in mouse brain but not in non-neuronal tissue.[15] Thereafter, SNAP-25 has been demonstrated to be widely but differently distributed both in the nervous system as well as in neuroendocrine and endocrine cells[16–20] and with an onset of gene expression that correlates to neuronal maturation.[21] SNAP-25 is a cytosolic protein and has two so-called SNARE motifs, which are 70-residue sequences present in all SNARE proteins; SNAP-25 thus contributes with two α-helices to the coiled–coiled structure that holds the core fusion trans-SNARE complex together.[3] A region of the amino terminus, residues 7–36, forms an amphipathic α-helix, and a central domain of the protein, positions 85–92, spans the only cysteine residues present in the polypeptide. The "cysteine quartet" is substrate for posttranslational palmitoylation and is believed to mediate membrane anchoring of the protein.[22–25] SNAP-25 is also a substrate for protein kinase A (PKA) phosphorylation at position 138 (Thr) and PKC phosphorylation at residue Ser187.[26–30]

When the gene for SNAP-25 was characterized, it was demonstrated that the polypeptide was encoded for by eight different exons spaced by quite large intron sequences, resulting in the gene encompassing more than 80 kb.[31] Surprisingly, the structure of the *Snap*25 gene revealed that two closely spaced and similar exon 5 sequences exist. Thus, an additional SNAP-25 variant, which only differed in nine amino acids from the originally described SNAP-25 protein, was identified and termed SNAP-25a.[31] *Snap*25 is a single copy gene, and SNAP-25a or SNAP-25b mRNAs are generated by obligate alternative splicing between these two similar but distinct exon five sequences, both encoding 39 amino acids. The nine amino acids that distinguish SNAP-25b from SNAP-25a include the region in the SNAP-25 protein that spans the last part of the N-terminal SNARE motif and the first part of the linker that separates the N- and C-terminal α-helices. Thus, the amino acid differences participating in forming the four α-helix coil–coil structure in the SNARE complex could possibly interfere with SNARE complex stability or ability to interact with accessory proteins. In addition, differences in the linker region distort the cystein quartet, and although all four cysteines are retained, they appear in a different organization and sequence context.

Studies of gene expression of these two alternative SNAP-25 variants in mouse brain have demonstrated that the two mRNAs are differently expressed during development and their neuroanatomical distributions are individual and distinct.[17–19,32,33] The SNAP-25a protein appears to represent a developmentally earlier and minor isoform, whereas the SNAP-25b protein is the absolute dominant species in adult brain. SNAP-25a is, although expressed at relatively lower levels than SNAP-25b, the dominant species in neuroendocrine cells, such as chromaffin cells derived from the adrenal gland, but also in selected brain structures in adult animals.[19,34,35] SNAP-25 is also expressed in the endocrine β-cell,[16,36] and in primary β-cells SNAP-25a also appears to be the dominant isoform in adulthood, although a more comprehensive quantitative study of mRNA expression is required to confirm this.[20]

In line with the hypothesis that the different isoforms of SNAP-25 confer different biological functions during neuronal growth and remodeling, several investigations have been performed. These include studies of developmental regulation in mouse and rat,[19,37] expression of SNAP-25a and SNAP-25b after nerve injury,[17,37,38] and involvement in long-term potentiation.[39] However, even if the SNAP-25a isoform was upregulated after nerve injury, no obvious functional differences between the two SNAP-25 isoforms were uncovered in these studies. Neither did *in vitro* studies in endocrine β-cells or insulinoma cell lines unravel a functional distinction between SNAP-25a and SNAP-25b in regulated exocytosis.[20,36] The studies in β-cells basically demonstrated that the two SNAP-25 protein variants are equally important for stimulated insulin secretion and, most unexpectedly, their function could be rescued by (over)expression of the cellular homologue to SNAP-25, SNAP-23.[20,36] SNAP-23 is ubiquitously expressed and believed to participate in late molecular steps preceding constitutive exocytosis. Are SNAP-25a and SNAP-25b functionally mutually exchangeable in regulated exocytosis or are *in vitro* experimental procedures usually too blunt to distinguish physiological differences between two similar SNAP-25 proteins?

Gene-targeted Mouse Mutants for SNAP-25

The SNAP-25 Knockout Mouse

The complete knockout of the *Snap*25 gene in mouse resolutely demonstrated the importance of SNAP-25 in neuroexocytosis.[40] Whereas spontaneous transmitter release was intact in SNAP-25 null mutants, evoked action potential-dependent exocytosis was completely abolished and homozygous mutants died at birth.[40] Interestingly, the targeted disruption of SNAP-25 did not affect normal nervous system development or spontaneous transmitter release. This was actually a finding in contradiction to early studies in PC12 cells that did suggest, based on an antisense approach, the importance of SNAP-25 in axonal outgrowth.[41] Surprisingly, heterozygous SNAP-25 null mutants expressing only half the amount of SNAP-25 mRNA and protein were experimentally indistinguishable from wild-type (WT) littermates (see also Table 1). Homozygous SNAP-25 null mutants demonstrated embryonic lethality, but nevertheless they proved to be critical tools for finally sorting out a biological difference between the SNAP-25a and SNAP-25b proteins. In an elegant study by Sørensen that is (2003),[42] fast Ca^{2+}-triggered exocytosis was rescued by viral introduction and overexpression of SNAP-25 in embryonic *Snap*25 null mutant chromaffin cells.[42] In addition, rescuing experiments when either SNAP-25a or SNAP-25b was introduced into these SNAP-25-deficient chromaffin cells demonstrated that SNAP-25b had a greater ability to maintain a larger pool of vesicles in a primed state, that is, ready for fusion.[42] Actually, SNAP-25b overexpression increased the size of the releasable vesicle pools about threefold without changing the fusion kinetics. Sørensen and colleagues suggested that the two SNAP-25 isoforms provide different stability to primed vesicles not by changing the forward rate of the priming process but rather that the rate of depriming was lower in cells overexpressing SNAP-25b. In a following study

TABLE 1. SNAP-25 Mouse Mutants

Mouse Mutant	Phenotype	SNAP-25 Isoform Gene Expression	Reference
Snap25 Knockout Mice			40
Homozygous Mutants	Die at birth, no evoked exocytosis	No SNAP-25 expression	
Heterozygous Mutants	No reported phenotype	50% of normal SNAP-25 expression levels, ratio between SNAP-25a/SNAP-25b normal	
Snap25$^{Tkneo/Tkneo}$ Mutants			44
Homozygous Mutants	Most mutants die between 3–5 weeks of age, impaired short-term synaptic plasticity, seizures	50% of normal SNAP-25 expression levels, ratio between SNAP-25a/SNAP-25b isoforms shifted towards more SNAP-25a	
Blind-drunk (Bdr) Mutants			45, 46
Heterozygous Mutants	Impaired sensorimotor gating and ataxia, impaired spontaneous and evoked release	Normal SNAP-25 expression levels	
SNAP-25b-deficient Mice			47
- Neo-containing Homozygous Mutants	Developmental defects, seizures, impaired short-term synaptic plasticity, lethal in the third postnatal week	50% of normal SNAP-25 expression levels, only SNAP-25a	
- Neo-excised Homozygous Mutants	Impaired short-term synaptic plasticity, progressive morphological changes in brain, learning deficits, seizures	Normal SNAP-25 expression levels but only SNAP-25a	

they also identified that much of the differences between SNAP-25a and SNAP-25b in secretion capacity were the amino acid differences located in their N-terminal SNARE domain.[43] Overexpression of SNAP-23 in *Snap*25-deficient chromaffin cells could partly rescue sustained secretion, but the "exocytotic burst," a hallmark of fast Ca^{2+}-triggered secretion, was still impaired. These data also support the idea that SNAP-23 only is important for slower modes of exocytosis and normally is not involved in fast modes of regulated membrane fusion. The finding that SNAP-25 isoforms control late steps prior to fusion by regulating the size of different vesicle pools indeed explains part of the biological importance for signaling cells to express either or both variants of SNAP-25. Still, as chromaffin cells normally predominantly express SNAP-25a in adulthood, the repertoire of accessory proteins matching the SNAP-25b protein might not be completely physiological when assayed in embryonic chromaffin cells.

The *Snap25*$^{Tkneo/Tkneo}$ Mini Transgene

Our own studies using a gene-targeted mouse model, designated *Snap*25$^{Tkneo/Tkneo}$, with a selection gene inserted in the intron sequence upstream of exon 6 in the *Snap*25 gene supported the role of the two SNAP-25 isoforms in synaptic plasticity.[44] This *Snap*25$^{Tkneo/Tkneo}$ mouse mutant, or "mini transgene," exhibited an impaired alternative splicing pattern between the two alternative exon 5 sequences, shifting the relative ratio toward more SNAP-25a versus SNAP-25b protein in brain (Table 1). The compromised alternative splicing pattern between SNAP-25a and SNAP-25b isoforms influenced short-term synaptic plasticity in neurotransmission by affecting facilitation but not the probability of release. The majority of homozygous *Snap*25$^{Tkneo/Tkneo}$ mutants die between 3 and 5 weeks of age, which coincides with the time when SNAP-25b expression in brain normally reaches mature levels and synapse formation is largely completed. Heterozygous mutants did not demonstrate an obvious phenotype. This suggests that the mechanisms controlling alternative splicing between SNAP-25 isoforms may serve as a molecular switch regulating synaptic plasticity by guiding the transition from immature to mature synaptic connections as well as synapse regrowth and remodeling after neural injury.[44] However, the total levels of SNAP-25 expression in these homozygous mouse mutants were down to approximately 50% of normal as the presence of the *Tkneo* transgene not only impaired alternative splicing but also suppressed total gene expression. Therefore, the phenotype observed in the *Snap*25$^{Tkneo/Tkneo}$ could be the result of a combination of an imbalance in splicing and reduced total expression levels of SNAP-25. *In vivo* excision of the *Tkneo* transgene in these mutants completely reversed the phenotype.

The Blind-drunk Mouse

It has previously been established that SNAP-25b containing SNARE complexes are more stable than those containing SNAP-25a.[3,42] Interestingly, from a large-scale mutagenesis screen, a mouse mutant with even more stable SNARE complexes was identified.[45] The blind-drunk (Bdr) mouse expresses a SNAP-25b protein with a dominant point mutation in the 5b specific exon, resulting in an even higher affinity for syntaxin than WT SNAP-25b.[45] This results in an impairment of both spontaneous and evoked release from cortical neurons and lack of facilitation during trains of high-frequency stimulation (Table 1). This could be because the higher stability of the SNARE complex interferes with normal SNARE complex dissociation and membrane recycling, which then also impair replenishment of releasable vesicle pools. The phenotype is mild, but mutants demonstrate ataxia and impaired sensorimotor gating. The Bdr mutant, as understood from the name, has poor eye sight or is blind, but that is merely a result of the background strains used in the screening and not because of the mutation in the SNAP-25b protein. The Bdr mutant was collected during a mutagenesis screen with ethylnitrosurea (ENU), which involved mutagenizing BALB/c males (albinos, known to have retinal abnormalities from the absence of melanin) and crossing them to C3H/HeH females, which carry the retinal degeneration mutation known as rd-1.[46]

A Gene-targeted SNAP-25b-deficient Mouse Mutant

To more directly address the specific function of the different SNAP-25 isoforms in a physiological setting, we recently developed SNAP-25b-deficient mouse mutants. The experimental idea was to manipulate the endogenous *Snap*25 gene as little as possible, and therefore we removed the downstream exon 5, 5b and simultaneously replaced it with an additional exon 5a equivalent without interfering with surrounding intron sequences. Thus, the organization of the *Snap*25 gene was kept intact, still allowing alternative splicing to take place, but splicing to the downstream exon 5 also produced SNAP-25a protein. The SNAP-25b-deficient mutants exist in two versions, one with a *Tkneo* gene still present (necessary for selection of positive embryonic stem cell clones and placed in the intron sequence upstream of exon 6) and a second mutant where the selection gene had been removed by Cre-mediated *in vivo* excision (Table 1). *Neo*-containing SNAP-25b-deficient mutants were viable at birth and appeared normal during the first postnatal (PN) week. Thereafter, they demonstrated a severe phenotype that progressed with age and included developmental defects, such as reduced bone growth, inability to gain normal weight, as well as delayed eye opening and hearing ability. The phenotype also included spontaneous seizure activity, circling behavior, tremors and hyperactive episodes, and reduced paired-pulse facilitation (PPF), a parameter of short-term synaptic plasticity. A closer investigation of the expression profile from these homozygous *neo*-containing mutants demonstrated that the *Tkneo* gene in these mice suppressed mRNA expression in brain down to 50% of WT littermates. However, at the protein level, mutants still had 80% of normal SNAP-25 levels but only the SNAP-25a isoform.

When the *Tkneo* gene was excised, homozygous SNAP-25b-deficient mutants escaped the critical period during early development and lived until adulthood. The levels of SNAP-25 mRNA and protein were as in WT littermates, but only the SNAP-25a isoform was expressed. However, *neo*-excised SNAP-25b-deficient mutants also demonstrated reduced short-term synaptic plasticity. The reduction of PPF in juvenile SNAP-25b-deficient mutants (PN12–16) appeared to differ from young adult *Snap*25$^{Tkneo/Tkneo}$ mice, as in the latter the relative proportion of SNAP-25a is increased but total SNAP-25 protein levels decreased compared to WT animals.[44] In *Snap*25$^{Tkneo/Tkneo}$ mutants, PPF was increased at shorter (50 ms) but not at longer interpulse intervals as measured around PN21. PPF has been demonstrated to be low in PN7–9 WT animals,[48] peaking around PN14 and thereafter decreasing.[49] A possibility is that the switch from SNAP-25a to SNAP-25b is instrumental in tuning presynaptic function from an immature to a mature state. Thus, it is possible that the conditions in *Snap*25$^{Tkneo/Tkneo}$ mutants resemble a synapse trapped at an immature state as a result of limited SNAP-25b expression, whereas both *neo*-containing and *neo*-excised PN12–16 SNAP-25b-deficient mutants correspond to an even earlier presynaptic developmental time point as a consequence of the complete absence of SNAP-25b.

In the *neo*-excised SNAP-25b-deficient animals, which live until adulthood, we could also address the question of the importance of SNAP-25b in behavioral functions related to long-term plasticity mechanisms in hippocampus and amygdala. The mice were tested in several behavioral tasks, including the Morris water maze and the elevated plus maze. The results showed that *neo*-excised SNAP-25b-deficient mice, irrespective of gender, demonstrate a higher anxiety index, a severe impairment in spatial learning with female mice being most severely affected, and a lower locomotor activity when compared to WT controls. To further address the severe cognitive impairment identified, we analyzed the morphology of the hippocampal structures more carefully with histochemistry and immunohistochemistry. In the CA3 region of adult *neo*-excised SNAP-25b-deficient mutants, SNAP-25-positive mossy fibers were expanded and had formed large bundles. By using synaptophysin as a synaptic marker, we could identify island-like structures that apparently were devoid of immunoreactivity. These morphological differences observed in *neo*-excised SNAP-25b-deficient mouse mutants appeared to progress with age. Thus, evidently the absence of SNAP-25b in the adult nervous system results in progressing pathological changes in brain areas important for cognitive functions.

Conclusions

SNARE-mediated regulated membrane fusion at the plasma membrane, that is, regulated exocytosis, determines complex animal homeostasis and behavior. Much has been learned about SNARE protein function and expression in *in vitro* systems, but investigations need to be combined with studies that address how variation of expression of SNAREs, or SNARE-regulating proteins, manifests in intact organisms *in vivo*. Even minor deviations from a normal secretion pattern of classical neurotransmitters and hormones are expected to contribute to changed phenotypes, such as neuropsychiatric and/or metabolic disorders. Recently, genome-wide scans and linkage analysis in humans have indicated an association between polymorphisms in the gene for SNAP-25 and vulnerability to develop attention-deficit hyperactivity disorder, ADHD.[50,51] In humans different *Snap*25 alleles have also demonstrated inheritance correlated to intelligence.[52,53] Thus, with this different SNAP-25 mouse mutant in hand, it will now be possible to address many new questions in a physiological setting and start to unravel why the expression of SNAP-25a and SNAP-25b during development and in different cellular populations are so strictly regulated. With the initial characterization of these SNAP-25 mouse mutants, it has already become increasingly clear that one key function of the different SNAP-25 proteins is to control the stability of the SNARE complex. This seems to affect vesicle trafficking, the size of different vesicle pools, and the kinetics of priming/depriming of fusion-competent vesicles. It is likely that SNAP-25a and SNAP-25b, by differently modulating regulated exocytosis, play a role in "adapting" the exocytotic machinery toward different modes of transmitter release.

Acknowledgments

I would like to thank the the Swedish Research Council, the Novo Nordisk Foundation, Magnus Bergvalls Foundation, Alzheimerfoundation Sweden, OE and Edla Johanssons Foundation, Längmanska Foundation, Gun and Bertil Stohnes Foundation, and the Karolinska Institutet for support.

Conflicts of Interest

The authors declare no conflicts of interest.

References

1. Söllner, T., S.W. Whiteheart, M. Brunner, *et al.* 1993. SNAP receptors implicated in vesicle targeting and fusion. *Nature* **362:** 318–324.
2. Söllner, T., M.K. Bennett, S.W. Whiteheart, *et al.* 1993. A protein assembly disassembly pathway in vitro that may correspond to sequential steps of synaptic vesicle docking, activation, and fusion. *Cell* **75:** 409–418.
3. Sutton, R.B., D. Fasshauer, R. Jahn & A.T. Brunger. 1998. Crystal structure of a SNARE complex involved in synaptic exocytosis at 2.4 A resolution. *Nature* **395:** 347–353.
4. Hökfelt, T. 1991. Neuropeptides in perspective: the last ten years. *Neuron* **7:** 867–879.
5. Hökfelt, T., C. Broberger, Z.Q. Xu, *et al.* 2000. Neuropeptides–an overview. *Neuropharmacology* **39:** 1337–1356.
6. Jahn, R., J. Hell & P.R. Maycox. 1990. Synaptic vesicles: key organelles involved in neurotransmission. *J. Physiol.* **84:** 128–133
7. Hökfelt, T., O. Johansson, A. Ljungdahl, *et al.* 1980. Peptidergic neurones. *Nature* **284:** 515–521.
8. Südhof, T.C. 2004. The synaptic vesicle cycle. *Annu. Rev. Neurosci.* **27:** 509–547.
9. Jackson, M.B. & E.R. Chapman. 2006. Fusion pores and fusion machines in Ca^{2+}-triggered exocytosis. *Annu. Rev. Biophys. Biomol. Struct.* **35:** 135–160.
10. Jahn, R. & R.H. Scheller. 2006. SNAREs – engines for membrane fusion. *Nat. Rev. Mol. Cell Biol.* **7:** 631–643.
11. Rizo, J. & C. Rosenmund. 2008. Synaptic vesicle fusion. *Nat. Struct. Mol. Biol.* **15:** 665–674.
12. Wickner, W. & R. Schekman. 2008. Membrane fusion. *Nat. Struct. Mol. Biol.* **15:** 658–664.
13. Giraudo, C.G., W.S. Eng, T.J. Melia & J.E. Rothman. 2006. A clamping mechanism involved in SNARE-dependent exocytosis. *Science* **313:** 676–680.
14. Pobbati, A.V., A. Stein & D. Fasshauer. 2006. N- to C-terminal SNARE complex assembly promotes rapid membrane fusion. *Science* **313:** 673–676.
15. Oyler, G.A., G.A. Higgins, R.A. Hart, *et al.* 1989. The identification of a novel synaptosomal-associated protein, SNAP-25, differentially expressed by neuronal subpopulations. *J. Cell Biol.* **109:** 3039–3052.
16. Jacobsson, G., A.J. Bean, R.H. Scheller, *et al.* 1994. Identification of synaptic proteins and their isoform mRNAs in compartments of pancreatic endocrine cells. *Proc. Natl. Acad. Sci. USA* **91:** 12487–12491.
17. Jacobsson, G., F. Piehl, I.C. Bark, *et al.* 1996. Differential subcellular localization of SNAP-25a and SNAP-25b RNA transcripts in spinal motoneurons and plasticity in expression after nerve injury. *Mol. Brain Res.* **37:** 49–62.
18. Jacobsson, G., C. Bark & B. Meister. 1999. Isoform-specific expression of SNAP-25a and SNAP-25b RNA transcripts in cranial nerve nuclei. *J. Comp. Neurol.* **4:** 591–600.
19. Bark, I.C., K.M. Hahn, A.E. Ryabinin & M.C. Wilson. 1995. Differential expression of SNAP-25 isoforms during divergent vesicle fusion events of

neural development. *Proc. Natl. Acad. Sci. USA* **92:** 1510–1514.

20. Gonelle-Gispert, C., P.A. Halban, H. Niemann, *et al.* 1999. SNAP-25a and -25b isoforms are both expressed in insulin-secreting cells and can function in insulin secretion. *Biochem. J.* **339:** 159–165.
21. Catsicas, S., D. Larhammar, A. Blomqvist, *et al.* 1991. Expression of a conserved cell-type-specific protein in nerve terminals coincides with synaptogenesis. *Proc. Natl. Acad. Sci. USA* **88:** 785–789.
22. Lane, S.R. & Y. Liu. 1997. Characterization of the palmitoylation domain of SNAP-25. *J. Neurochem.* **69:** 1864–1869.
23. Gonzalo, S. & M.E. Linder. 1998. SNAP-25 palmitoylation and plasma membrane targeting require a functional secretory pathway. *Mol. Biol. Cell* **9:** 585–597.
24. Gonzalo, S., W.K. Greentree & M.E. Linder. 1999. SNAP-25 is targeted to the plasma membrane through a novel membrane-binding domain. *J. Biol. Chem.* **274:** 21313–21318.
25. Koticha, D.K., S.J. Huddleston, J.W. Witkin & G. Baldini. 1999. Role of the cysteine-rich domain of the t-SNARE component, SYNDET, in membrane binding and subcellular localization. *J. Biol. Chem.* **274:** 9053–9060.
26. Shimazaki, Y., T. Nishiki, A. Omori, *et al.* 1996. Phosphorylation of 25-kDa synaptosome-associated protein. Possible involvement in protein kinase C-mediated regulation of neurotransmitter release. *J. Biol. Chem.* **271:** 14548–14553.
27. Genoud, S., W. Pralong, B.M. Riederer, *et al.* 1999. Activity-dependent phosphorylation of SNAP-25 in hippocampal organotypic cultures. *J. Neurochem.* **72:** 1699–1706.
28. Risinger, C. & M.K. Bennett. 1999. Differential phosphorylation of syntaxin and synaptosome-associated protein of 25 kDa (SNAP-25) isoforms. *J. Neurochem.* **72:** 614–624.
29. Nagy, G., U. Matti, R.B. Nehring, *et al.* 2002. Protein kinase C-dependent phosphorylation of synaptosome-associated protein of 25 kDa at Ser187 potentiates vesicle recruitment. *J. Neurosci.* **22:** 9278–9286.
30. Nagy, G., K. Reim, U. Matti, *et al.* 2004. Regulation of releasable vesicle pool sizes by protein kinase A-dependent phosphorylation of SNAP-25. *Neuron* **41:** 351–365.
31. Bark, I.C. 1993. Structure of the chicken gene for SNAP-25 reveals duplicated exons encoding distinct isoforms of the protein. *J. Mol. Biol.* **233:** 67–76.
32. Bark, I.C. & M.C. Wilson. 1994a. Cloning of human cDNA clones encoding two different isoforms of the nerve terminal protein SNAP-25. *Gene* **139:** 291–292.
33. Bark, I.C. & M.C. Wilson. 1994b. Regulated vesicular fusion in neurons – snapping together the details. *Proc. Natl. Acad. Sci. USA* **91:** 4621–4624.
34. Grant, N.J., R. Hepp, W. Krause, *et al.* 1999. Differential Expression of SNAP-25 isoforms and SNAP-23 in the adrenal gland. *J. Neurochem.* **72:** 363–372.
35. Langley, K., R. Hepp, N.J. Grant, *et al.* 2002. Thyroid hormones regulate adrenal chromaffin cell SNAP-25. *Ann. N.Y. Acad. Sci.* **971:** 277–280.
36. Sadoul, K., J. Lang, C. Montecucco, *et al.* 1995. SNAP-25 is expressed in islets of Langerhans and is involved in insulin release. *J. Cell Biol.* **128:** 1019–1028.
37. Boschert, U., C. O'Shaughnessy, R. Dickinson, *et al.* 1996. Developmental and plasticity-related differential expression of two SNAP-25 isoforms in the rat brain. *J. Comp. Neurol.* **367:** 177–193.
38. Patanow, C.M., J.R. Day & M.L. Billingsley. 1997. Alterations in hippocampal expression of SNAP-25, GAP-43, stannin and glial fibrillary acidic protein following mechanical and trimethyltin-induced injury in the rat. *Neuroscience* **76:** 187–202.
39. Roberts, L.A., B.J. Morris & C.T. O'Shaughnessy. 1998. Involvement of two isoforms of SNAP-25 in the expression of long-term potentiation in the rat hippocampus. *Neuroreport* **9:** 33–36.
40. Washbourne, P., P.M. Thompson, M. Carta, *et al.* 2002. Genetic ablation of the t-SNARE SNAP-25 distinguishes mechanisms of neuroexocytosis. *Nat. Neurosci.* **5:** 19–26.
41. Osen-Sand, A., M. Catsicas, J.K. Staple, *et al.* 1993. Inhibition of axonal growth by SNAP-25 antisense oligonucleotides in vitro and in vivo. *Nature* **364:** 445–448.
42. Sørensen, J.B., G. Nagy, F. Varoqueaux, *et al.* 2003. Differential control of the releasable vesicle pools by SNAP-25 splice variants and SNAP-23. *Cell* **114:** 75–86.
43. Nagy, G., I. Milosevic, D. Fasshauer, *et al.* 2005. Alternative splicing of SNAP-25 regulates secretion through nonconservative substitutions in the SNARE domain. *Mol. Biol. Cell* **16:** 5675–5685.
44. Bark, C., F.P. Bellinger, A. Kaushal, *et al.* 2004. Developmentally regulated switch in alternatively spliced SNAP-25 isoforms alters facilitation of synaptic transmission. *J. Neurosci.* **24:** 8796–8805.
45. Jeans, A.F., P.L. Oliver, R. Johnson, *et al.* 2007. A dominant mutation in *Snap25* causes impaired vesicle trafficking, sensorimotor gating, and ataxia in the blind-drunk mouse. *Proc. Natl. Acad. Sci. USA* 104: 2431–2436.
46. Nolan, P.M., J. Peters, M. Strivens, *et al.* 2000. A systematic, genome-wide, phenotype-driven mutagenesis programme for gene function studies in the mouse. *Nat. Genet.* **25:** 440–443.

47. Johansson, J.U., J. Ericsson, J, Janson, *et al*. 2008. An Ancient Duplication of Exon 5 in the *Snap*25 Gene is Required for Complex Neuronal Development/Function. *PLoS Genetics* **4:** [Epub Nov. 28, 2008].
48. Müller, D., M. Olivier & G. Lynch. 1989. Developmental changes in synaptic properties in hippocampus of neonatal rats. *Brain Res. Dev. Brain Res.* **49:** 105–114.
49. deKay, J.G.T., T.C. Chang, N. Mills, *et al*. 2006. Responses of excitatory hippocampal synapses to natural stimulus patterns reveal a decrease in short-term facilitation and increase in short-term depression during postnatal development. *Hippocampus* **16:** 66–79.
50. Faraone, S.V., R.H. Perlis, A.E. Doyle, *et al*. 2005. Molecular genetics of attention-deficit/hyperactivity disorder. *Biol. Psychiatry* **57:** 1313–1323.
51. Thapar, A., M. O'Donovan & M.J. Owen. 2005. The genetics of attention deficit hyperactivity disorder. *Hum. Mol. Genet.* **14**(Spec No. 2): R275–282.
52. Gosso, M.F., E.J. de Geus, M.J. van Belzen, *et al*. 2006. The SNAP-25 gene is associated with cognitive ability: evidence from a family-based study in two independent Dutch cohorts. *Mol. Psychiatry* **11:** 878–886.
53. Gosso, M.F., E.J. de Geus, T.J. Polderman, *et al*. 2008. Common variants underlying cognitive ability: further evidence for association between the SNAP-25 gene and cognition using a family-based study in two independent Dutch cohorts. *Genes Brain Behav.* **7:** 355–364.

Rescuing the Subprime Meltdown in Insulin Exocytosis in Diabetes

Edwin P. Kwan[a,b] **and Herbert Y. Gaisano**[a]

[a] *Departments of Physiology and Medicine, University of Toronto, Toronto, Ontario, Canada*

[b] *Division of Endocrinology, Beth Israel Deaconess Medical Center and Harvard Medical School, Boston, Massachusetts, USA*

Neuroendocrine pancreatic islet β-cells secrete the hormone insulin in response to glucose stimulation and adapt efficiently to increased demand by peripheral tissues to maintain glucose homeostasis. Insulin is packed within dense-core granules, which traffic and dock onto the plasma membrane whereby a Ca^{2+} stimulus evokes exocytosis by soluble *N*-ethylmaleimide-sensitive factor attachment protein receptor (SNARE), complex-mediated, membrane fusion. Recent studies have unveiled postdocking steps mediated by "priming" factors that influence SNARE complex assembly to confer fusion readiness to the docked granules. This review will summarize recent insights into the priming role for Munc13 in the exocytosis of insulin granules. We present evidence for the interaction of Munc13-1 with exocytotic substrates involved in cAMP-mediated potentiation of insulin release, the latter we show to mediate enhanced granule-to-granule fusion events underlying compound exocytosis. We thus also further review the current understanding of granule-to-granule fusion. As agents acting on cAMP signaling are clinically used to augment insulin release in diabetes, this better understanding of priming steps may reveal additional novel therapeutic strategies to increase the capacity for insulin release to improve the treatment of diabetes.

Key words: **Munc13; insulin secretory granules; priming; exocytosis; SNARE proteins; diabetes; glucagon-like peptide-1; cAMP; compound exocytosis**

Introduction

In pancreatic islet β-cells, a small number of insulin secretory granules (SGs; <10%) are docked or tethered to the plasma membrane (PM) and only a fraction (<2%) of docked SGs are primed for immediate release (exocytosis).[1] Each of these sequential steps leading to secretion-insulin SG docking and tethering to the PM, priming, and exocytotic membrane fusion are mediated by distinct sets of molecules. We have found reduced expression of these exocytotic molecules, including soluble *N*-ethylmaleimide-sensitive factor attachment protein receptor (SNARE) complex proteins and priming factors in type 2 diabetic human islets[2] and rodent [prediabetic Zucker fa/fa and mildly diabetic Goto-Kakisaki (GK) rat] islets.[3,4] This should contribute to the insulin secretory deficiency in diabetes, including reduction of glucose-stimulated first phase insulin secretion and reduced efficiency of subsequent second phase insulin release, leading to poor glycemic control. Much is known about the molecules mediating exocytotic fusion *per se* (SNARE proteins). However, less is known about those molecules that mediate insulin SG priming (e.g., Munc13, Munc18), which we discuss in this review.

Address for correspondence: Edwin P. Kwan, PhD, Research North, Rm 390, 99 Brookline Avenue, Boston, MA, 02215. Voice: +617-667-1979; fax: +617-667-2927. ekwan@bidmc.harvard.edu

Mechanisms of Exocytosis: Ann. N.Y. Acad. Sci. 1152: 154–164 (2009).
doi: 10.1111/j.1749-6632.2008.03990.x

Steps in Exocytosis: Docking, Fusion, and Priming

Docking

Only a small subset of insulin SGs are docked onto the PM, and the mechanisms whereby this subset is selected to tether and dock with the PM are poorly understood. Munc18-1, a Sec1/Munc18-like (SM) protein known to regulate the target (t-) SNARE syntaxin-1 (stx-1) stability and localization, is a major molecule that directly targets catecholamine-containing secretory SGs to dock to the PM.[5] Munc18-1 action in the docking of SGs is generally viewed to require Munc18-1 binding to the closed conformation of stx-1, but this view remains under debate.[6] The docking of synaptic vesicles (SVs) in *Caenorhabditis elegans*, by contrast, requires stx to assume an open conformation.[7] Munc18-1 was recently reported to directly mediate the docking of insulin SGs via an additional interaction with granuphilin/Slp-4.[8]

Fusion

The SNARE complex, generally accepted as the putative machinery for exocytotic fusion of SVs/SGs, is assembled as a parallel bundle of four α-helices, with one helix from PM stx, another two helices from another PM t-SNARE (SNAP), and the fourth helix from vesicle(v)-SNARE (VAMP).[9,10] There are multiple isoforms of each of the SNAREs in the β-cells, of which the roles of stx-1, SNAP-25, and VAMP-2 are similar in insulin exocytosis as with other regulated secretory systems and have been better characterized[11,12] than those of stx-2, stx-3, stx-4, and SNAP-25.[13,14] Ca^{2+} is the primary trigger for fusion of insulin SGs. The rise in $[Ca^{2+}]_i$ is most likely sensed by synaptotagmin(s) (syt), which subsequently elicits fusion pore opening.[15] Multiple syt isoforms (syt-1, -2, -3, -5, -7, and -9) localized to either the PM or SG membrane are expressed in the β-cells.[16] However, which of these syts and what exact role they have in triggering insulin exocytosis remain unclear, although syt-7 was recently reported to be the likely lead syt candidate in β-cells.[17]

Priming

The priming of insulin SGs is generally regarded as the step in which docked SGs are rendered fusion competent. Extensive studies on SVs and/or SGs in different cell types have suggested that many factors, such as Ca^{2+}, Cl^-, ATP, GTP, *N*-ethylmaleimide (NEM)-sensitive factor (NSF), actin-microfilament and membrane-actin cytoskeleton interactions, and phosphoinositides, are involved in the priming step.[1,18,19] These factors indeed play a crucial role in regulated exocytosis, but nonetheless the molecular substrates that these signals act on to confer fusion competence remain largely unclear. Moreover, investigation of the priming step has continuously been hampered by the lack of real-time morphological assays that can precisely distinguish docking, priming, and fusion. More insights into these steps have now become possible with the use of powerful imaging techniques, such as total internal reflection microscopy (TIRFM) to resolve the exocytotic behavior of single secretory SVs/SGs,[20] particularly when used with genetic manipulations.

Munc13-1 as a Priming Factor in Insulin Exocytosis

It has been about a decade since Munc13 was first suggested to play a crucial role in the priming of SVs (reviewed in ref. 21).[21] Mammalian Munc13-1, 2, 3, and 4 are encoded by four separate genes and display both overlapping and distinct tissue-, cell-, and subcellular localizations.[21,22] Hippocampal neurons of mice carrying a null deletion of *munc13-1* have almost completely abolished release of glutamate neurotransmitters. Using electrophysiological analyses, the secretory deficiency was shown not to be caused by a defect in fusion *per se* or alteration in the sensitivity of the exocytotic machinery to Ca^{2+} but is attributed to aberration in a postdocking step. Ablation of Munc13-1 reduced the size of the readily-releasable pool (RRP) of SVs, confirming the priming role of Munc13-1 in rendering SV fusion competence. Subsequent studies showed that Munc13 isoforms in the brain regulate SV priming in a neuron- and activity-dependent manner to regulate synaptic plasticity. Munc13-1 also facilitates the priming of dense-core SGs in chromaffin cells. Orthologues Dunc-13 and Unc-13 were also similarly demonstrated to regulate SV priming in *Drosophila* and *C. elegans*, respectively.

We and others had reported that Munc13-1 is expressed in pancreatic β-cells and insulin-secreting cell lines in the cytosol and PM, whereas other isoforms could not be reliably detected.[23,24] Heterozygous Munc13-1 ablation (Munc13-$1^{+/-}$) mice exhibit glucose intolerance after a glucose challenge but normal insulin sensitivity, indicating insufficient release of insulin to lower blood glucose levels.[25] Indeed, insulin release from pancreatic islets of Munc13-1 null mice is severely impaired.[23,25] Importantly, Munc13-1 levels are severely reduced in islets of type 2 diabetic patients,[2] suggesting that a priming defect from Munc13-1 deficiency may contribute to the deficient insulin secretion in these diabetic patients. In fact, using membrane capacitance (C_m) measurements, Munc13-$1^{+/-}$ β-cells were shown to have both a reduction in RRP size and RRP replenishment rate, consistent with the observed respective reduction in first phase and second phase insulin release from isolated islets.[23,25]

Diacyglycerol (DAG) second messenger signaling activated by glucose causes marked potentiation of insulin release in part by increasing the size and refilling of RRP of insulin SGs. Most Munc13 homologues contain a phorbol ester (PE)/DAG-binding C_1 domain similar to that of protein kinase C (PKC) and which assists in recruiting Munc13 to the PM,[21] as was also observed in β-cells.[24] Homozygous mice harboring a histidine-to-lysine mutation in the C_1 domain (H597K) reduced DAG binding and consequently PE/DAG-mediated potentiation of neurotransmitter release, which indicate that both Munc13 and PKC[26] integrate DAG-mediating signaling to facilitate the potentiation of exocytosis. Characterization of Munc13-1^{H597K} mutant mouse islets and β-cells demonstrated that this C_1 domain is indeed important for DAG-mediated enhancement of the size, release probability, and replenishment of RRP, which collectively account for the potentiation of insulin release.[11,27] The precise molecular mechanism whereby Munc13-1 enhances PE/DAG-mediated potentiation of exocytosis of insulin SGs and SVs require further investigation but is suggested to involve the interaction of Munc13-1 with other substrates outside its C_1 domain [e.g., C_2 domains, calmodulin-binding domain, and Rab3-interacting molecule (RIM)-binding domains].[21] Munc13-1 also acts in concert with PKCs to effect the maximal potentiating effect of DAG on insulin release.[28]

Proposed Mechanism of the Priming Action of Munc13-1

The molecular basis for the priming activity of Munc13-1 remains a hotly debated topic. Munc18-1 binds to and stabilizes the closed conformation of stx-1 where its N-terminal H_{ABC} domain folds back onto its C-terminal H_3 SNARE motif.[5] It was originally believed that Munc18-1-bound closed stx-1 is unable to interact with SNAP-25 and VAMP-2 to form SNARE complex and that Munc13-1 primes SVs by "unfolding" stx-1 to an "open" conformation. Munc13-1 binds the H_{ABC} domain of stx-1 via its C-terminal domain, as shown by *in vitro* biochemical assays; overexpression of open stx-1 restores SV priming in *unc-13* mutants; and the activity of Munc13-1/Unc-13 was suggested to rely on its interaction with stx-1.[21,29,30] However, this reported interaction between Munc13-1 and stx-1 has been questioned and argued to be a result of misfolded stx-1, since such interaction could not be detected in solution by nuclear magnetic resonance (NMR) experiments.[31] The purported Munc13-1 unfolding action on stx-1 has not been entirely refuted as Munc13-1 was reported to bind assembled, native, membrane-anchored SNARE complexes.[31] MUN domain, the Munc13-1 C-terminal minimal priming domain, only partially rescued SV release in Unc-13/Munc13-1/-2 double-null mutants, which supports a post-priming role for Munc13-1 that is facilitated by other Munc13-1 domains (e.g., calmodulin-binding and RIM-binding domains[11] (see below).

Other Priming Factors in Insulin Exocytosis

Munc18

Munc18-1 binding to and stabilization of closed stx-1 was once thought to suggest an inhibitory role for Munc18-1 in exocytosis. Indeed, initial studies showed that overexpressed Munc18-1 negatively regulated insulin release.[32] This thinking is now refined wherein the inhibitory/proofreading role for Munc18 in exocytosis is to prevent promiscuous pairings of noncognate SNAREs, pairings in non-3Q:1R topologies, and premature exocytotic events during membrane trafficking and fusion.[33–35] However, null mutation studies showed severely abolished neurotransmitter release (in Munc18-1 null mouse) and abnormal glucose intolerance (in Munc18-3 null mouse),[5,36] indicating a positive role for Munc18 in exocytosis. Munc18-1 facilitates SNARE complex formation and drives membrane fusion.[34,37] Since the priming step lies between SV/SG fusion and docking, which Munc18-1 also facilitates, Munc18-1 is thought to also play a role in priming. In support, Munc18-1 binds lipid-anchored stx-1/SNAP-25 heterodimer that serves as an acceptor complex for VAMP-2.[38] Munc13-1 MUN domain and Munc18-1 can simultaneously bind this heterodimeric acceptor complex but compete for binding with the assembled SNARE complex;[31] the latter raises the possibility that the competitive displacement of Munc13 by Munc18-1 might serve to trigger membrane merger. The exact nature of these binding interactions remains to be fully elucidated. Since Munc18-1 facilitates SV/SG docking by binding to closed stx-1[5,6] and stx-1 is required to be in an unfolded open conformation to form SNARE complex,[34] then how is Munc18-1 displaced from closed stx-1? It is also unknown what re-enables the displaced Munc18-1 to bind stx-1/SNAP-25 dimer to initiate the subsequent round of priming for SNARE complex formation. Perhaps Munc18-1 is only temporarily displaced from assembled SNARE complex and returns to drive membrane fusion. Adding to the complexity, Munc18-1 could also bind open form stx-1.[39] A labile half-closed conformation of stx-1 appears to support the transient binding modes of Munc18-1 in mediating priming and fusion.[38] Synaptotagmin-1 and complexin also assist in stabilizing the stx-1/SNAP-25 acceptor complex, suggesting their role in priming, which is separate from their more downstream role in membrane fusion.[31,40]

β-cells also express Munc18-2 and Munc18-3. Munc18-3 was shown to interact with stx-4 to influence insulin exocytosis, and Munc18-3 or stx-4 deletion reduced insulin secretion.[13,36] However, their direct role in priming insulin SGs has not been demonstrated. Munc18-2, like Munc18-1, could potentially also exert priming action by binding closed form stx-1 at a postdocking step, supporting this mode of action as a general mechanism by which Munc18 proteins mediate priming.[6] Munc18-2 could thus compete with (or compensate for) Munc18-1 in priming SGs. The role of Munc18-2 in insulin exocytosis remains to be examined.

Tomosyn

Tomosyn has been shown to negatively regulate exocytosis of insulin SGs and other neuropeptide-containing SGs as well as SVs,[41–43] primarily by acting antagonistically to Munc13. Tomosyn binds the SNARE motif of stx-1 via its VAMP homology domain, which reduces stx-1 availability to assemble with its partners to form functional SNARE complex; and disassembly of tomosyn/stx complex is regulated by NSF.[44] Deletion of tomosyn gene (*tom-1*) in *C. elegans* increased the RRP size of primed SVs, and, like open form stx-1, *tom-1* deletion partially rescued the defect of SV priming in *unc-13* null *C. elegans*.[45,46] In apparent contrast, overexpression of a tomosyn mutant defective in stx binding is almost as efficient as wild-type tomosyn in inhibiting secretion in neuroendocrine cells, suggesting the presence of other stx-1-independent inhibitory domains.[47] This study, however, used an overexpression system wherein endogenous tomosyn is present and should thus be interpreted with caution.

CAPS

Ca^{2+}-dependent activator protein for secretion (CAPS) (or Unc-31 in *C. elegans*) plays a role in docking and priming of neuropeptide-containing SGs and insulin SGs.[48–52] Knockout mice of both CAPS1 and CAPS2 isoforms display impaired glucose tolerance, insulin SG priming, and impaired exocytosis of insulin SGs.[51] The mechanism by which CAPS primes SVs or SGs remains unclear. The fact that CAPS contains the MUN domain has implicated that it functions like Munc13-1. Of note, priming of SVs and SGs in neurons of *C. elegans* seems to rely exclusively on Unc-13 and Unc-31, respectively[50] (but see Sieburth *et al.*[49]), whereas priming of both SVs and SGs require tomosyn action.[53] In contrast, priming of insulin SGs are dependent on both Munc13-1 and CAPS, suggesting that both proteins and tomosyn interact in priming insulin SGs.[11,51] The nature of this interaction is, however, unclear.

cAMP Potentiation of Insulin Exocytosis

Gut peptide hormone glucagon-like peptide-1 (GLP-1) greatly enhances insulin exocytosis by increasing the size of docked SG pool and the size and replenishment of RRP of insulin SGs, which account for the rescue of first and second phase insulin release in patients with type 2 diabetes.[54] These actions of GLP-1 on insulin exocytosis are via GLP-1 receptor-mediated production of cAMP, which acts on two receptors, PKA and Epac.[55] Both Epac and PKA can alter the expression and activities of ion channels and Ca^{2+} release pathways involved in insulin secretion.[55,56] PKA phosphorylates a number of exocytotic proteins, such as RIM, which our work has focused on.

GLP-1 Can Bypass the Requirement for Munc13-1 in the Priming of Insulin SGs

We initially hypothesized that Munc13-1 forms a functional complex with cAMP, Epac, PKA, and RIM, and that Munc13-1 deficiency would therefore cause impairment of cAMP/GLP-1-induced potentiation of insulin release. Surprisingly, we showed that cAMP/GLP-1 stimulation largely restored defects in the priming and refilling of RRP of insulin SGs in Munc13-1$^{+/-}$ islets and β-cells, culminating in normal GLP-1-mediated potentiation of first and second phase insulin release.[57] Using C_m measurements, we found that the ability of Epac to fully potentiate insulin exocytosis was restricted by Munc13-1 abundance. In contrast, PKA signaling, in part via RIM, was more efficacious in overcoming the Munc13-1 deficiency-induced defects in the priming and replenishment of RRP of insulin SGs. Munc13-1 interaction with RIM appeared to be essential for the cAMP rescue, consistent with the report of RIM-mediated partial normalization of priming defects in Munc13-1-deficient synapses.[58] This rescue may be attributed to direct actions of PKA-mediated phosphorylation of exocytotic proteins besides RIM, including α-SNAP,[59] synapsin I,[60] snapin,[61] cysteine string protein,[62] tomosyn,[63] and syntaphilin.[64] For instance, PKA-mediated phosphorylation has been suggested to reduce the inhibitory effect of tomosyn on stx.[63] Interestingly, PKA activation similarly bypasses the exocytotic defect of SGs in *C. elegans* neurons caused by the ablation of priming protein Unc-31 (CAPS).[52]

SG-to-SG Fusion Underlying Compound and Sequential Insulin Exocytosis

We previously demonstrated using FM1–43 fluorescence imaging and electron microscopy that GLP-1/cAMP stimulation promoted not only additional *de novo* fusion sites on the PM[65–67] but also recruited fusion events at pre-established fusion sites in the form of sequential and compound exocytosis of insulin SGs.[65,66,68] Sequential exocytosis is defined as an initial fusion of an SG with the PM followed by an orderly stepwise fusion of oncoming single SGs with the exocytosing SG. Compound exocytosis is defined as the fusion of multiple SGs frequently prefused in the cytosol (homotypic fusion) into a single fusion pore opening on the PM. Sequential and compound exocytoses were reported in a number of secretory cells, including lactotrophs, eosinophils, mast cells, neutrophils, and pancreatic acinar cells.[69–72] We reported that cAMP-activated Epac promoted the recruitment of *de novo* primary fusion sites on the PM,[66] which would account for cAMP/Epac potentiation of first phase insulin release.[55] PKA seems to have a preferential effect in inducing SG-to-SG fusion underlying sequential and compound exocytoses, which would temporally influence second phase insulin secretion. Here, PKA-mediated SG-to-SG fusion could be enhanced by Epac,[66] consistent with the report that knockdown of Epac expression inhibited half of the potentiating effect of cAMP on insulin release from pancreatic islets, attributable mainly to reduction in second phase insulin release.[55] Foxa2, a member of forkhead transcription factor known to regulate fetal development, was shown to regulate docking and balance of primary fusion and compound fusion of insulin SGs by controlling cAMP signaling.[73] Intriguingly, using 2-photon imaging technique, H. Kasai's group reported that potentiation of insulin exocytosis by cAMP was solely attributed to PKA signaling, whereas Epac signaling enhanced only the exocytosis of SVs in the β-cells.[74]

Different Isoforms of SNARE Proteins Regulate Distinct Insulin Exocytotic Events—Primary PM Fusion Sites and SG-to-SG Fusion

Whereas Epac and PKA can greatly potentiate insulin SG-to-SG fusion, they are not the primary mediators of SG-to-SG fusion *per se*, which are in fact essentially unknown. We discuss below several issues for consideration when exploring these mechanisms.

Stx-1, -2, -3, and -4 isoforms are present in the β-cells and localized predominantly to the PM, with the exception of stx-3, which is also present in the SG membrane.[14] Stx-1 is generally accepted to positively affect fusion of insulin SGs at the PM by forming the SNARE complex with its cognate partners, SNAP-25 and VAMP-2.[15] The roles of other stx isoforms in insulin exocytosis are less well understood. Stx-4 has been suggested to regulate insulin release,[13] but the precise exocytotic step remains unclear. These stx isoforms may contribute to the formation of the SNARE complexes via interaction with distinct isoforms of VAMP, SNAP, and Munc18. Much that is known about the exocytotic function of stx-4 is from work in glucose-transporter-4 (GLUT4)-containing vesicles in muscles and adipocytes, where stx-4 interacts with VAMP-2 and SNAP-23.[75] Munc18-3 binds with stx-4 to regulate SNARE complex assembly in a similar fashion as Munc18-1–stx-1 interactions. Munc18-3 is also present in β-cells to regulate insulin secretion[36] but could also bind stx-2,[76] which is present in the PM of β-cells. Both Munc18-1 and Munc18-2, known to be expressed in β-cells, have been shown to bind stx-1 and stx-2 on the PM and stx-3 on the PM and SG membrane.[76] More effort should be put into the investigation of the common and distinct roles of stx isoforms and Munc18 isoforms in regulating distinct insulin SG exocytotic events.

Moreover, the actions of distinct stx isoforms by their interaction with cognate SNARE partners may dictate the compartmental and temporal specificity and sensitivity of SG-to-PM fusion versus SG-to-SG fusion. For instance, what dictates the subset of SGs that become docked onto the PM versus the subset of insulin SGs (called *newcomers*) that are recruited to the PM to undergo fusion with minimal docking time?[77] In fact, stx-1 deletion blocked the exocytosis of previously docked insulin SGs but did not seem to affect the fusion of the newcomer insulin SGs,[77] suggesting that exocytosis of newcomer SGs must be mediated by a stx other than stx-1. For SG-to-SG fusion, what dictates the fusion of an oncoming insulin SG to a previous PM exocytic site versus a more rapid by orderly stepwise fusion of single insulin SGs with an SG just undergoing exocytotic fusion (sequential exocytosis) versus insulin SG-to-SG already occurring deep inside the cytosol (homotypic fusion) before exiting at a single PM site (compound exocytosis)? Insulin SG-to-SG fusion likely resembles homotypic fusion of yeast vacuoles or mammalian endosomes[9] and is therefore likely mediated by SG membrane-anchored SNARE isoforms. Stx-3 appears to be the only stx-3 isoform localized to insulin SGs and is thus the best stx candidate for SG-to-SG fusion. Interestingly, proteome analysis of SVs suggested that the copy number of SNARE proteins is sufficiently high for simultaneous formation of multiple SNARE complexes,[78] indicating that the disassembly of SNARE complexes at the PM and the subsequent redistribution of SNARE proteins are not necessary to "recycle" SNARE proteins for homotypic fusion. Alternatively, can PM stx-1, -2-, and -4 also participate in SG-to-SG fusion, perhaps by having their activation and cellular location dictated by the different Munc18 isoforms [stx-1, -2, and -3 are bound by Munc18-1 and Munc18-2; stx-4 and stx-2 (weakly) bind Munc18-3]? Indeed, the pairings between cognate and noncognate SNARE partners are highly promiscuous and are quality proofread by specific Munc18 isoforms.[33,35] The identities of stx(s) cognate partners involved in homotypic fusion of insulin SGs are also unknown. SNAP-25 is present in β-cells and known to mediate insulin exocytosis;[79] but SNAP-23, which forms SNARE complexes with stx-4 and VAMP-2 in muscles and adipocytes, appears to be absent in β-cells. SNAP-25, like SNAP-23 involvement in homotypic granule fusion in mast cells,[80] redistributes from the PM to sequentially fused insulin SGs during cAMP-induced sequential exocytosis.[72] The redistributed SNAP proteins would bind cognate partners in SGs to form SNARE complexes that mediate SG-to-SG fusion.[80] Taken together, these reports suggest that homotypic fusion between insulin SGs is likely mediated by SNARE complexes assembled from stx-3, VAMP-2, and SNAP-25. Stx-2 shown to undergo a similar redistribution from PM to sequentially fused SGs in pancreatic acinar cells[81] might also have a role in insulin SG-to-SG fusion.

The topology of SNARE formation, in which one SNARE on the SG donor and three SNAREs on the acceptor, such as the PM (3Q:1R), was once thought to be the only topology allowed for heterotypic fusion.[82] However, recent studies have suggested that in homotypic fusion of early endosomes and yeast vacuoles, several topologies of SNARE complexes are possible.[35,83] The contributing SNAREs do not have to belong to the same subclass defined on the basis of similarity between the zero-layer residue in the SNARE motifs. Furthermore, two SNAREs from each of the donor SG and acceptor SG may contribute to form SNARE complexes. Insights on homotypic insulin SG-to-SG fusion and heterotypic insulin SGs-PM fusion could be gained from studies using cell-free assays of fusion of liposomes reconstituted with the different combinations of SNARE proteins (proteoliposomes) found in β-cells that could be configured into different topologies under the regulation of different Munc18 isoforms.[84] Moreover, imaging techniques, such as fluorescence resonance energy transfer, to verify SNARE protein interactions in intact cells have emerged as a useful approach to substantiate the SNARE interactions detected by immunoprecipitations from previous studies to be functionally relevant.[47]

Full fusion mode was believed to be the predominant mode of fusion event for insulin SGs, wherein SG fusion pore dilates completely and the SG membrane collapses onto the PM, subsequently causing the complete emptying of the SG cargo. By contrast, electron micrographs of SG-to-SG fusion events at primary PM exocytotic fusion sites or deep within the β-cells showed that SGs involved in SG-to-SG fusion maintain a narrow and stable fusion pore.[65,85] This observation suggests that additional undefined factors differentially regulate the fusion pore opening dynamics for SG-to-PM fusion and SG-to-SG fusion. In fact, it has been suggested that most insulin SGs undergo partial and transient fusion ("kiss and run").[86] Stx binding to lipid phosphatidic acid has been implicated to tip the balance between negative-curvature favoring membranes that promote fusion pore dilation and positive-curvature favoring membranes that prevent the dilation.[87] Moreover, Munc18 has also been suggested to regulate fusion pore dynamics, but its role is still under debate.[6,88]

Acknowledgments

This work was supported by grants to HYG from the Canadian Institutes of Health Research (CIHR MOP-86544). EPK was supported by a Canada Graduate Scholarships Doctoral Award from CIHR (CGD 76317) and a Banting and Best Diabetes Centre Novo-Nordisk Studentship Award from the University of Toronto.

Conflicts of Interest

The authors declare no conflicts of interest.

References

1. Rorsman, P. & E. Renstrom. 2003. Insulin granule dynamics in pancreatic beta cells. *Diabetologia* **46:** 1029–1045.
2. Ostenson, C., H. Gaisano, *et al.* 2006. Impaired gene and protein expression of exocytotic soluble N-ethylmaleimide attachment protein receptor complex proteins in pancreatic islets of type 2 diabetic patients. *Diabetes* **55:** 435–440.
3. Chan, C.B., R.M. Macphail, *et al.* 1999. Beta-cell hypertrophy in fa/fa rats is associated with basal glucose hypersensitivity and reduced SNARE protein expression. *Diabetes* **48:** 997–1005.
4. Gaisano, H.Y., C.G. Ostenson, *et al.* 2002. Abnormal expression of pancreatic islet exocytotic soluble N-ethylmaleimide-sensitive factor attachment protein receptors in Goto-Kakizaki rats is partially restored by phlorizin treatment and accentuated by high glucose treatment. *Endocrinology* **143:** 4218–4226.
5. Toonen, R.F. & M. Verhage. 2007. Munc18-1 in secretion: lonely Munc joins SNARE team and takes control. *Trends Neurosci.* **30:** 564–572.
6. Gulyas-Kovacs, A., H. De Wit, *et al.* 2007. Munc18-1: sequential interactions with the fusion machinery stimulate vesicle docking and priming. *J. Neurosci.* **27:** 8676–8686.
7. Hammarlund, M., M.T. Palfreyman, *et al.* 2007. Open syntaxin docks synaptic vesicles. *PLoS. Biol.* **5:** e198.
8. Tomas, A., P. Meda, *et al.* 2008. Munc 18-1 and granuphilin collaborate during insulin granule exocytosis. *Traffic* **9:** 813–832.
9. Hong, W. 2005. SNAREs and traffic. *Biochim. Biophys. Acta* **1744:** 493–517.
10. Jahn, R. & R.H. Scheller. 2006. SNAREs–engines for membrane fusion. *Nat. Rev. Mol. Cell Biol.* **7:** 631–643.
11. Kwan, E.P. & H.Y. Gaisano. 2007. New insights into the molecular mechanisms of priming of insulin exocytosis. *Diabetes Obes. Metab.* **9**(Suppl 2): 99–108.
12. Leung, Y.M., E.P. Kwan, *et al.* 2007. SNAREing voltage-gated K+ and ATP-sensitive K+ channels: tuning beta-cell excitability with syntaxin-1A and other exocytotic proteins. *Endocr. Rev.* **28:** 653–663.
13. Spurlin, B.A. & D.C. Thurmond. 2006. Syntaxin 4 facilitates biphasic glucose-stimulated insulin secretion from pancreatic beta-cells *Mol. Endocrinol.* **20:** 183–193.
14. Wheeler, M.B., L. Sheu, *et al.* 1996. Characterization of SNARE protein expression in beta cell lines and pancreatic islets. *Endocrinology* **137:** 1340–1348.
15. Gerber, S.H. & T.C. Sudhof. 2002. Molecular determinants of regulated exocytosis. *Diabetes* **51**(Suppl 1): S3–S11.

16. Iezzi, M., G. Kouri, *et al.* 2004. Synaptotagmin V and IX isoforms control Ca^{2+}-dependent insulin exocytosis *J. Cell Sci.* **117:** 3119–3127.
17. Gustavsson, N., Y. Lao, *et al.* 2008. Impaired insulin secretion and glucose intolerance in synaptotagmin-7 null mutant mice. *Proc. Natl. Acad. Sci. USA* **105:** 3992–3997.
18. Klenchin, V.A. & T.F. Martin. 2000. Priming in exocytosis: attaining fusion-competence after vesicle docking. *Biochimie* **82:** 399–407.
19. Cremona, O. & P. De Camilli. 2001. Phosphoinositides in membrane traffic at the synapse. *J. Cell Sci.* **114:** 1041–1052.
20. Bai, L., Y. Wang, *et al.* 2007. Dissecting multiple steps of GLUT4 trafficking and identifying the sites of insulin action. *Cell Metab.* **5:** 47–57.
21. Rosenmund, C., J. Rettig, *et al.* 2003. Molecular mechanisms of active zone function. *Curr. Opin. Neurobiol.* **13:** 509–519.
22. Feldmann, J., I. Callebaut, *et al.* 2003. Munc13-4 is essential for cytolytic granules fusion and is mutated in a form of familial hemophagocytic lymphohistiocytosis (FHL3). *Cell* **115:** 461–473.
23. Kang, L., Z. He, *et al.* 2006. Munc13-1 is required for the sustained release of insulin from pancreatic beta cells. *Cell Metab.* **3:** 463–468.
24. Sheu, L., E.A. Pasyk, *et al.* 2003. Regulation of insulin exocytosis by Munc13-1. *J. Biol. Chem.* **278:** 27556–27563.
25. Kwan, E.P., L. Xie, *et al.* 2006. Munc13-1 deficiency reduces insulin secretion and causes abnormal glucose tolerance. *Diabetes* **55:** 1421–1429.
26. Wierda, K.D., R.T. Toonen, *et al.* 2007. Interdependence of PKC-dependent and PKC-independent pathways for presynaptic plasticity. *Neuron* **54:** 275–290.
27. Basu, J., A. Betz, *et al.* 2007. Munc13-1 C1 domain activation lowers the energy barrier for synaptic vesicle fusion. *J. Neurosci.* **27:** 1200–1210.
28. Nesher, R., E. Anteby, *et al.* 2002. Beta-cell protein kinases and the dynamics of the insulin response to glucose. *Diabetes* **51**(Suppl 1): S68–S73.
29. Madison, J.M., S. Nurrish, *et al.* 2005. UNC-13 interaction with syntaxin is required for synaptic transmission. *Curr. Biol.* **15:** 2236–2242.
30. Stevens, D.R., Z.X. Wu, *et al.* 2005. Identification of the minimal protein domain required for priming activity of Munc13-1. *Curr. Biol.* **15:** 2243–2248.
31. Guan, R., H. Dai, *et al.* 2008. Binding of the Munc13-1 MUN domain to membrane-anchored SNARE complexes. *Biochemistry* **47:** 1474–1481.
32. Zhang, W., A. Efanov, *et al.* 2000. Munc-18 associates with syntaxin and serves as a negative regulator of exocytosis in the pancreatic beta -cell. *J. Biol. Chem.* **275:** 41521–41527.
33. Bethani, I., T. Lang, *et al.* 2007. The specificity of SNARE pairing in biological membranes is mediated by both proof-reading and spatial segregation. *EMBO J.* **26:** 3981–3992.
34. Shen, J., D.C. Tareste, *et al.* 2007. Selective activation of cognate SNAREpins by Sec1/Munc18 proteins. *Cell* **128:** 183–195.
35. Starai, V.J., C.M. Hickey, *et al.* 2008. HOPS proofreads the trans-SNARE complex for yeast vacuole fusion. *Mol. Biol. Cell.* **19:** 2500–2508.
36. Oh, E., B.A. Spurlin, *et al.* 2005. Munc18c heterozygous knockout mice display increased susceptibility for severe glucose intolerance. *Diabetes* **54:** 638–647.
37. Dulubova, I., M. Khvotchev, *et al.* 2007. Munc18-1 binds directly to the neuronal SNARE complex. *Proc. Natl. Acad. Sci. USA* **104:** 2697–2702.
38. Zilly, F.E., J.B. Sorensen, *et al.* 2006. Munc18-bound syntaxin readily forms SNARE complexes with synaptobrevin in native plasma membranes. *PLoS. Biol.* **4:** e330.
39. Rickman, C., C.N. Medine, *et al.* 2007. Functionally and spatially distinct modes of MUNC18-syntaxin 1 interaction. *J. Biol. Chem.* **282:** 12097–12103.
40. Weninger, K., M.E. Bowen, *et al.* 2008. Accessory proteins stabilize the acceptor complex for synaptobrevin, the 1:1 syntaxin/SNAP-25 complex. *Structure* **16:** 308–320.
41. Fujita, Y., H. Shirataki, *et al.* 1998. Tomosyn: a syntaxin-1-binding protein that forms a novel complex in the neurotransmitter release process. *Neuron* **20:** 905–915.
42. Gracheva, E.O., A.O. Burdina, *et al.* 2006. Tomosyn inhibits synaptic vesicle priming in Caenorhabditis elegans. *PLoS. Biol.* **4:** e261.
43. Zhang, W., L. Lilja, *et al.* 2006. Tomosyn is expressed in beta-cells and negatively regulates insulin exocytosis. *Diabetes* **55:** 574–581.
44. Hatsuzawa, K., T. Lang, *et al.* 2003. The R-SNARE motif of tomosyn forms SNARE core complexes with syntaxin 1 and SNAP-25 and down-regulates exocytosis. *J. Biol. Chem.* **278:** 31159–31166.
45. Gracheva, E.O., A.O. Burdina, *et al.* 2007. Tomosyn negatively regulates CAPS-dependent peptide release at Caenorhabditis elegans synapses. *J. Neurosci.* **27:** 10176–10184.
46. McEwen, J.M., J.M. Madison, *et al.* 2006. Antagonistic regulation of synaptic vesicle priming by Tomosyn and UNC-13. *Neuron* **51:** 303–315.
47. Gladycheva, S.E., A.D. Lam, *et al.* 2007. Receptor-mediated regulation of tomosyn-syntaxin 1A interactions in bovine adrenal chromaffin cells. *J. Biol. Chem.* **282:** 22887–22899.
48. Fujita, Y., A. Xu, *et al.* 2007. Ca^{2+}-dependent activator protein for secretion 1 is critical for constitutive and regulated exocytosis but not for loading of

transmitters into dense core vesicles. *J. Biol. Chem.* **282:** 21392–21403.

49. Sieburth, D., J.M. Madison, *et al.* 2007. PKC-1 regulates secretion of neuropeptides. *Nat. Neurosci.* **10:** 49–57.
50. Speese, S., M. Petrie, *et al.* 2007. UNC-31 (CAPS) is required for dense-core vesicle but not synaptic vesicle exocytosis in Caenorhabditis elegans. *J. Neurosci.* **27:** 6150–6162.
51. Speidel, D., A. Salehi, *et al.* 2008. CAPS1 and CAPS2 regulate stability and recruitment of insulin granules in mouse pancreatic beta cells. *Cell Metab.* **7:** 57–67.
52. Zhou, K.M., Y.M. Dong, *et al.* 2007. PKA activation bypasses the requirement for UNC-31 in the docking of dense core vesicles from C. elegans neurons. *Neuron* **56:** 657–669.
53. Gracheva, E.O., A.O. Burdina, *et al.* 2007. Tomosyn negatively regulates both synaptic transmitter and neuropeptide release at the C. elegans neuromuscular junction. *J. Physiol.* **585:** 705–709.
54. Drucker, D.J. 2006. The biology of incretin hormones. *Cell Metab.* **3:** 153–165.
55. Seino, S. & T. Shibasaki. 2005. PKA-dependent and PKA-independent pathways for cAMP-regulated exocytosis. *Physiol. Rev.* **85:** 1303–1342.
56. Holz, G.G., G. Kang, *et al.* 2006. Cell physiology of cAMP sensor. *Epac J. Physiol.* **577:** 5–15.
57. Kwan, E.P., L. Xie, *et al.* 2007. Interaction between Munc13-1 and RIM is critical for glucagon-like peptide-1 mediated rescue of exocytotic defects in Munc13-1 deficient pancreatic beta-cells. *Diabetes* **56:** 2579–2588.
58. Kaeser, P.S. & T.C. Sudhof. 2005. RIM function in short- and long-term synaptic plasticity. *Biochem. Soc. Trans.* **33:** 1345–1349.
59. Hirling, H. & R.H.S. cheller. 1996. Phosphorylation of synaptic vesicle proteins: modulation of the alpha SNAP interaction with the core complex. *Proc. Natl. Acad. Sci. USA* **93:** 11945–11949.
60. Jovanovic, J.N., T.S. Sihra, *et al.* 2001. Opposing changes in phosphorylation of specific sites in synapsin I during Ca2+-dependent glutamate release in isolated nerve terminals. *J. Neurosci.* **21:** 7944–7953.
61. Chheda, M.G., U. Ashery, *et al.* 2001. Phosphorylation of Snapin by PKA modulates its interaction with the SNARE complex. *Nat. Cell Biol.* **3:** 331–338.
62. Evans, G.J., M.C. Wilkinson, *et al.* 2001. Phosphorylation of cysteine string protein by protein kinase A. Implications for the modulation of exocytosis. *J. Biol. Chem.* **276:** 47877–47885.
63. Baba, T., T. Sakisaka, *et al.* 2005. PKA-catalyzed phosphorylation of tomosyn and its implication in Ca2+-dependent exocytosis of neurotransmitter. *J. Cell Biol.* **170:** 1113–1125.
64. Boczan, J., A.G. Leenders, *et al.* 2004. Phosphorylation of syntaphilin by cAMP-dependent protein kinase modulates its interaction with syntaxin-1 and annuls its inhibitory effect on vesicle exocytosis. *J. Biol. Chem.* **279:** 18911–18919.
65. Kwan, E.P. & H.Y. Gaisano. 2005. Glucagon-like peptide 1 regulates sequential and compound exocytosis in pancreatic islet beta-cells. *Diabetes* **54:** 2734–2743.
66. Kwan, E.P., X. Gao, *et al.* 2007. Activation of exchange protein directly activated by cyclic adenosine monophosphate and protein kinase A regulate common and distinct steps in promoting plasma membrane exocytic and granule-to-granule fusions in rat islet beta cells. *Pancreas* **35:** e45–e54.
67. Takahashi, N., T. Kishimoto, *et al.* 2002. Fusion pore dynamics and insulin granule exocytosis in the pancreatic islet. *Science* **297:** 1349–1352.
68. Leung, Y.M., L. Sheu, *et al.* 2002. Visualization of sequential exocytosis in rat pancreatic islet beta cells. *Biochem. Biophys. Res. Commun.* **292:** 980–986.
69. Cochilla, A.J., J.K. Angleson, *et al.* 2000. Differential regulation of granule-to-granule and granule-to-plasma membrane fusion during secretion from rat pituitary lactotrophs. *J. Cell Biol.* **150:** 839–848.
70. Nemoto, T., R. Kimura, *et al.* 2001. Sequential-replenishment mechanism of exocytosis in pancreatic acini. *Nat. Cell Biol.* **3:** 253–258.
71. Scepek, S. & M. Lindau. 1993. Focal exocytosis by eosinophils–compound exocytosis and cumulative fusion. *EMBO J.* **12:** 1811–1817.
72. Takahashi, N., H. Hatakeyama, *et al.* 2004. Sequential exocytosis of insulin granules is associated with redistribution of SNAP25. *J. Cell Biol.* **165:** 255–262.
73. Gao, N., P. White, *et al.* 2007. Foxa2 controls vesicle docking and insulin secretion in mature. *Beta Cells Cell Metab.* **6:** 267–279.
74. Hatakeyama, H., N. Takahashi, *et al.* 2007. Two cAMP-dependent pathways differentially regulate exocytosis of large dense-core and small vesicles in mouse {beta} cells. *J. Physiol.* **582:** 1087–1098.
75. Latham, C.F., J.A. Lopez, *et al.* 2006. Molecular dissection of the Munc18c/syntaxin4 interaction: implications for regulation of membrane trafficking. *Traffic* **7:** 1408–1419.
76. Tellam, J.T., S.L. Macaulay, *et al.* 1997. Characterization of Munc-18c and syntaxin-4 in 3T3-L1 adipocytes. Putative role in insulin-dependent movement of GLUT-4. *J. Biol. Chem.* **272:** 6179–6186.
77. Ohara-Imaizumi, M., T. Fujiwara, *et al.* 2007. Imaging analysis reveals mechanistic differences between first- and second-phase insulin exocytosis. *J. Cell Biol.* **177:** 695–705.

78. Takamori, S., M. Holt, *et al.* 2006. Molecular anatomy of a trafficking organelle. *Cell* **127:** 831–846.
79. Huang, X., Y.H. Kang, *et al.* 2001. Ca(2+) influx and cAMP elevation overcame botulinum toxin A but not tetanus toxin inhibition of insulin exocytosis. *Am. J. Physiol. Cell Physiol.* **281:** C740–C750.
80. Guo, Z., C. Turner, *et al.* 1998. Relocation of the t-SNARE SNAP-23 from lamellipodia-like cell surface projections regulates compound exocytosis in mast cells. *Cell* **94:** 537–548.
81. Pickett, J.A., P. Thorn, *et al.* 2005. The plasma membrane Q-SNARE syntaxin 2 enters the zymogen granule membrane during exocytosis in the pancreatic acinar cell. *J. Biol. Chem.* **280:** 1506–1511.
82. Parlati, F., O. Varlamov, *et al.* 2002. Distinct SNARE complexes mediating membrane fusion in Golgi transport based on combinatorial specificity. *Proc. Natl. Acad. Sci. USA* **99:** 5424–5429.
83. Zwilling, D., A. Cypionka, *et al.* 2007. Early endosomal SNAREs form a structurally conserved SNARE complex and fuse liposomes with multiple topologies. *EMBO J.* **26:** 9–18.
84. Kreye, S., J. Malsam, *et al.* 2008. In vitro assays to measure SNARE-mediated vesicle fusion. *Methods Mol. Biol.* **440:** 37–50.
85. Orci, L. & W. Malaisse. 1980. Hypothesis: single and chain release of insulin secretory granules is related to anionic transport at exocytotic sites. *Diabetes* **29:** 943–944.
86. Tsuboi, T. & G.A. Rutter. 2003. Multiple forms of "kiss-and-run" exocytosis revealed by evanescent wave microscopy. *Curr. Biol.* **13:** 563–567.
87. Lam, A.D., P. Tryoen-Toth, *et al.* 2008. SNARE-catalyzed fusion events are regulated by syntaxin1A-lipid interactions. *Mol. Biol. Cell* **19:** 485–497.
88. Fisher, R.J., J. Pevsner, *et al.* 2001. Control of fusion pore dynamics during exocytosis by Munc18. *Science* **291:** 875–878.

Phosphatidylinositol-4,5-bisphosphate-dependent Facilitation of the ATP-dependent Secretory Activity in Mouse Pituitary Cells

Simon Sedej,[a,b] Iman Singh Gurung,[a,c] Thomas Binz,[d] and Marjan Rupnik[a,e]

[a]*European Neuroscience Institute-Göttingen, Göttingen, Germany*

[b]*Medical University Graz, Graz, Austria*

[c]*Department of Biochemistry, University of Cambridge, Cambridge, United Kingdom*

[d]*Institute of Biochemistry, Medizinische Hochschule Hannover, Hannover, Germany*

[e]*Institute of Physiology, Faculty of Medicine, University of Maribor, Maribor, Slovenia*

Phosphatidylinositol-4, 5-bisphosphate [PI(4,5)P_2] has been implicated in the priming of large dense-core vesicles in many secretory cells; however, its role in the Ca^{2+}-dependent secretory activity in pituitary cells remains elusive. We assessed the effect of elevated intracellular PI(4,5)P_2 on the kinetics of Ca^{2+}-dependent exocytosis, using a whole-cell patch-clamp technique in wild-type mouse melanotrophs from fresh pituitary tissue slices. We found that 1 μmol/L PI(4,5)P_2 significantly increased Ca^{2+}-dependent exocytosis of vesicles that need to go through ATP-dependent reactions; however, the exocytosis of release-ready vesicles (ATP-independent release) and voltage-activated Ca^{2+} currents remained unaffected. We suggest that PI(4,5)P_2 increases the size of the readily releasable vesicle pool by regulating the effectiveness of vesicular mobilization and fusion in an ATP-dependent manner.

Key words: **phosphatidylinositol-4; 5-biphosphate; release-ready vesicles; Ca^{2+}-dependent exocytosis; priming; capacitance measurements; ATP-independent pathway; ATP-dependent pathway; acute tissue slices; pituitary; melanotrophs**

Introduction

Phosphatidylinositol 4,5-biphosphate [PI(4, 5)P_2] is a key phospholipid molecule acting as a regulator of secretion.[1] PI(4,5)P_2 is synthesized from phosphatydilinositol (PI) in the presence of ATP by phosphatidylinositol-4-phosphate-5-kinase and a phosphatidylinositol transfer protein and is involved in the ATP-dependent priming reaction.[2–4]

Melanotrophs, neuroendocrine cells from the intermediate lobe of the pituitary, secrete α-melanocyte-stimulating hormone (α-MSH) and β-endorphin in response to an electrical activity.[5] Recently, we showed that mouse melanotrophs from tissue slices display a clear separation between the fusion of release-ready vesicles within the Ca^{2+}-microdomain (approximately 15 vesicles) from the release of secretory vesicles outside this microdomain.[6] The exocytotic process outside the microdomain can be readily separated into two distinct kinetic components: the linear component (ATP independent) and the threshold component (ATP dependent) assigned to the exocytosis of release-ready vesicles and exocytosis of secretory vesicles that required ATP-dependent reaction(s), respectively.[6] Both components can be selectively controlled by cAMP levels;[6] however, the effects of

Address for correspondence: Dr. Marjan Rupnik, Institute of Physiology, University of Maribor, Slomškov trg 15, 2000 Maribor, Slovenia. Voice: +386 2 330 58 54; fax: +386 2 330 58 53. marjan.rupnik @uni-mb.si

Mechanisms of Exocytosis: Ann. N.Y. Acad. Sci. 1152: 165–173 (2009).
doi: 10.1111/j.1749-6632.2008.04002.x

PI(4,5)P2 on the kinetics of Ca^{2+}-dependent exocytosis in pituitary cells still remains to be elucidated.

MgATP dependence of different exocytotic components has already been tested in cultured cells.[7–11] Although there are some reports showing that Ca^{2+} triggers exocytosis in the absence of MgATP,[8,9] such studies were not consequently reproduced.[11] There are two possible candidates for the last ATP-requiring reactions: ATP hydrolysis by N-ethylmaleimide-sensitive factor (NSF)[12,13] and the ATP-dependent formation of PI(4,5)P_2.[3,4,14] It is not yet clear at which stages these reactions take place.

Many molecular components of the secretory apparatus have been identified; however, their precise role and order of involvement in the different stages of Ca^{2+}-dependent secretory activity is still controversial. As in several other neuroendocrine systems, both protein kinase A (PKA)- and cAMP-Guaniding Nucleotide Exchange Factor II (GEFII)/Epac2-dependent pathways selectively control the Ca^{2+}-dependent secretory activity at resting and elevated cAMP levels, respectively.[6,15–20] Neurotoxins have been crucial to dissect the role of some proteins involved.[11] It is quite intriguing that both the exocytotic burst and slower component elicited by the flash photolysis experiment are sensitive to neurotoxins, implying all of synaptosome-associated protein of 25 Kda (SNAP-25) or syntaxin or synaptobrevin must be assembling/disassembling during the action of the neurotoxin.[11,21]

Our results show that a prolonged depolarization train can readily reveal the secretory component resulting primarily from the mobilization of undocked vesicles and which is highly ATP dependent and clostridial neurotoxin sensitive. We demonstrate that PI(4,5)P_2 augmented the ATP-dependent secretory component while having no effect on the density of voltage-activated Ca^{2+} channels. Washout of ATP from the cell during dialysis limited the PI(4,5)P_2 action, implying that ATP is indispensable for the last steps prior to fusion. We suggest that PI(4,5)P_2 promotes ATP-dependent exocytosis by increasing the releasable vesicle pool size.

Materials and Methods

Pituitary Slice Preparation

Adult Naval Medical Research Institute (NMRI) mice (8–12 weeks) of either sex were sacrificed by cervical dislocation prior to decapitation. Animal work was performed according to the regulations of the State of Lower Saxony, Germany. Acute pituitary slices were prepared as described elsewhere.[22] Fresh slices of 80-μm thickness were kept at 32°C for up to 8 h.

Electrophysiology

Changes in cell membrane capacitance (ΔC_m) and voltage-activated Ca^{2+} currents were measured in single melanotrophs within intact clusters of the intermediate lobe of the pituitary gland by the whole-cell patch-clamp technique.[23] During the measurements, the slice was perfused continuously at the rate of approximately 2 mL/min with external solution composed of (in mmol/L): 125 NaCl, 2.5 KCl, 1.25 NaH_2PO_4, 2 Na-pyruvate, 3 myo-inositol, 0.5 ascorbic acid, 10 glucose, 26 $NaHCO_3$, 1 $MgCl_2$, 2 $CaCl_2$, 6 lactic acid. External solution was continuously bubbled with 95% O_2 and 5% CO_2 to keep the pH stable at 7.3.

The intracellular solution was (in mmol/L): 140 CsCl, 10 HEPES, 2 $MgCl_2$, 20 tetraethylammonium (TEA)-Cl, 2 MgATP, 0.05 EGTA; pH was adjusted to 7.2 with CsOH. In the pipette-filling solution 1 μmol/L PI(4,5)P_2 was dissolved by sonication for 30 min on ice prior to the experiments.[24] At least 5 and 10 min of dialysis were allowed before elevation of Ca^{2+} to ensure a complete action of both 1 μmol/L PI(4,5)P_2 or 100 μmol/L neomycin, a PI(4,5)P_2-masking agent, and 400 nmol/L botulinum toxin E, respectively. All chemicals were obtained from Sigma (St. Louis, MO) unless otherwise stated.

Cells were visualized with an upright Nikon Eclipse E600 FN microscope (Nikon, Tokyo, Japan) and a mounted charge-coupled device (CCD) camera (Cohu, San Diego, CA). Pipettes had a resistance of 2–3 MΩ in CsCl-based solution. Whole-cell currents and capacitance changes were measured with a lock-in patch-clamp amplifier (SWAM IIC, Celica, Ljubljana, Slovenia), low-pass filtered (3 kHz, −3 dB), and stored on a standard PC. The cells were voltage clamped at −80 mV. Ca^{2+} currents were leak corrected. Cells that showed less than −30 pA peak amplitude Ca^{2+} current (approximately 10% of cells) were discarded from the analysis. WinWCP V3.3.9. software from J. Dempster (Strathclyde University, Glasgow, UK) was used for pulse generation, data acquisition, and basic analysis. High-resolution capacitance measurements (in the fF range) were made by using the compensated technique as described previously,[25] using a sinusoidal voltage (1600 Hz, 10 mV peak-to-peak). To determine ΔC_m values, membrane capacitance was first averaged over 30 ms preceding the depolarization to obtain a baseline value, which was subtracted from the value estimated after the depolarization averaged over a 40-ms window. The first 30 ms after the depolarization were excluded from the C_m measurement to avoid contamination by nonsecretory capacitative transients related to gating charge movement.[26] The use of tetrodotoxin was avoided because it prolongs a transient nonsecretory capacitance change.[27] All electrophysiological experiments were performed at 30–32°C. Signal processing and curve fitting was done using Sigmaplot (SPSS, Chicago, IL) and Matlab (Mathworks, Novi, MI). Data and results are reported as mean ± SEM for the indicated number of experiments. Estimates of the number of vesicles released are based on a value of 1 fF per vesicle for adult melanotrophs.[10] Differences between samples were tested using Student's *t*-test, one-way ANOVA, or one-way ANOVA repeated measures (RM) as indicated (Tukey's post-hoc test; SigmaSTAT, SPSS). The significance level was chosen at $P < 0.05$.

Results

ATP Dependence of Secretory Vesicle Fusion

Nonbound Mg-ATP is essentially completely lost from the cytosol within 2 min of patch break, and 95% of total Mg-ATP, including bound ATP, is lost after 6 min of dialysis with MgATP-free internal solution.[6,28] Importantly, under such experimental conditions peak voltage-activated Ca^{2+} channel (VACC) amplitudes ran down only about 1% per minute during the first 8 min of dialysis (not shown). Considering this, melanotrophs were dialyzed 6 min under Mg-ATP (2 mmol/L) or Mg-ATP-free condition followed by a train of 75 pulses from −80 mV to 0 mV at 10 Hz (40-ms duration) (Fig. 1A). Secretory responses were measured as the cumulative increase in membrane capacitance. The membrane capacitance in Mg-ATP-free cells increased linearly, and the linear growth subsided after 6–7s of stimulation (Fig. 1A). The second depolarization train applied 2 min after the first depolarization train failed to produce any further increase in capacitance. Lowering of pH to 6.6 reduced the slope of the linear component (Fig. 1A). Inclusion of Mg-ATP in the pipette revealed an additional threshold component, which considerably increased a secretory response as a result of the fusion of vesicles that required Mg-ATP-dependent reaction (Fig. 1A).[6]

The linear and threshold kinetic components were entirely suppressed after 12 min of dialysis with botulinum neurotoxin E, which cleaves SNAP-25 protein.[29] Importantly, throughout the experiment neither changes in membrane capacitance nor Ca^{2+} current rundown were observed (Fig. 1B inset).

PI(4,5)P$_2$ and Ca^{2+}-dependent Secretory Activity

Next, we tested how PI(4,5)P$_2$ may affect the Ca^{2+}-dependent secretory activity in mouse melanotrophs. For this purpose, we delivered

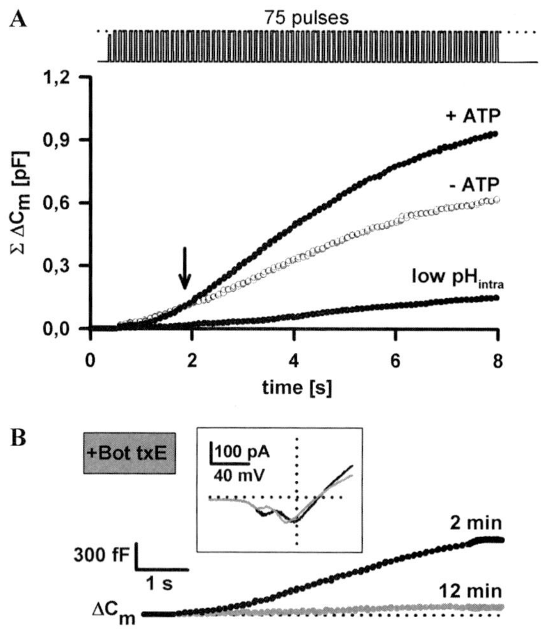

Figure 1. The effect of ATP and low pH on the Ca^{2+}-dependent exocytosis in mouse melanotrophs. (**A**) Typical depolarization-induced secretory response (75 pulses, from −80 mV to 0 mV, 40 ms, at 10 Hz) after the washout of ATP was linear ($n = 3$; the linear ATP-independent component). Note that the inclusion of intracellular ATP separates the linear component from the threshold component ($n = 8$; the onset of ATP-dependent component is shown by arrow). Low intracellular pH (6.6) dramatically reduced the linear component ($n = 6$). (**B**) The representative C_m recording from a melanotroph dialyzed with Botulinum toxin E (Bot txE)-containing pipette solution in the presence of 2 mmol/L ATP. Bot txE suppressed the secretory response after at least 12 min (gray trace) without affecting peak Ca^{2+} channel amplitude (inset).

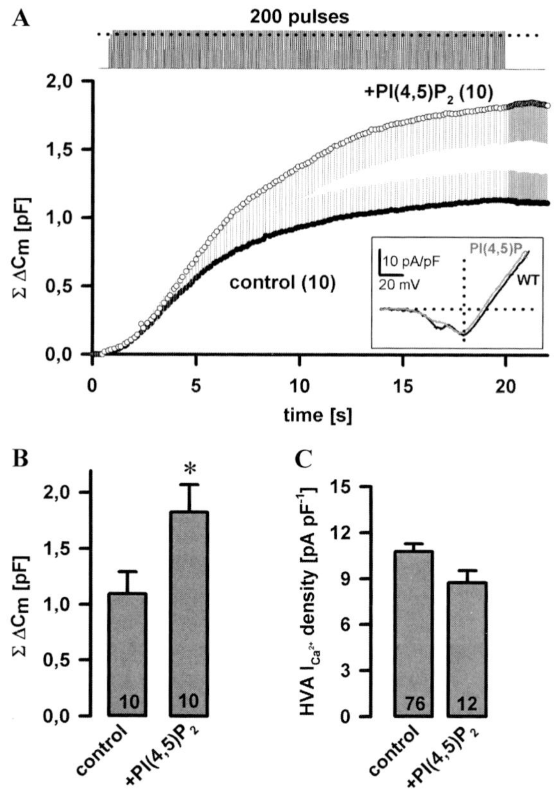

Figure 2. The effect of phosphatidylinositol 4, 5-bisphosphate [PI(4,5)P_2] on Ca^{2+}-dependent exocytosis and Ca^{2+} channel density. (**A**) The secretory response was measured as a cumulative membrane capacitance change (ΔC_m) during 200 depolarization pulses (40 ms, from −80 mV to 10 mV) at 10 Hz. The ΔC_m traces are shown as mean and SEM for each pulse. Note that I-V relationship in PI(4,5)P_2-treated and control cells remained unaltered (inset). (**B**) PI(4,5)P_2 significantly increased the cumulative ΔC_m after 200 pulses compared to control cells as shown by the asterisk ($P < 0.03$; Student's *t*-test) but not (**C**) High voltage activated (HVA) Ca^{2+} current density (see also inset). Numbers on bars or in the brackets indicate the number of cells tested.

a family of 200 depolarization pulses from −80 mV to 10 mV at 10 Hz (40-ms duration) to evoke Ca^{2+}-dependent secretion. As shown in Fig. 2A-B, 1 μmol/L PI(4,5)P_2 significantly increased the cumulative membrane capacitance from 1097 ± 197 fF ($n = 10$; control) to 1827 ± 245 fF ($n = 10$; $P < 0.03$; Fig. 2B). Importantly, we did not detect any alterations in the VACC density (Fig. 2C; control, 10.8 ± 0.5 pA/pF, $n = 76$ versus PI(4,5)P_2-treated, 8.8 ± 0.8 pA/pF), cytosolic Ca^{2+} change, and resting membrane capacitance (not shown).

To determine, which of the two kinetic components[6] (linear or threshold) was affected by PI(4,5)P_2, the analysis of the first 50 depolarization pulses was performed. Precisely, during the first second of stimulation only the linear component is present and its rate of membrane capacitance change (dC_m/dt) was determined. In addition, the average slope of the threshold component was estimated as the maximal

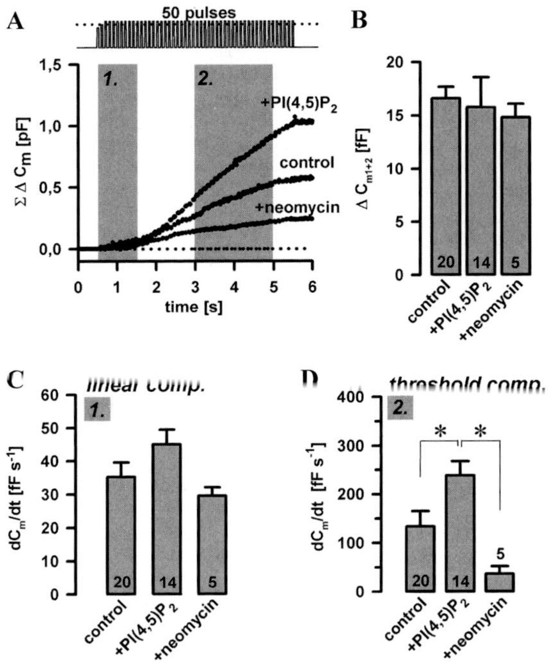

Figure 3. The effect of PI(4,5)P$_2$ on the ATP-independent and ATP-dependent component and release-ready pool size. (**A**) The average cumulative capacitance changes in PI(4,5)P$_2$-treated, neomycin-treated, and control melanotrophs after 50 depolarization pulses. Gray shaded areas #1 and #2 represent the time interval at which the dC_m/dt for the linear (ATP-independent) and the threshold component (ATP-dependent) was determined, respectively.[6] (**B**) Dual-pulse protocol failed to detect changes in immediately releasable pool (IRP) size (ΔC_{m1+2}) in the presence of 1 μmol/L PI(4,5)P$_2$- and 100 μmol/L neomycin. (**C**) The mean ΔC_m rate of the linear component (#1). (**D**) The average ΔC_m rate of threshold component (#2); PI(4,5)P$_2$-treated melanotrophs significantly differed with respect to control ($P < 0.003$; one-way ANOVA) and neomycin-treated cells ($P < 0.008$). Numbers on bars indicate the number of cells tested.

rate of capacitance change during a 2-s interval (measured between 2.5 s and 4.5 s of the depolarization train) and corrected by the rate of linear component (Fig. 3A). dC_m/dt of the linear component was similar in PI(4,5)P$_2$- and in neomycin-treated pituitary cells with respect to control melanotrophs (Fig. 3A gray area #1, Fig. 3C). In contrast, PI(4,5)P$_2$ significantly increased the average dC_m/dt of the threshold component (Fig. 3A gray area #2) from 134 ± 32 fF/s ($n = 20$; control) to 222 ± 39 fF/s ($P < 0.03$; one-way ANOVA; $n = 14$; Fig. 3D). On the other hand, the mean dC_m/dt rate of threshold component was significantly reduced in the presence of neomycin (37 ± 15 fF/s; $P < 0.008$; $n = 5$; Fig. 3D).

Further, to investigate the effect of PI(4,5)P$_2$ on the immediately releasable pool (IRP), which is a population of granules residing within the Ca^{2+} microdomain,[30] we applied a dual pulse protocol as previously described.[6,31] Briefly, the first two 40-ms depolarizations of the train separated with a 60-ms interval were chosen to elicit secretory depression. The first and second pulses were set at −10 mV and 0 mV to inject two comparable Ca^{2+} currents during both pulses, respectively.[6,31] The sum of the first two capacitance changes ($S = \Delta C_{m1} + \Delta C_{m2}$) was calculated to estimate the number of fused IRP vesicles. We did not detect any PI(4,5)P$_2$- and neomycin-dependent changes on the S value compared to the control (Fig. 3B).

PI(4,5)P$_2$ and Activity-dependent Vesicle Mobilization

To address whether PI(4,5)P$_2$ suffices for the refilling of release-ready vesicle pool, we dialyzed the cytosol of melanotrophs with PI(4,5)P$_2$-containing intracellular solution with-/out MgATP for 5 min prior to the delivery of depolarization trains. Two consecutive trains applied in 2-min intervals were utilized to stimulate the same cell. As shown in Fig. 4Aa, the total membrane capacitance increase evoked by the second train was of similar magnitude as that elicited by the first depolarization train in the presence of MgATP (control). When the same experiment was repeated after the inclusion of 1 μmol/L PI(4,5)P$_2$ in the pipette, both depolarization-induced C_m responses were about twofold higher (Fig. 4Ab). The removal of MgATP induced C_m changes after the first train similar to the control even though Mg^{2+} and PI(4;5)P$_2$ were present

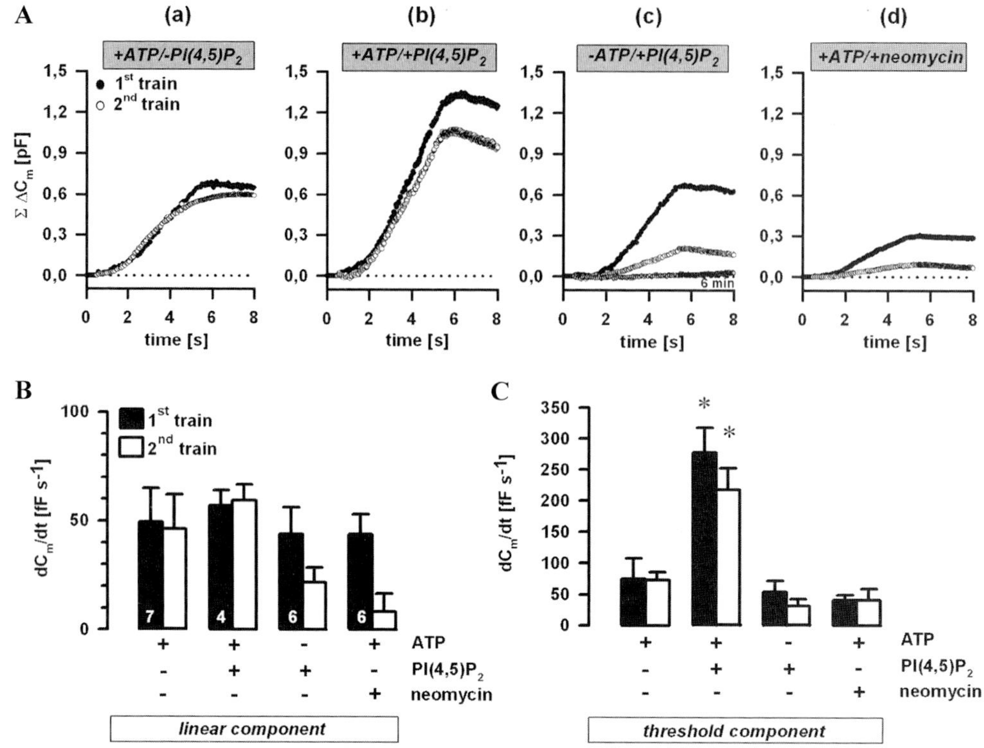

Figure 4. The effect of PI(4,5)P_2 on activity-dependent mobilization of vesicles. (**A**) The typical cumulative membrane capacitance recordings triggered by first (black circles) and second (empty circles) depolarization trains of 50 pulses at 10 Hz delivered in 2-min intervals. Melanotrophs were dialyzed with (a) 2 mmol/L MgATP (control), (b) 2 mmol/L MgATP and 1 μmol/L PI(4,5)P_2, (c) 1 μmol/L PI(4,5)P_2, and (d) 2 mmol/L MgATP and 100 μmol/L neomycin. (**B**) Mean dC_m/dt values for the linear component and (**C**) the threshold component. Note the significant difference of PI(4,5)P_2 effect on the threshold component with respect to the control ($P < 0.05$; one-way ANOVA).

(Fig. 4Ac). C_m increase evoked by the second train, however, was almost abolished and fully suppressed following the third train (Fig. 4Ac). The inclusion of neomycin and MgATP in the pipette solution halved the secretory response (Fig. 4Ad). We further dissected the secretory responses to assess whether the linear or the threshold component was affected by any of the interventions. As shown in Fig. 4B, the dC_m/dt of the linear component was not influenced by any of the compounds tested. In contrast, PI(4,5)P_2 together with MgATP significantly increased the threshold component with respect to MgATP or when only PI(4,5)P_2 was used (Fig. 4C; $P < 0.002$, one-way ANOVA RM).

Discussion

In this work we sought to characterize the effect of PI(4,5)P_2 on Ca^{2+}-dependent exocytosis in wild-type pituitary cells, using a whole-cell patch-clamp technique. We found that PI(4,5)P_2 augmented the ATP-dependent (threshold) component while leaving the ATP-independent (linear) component intact. The overall secretory increment was not accompanied by an obvious increase of the VACC function as reported by Wu *et al.*[32] Instead, enhanced stimulus-secretion coupling was assigned to the increased size of the readily releasable pool. This is consistent with previous studies in pancreatic beta cells[24] and chromaffin cells[33] where $PI(4,5)P_2$ enhanced stimulus-secretion coupling by increasing the releasable vesicle pool. At least two mechanisms can be responsible to achieve such an increase in the vesicle pool size. First, the capability of $PI(4,5)P_2$ to recruit several exocytotic proteins [e.g., synaptotagmin, Rabphillin 3a, Ca^{2+}-dependent activator protein for secretion (CAPS)] to their site of action[34,35] and, second, the rapid regulation of actin dynamics.[36,37]

Given that no significant endocytotic activity was detected after the end of the depolarization train, we excluded the possibility that increased endocytosis accounted for the reduced rate of capacitance increase[38] during the last pulses of the stimulation. Instead, the cessation of exocytosis was likely attributable either to the exhaustion of vesicles from the release-ready pool or to increased unpriming (and decreased priming).

The use of neurotoxins is a valuable tool to dissect the role of soluble N-ethylmaleimide-sensitive factor attachment protein receptor (SNARE) proteins involved in the Ca^{2+}-dependent exocytosis.[11] Both the exocytotic burst and slower component elicited by the flash photolysis experiment are sensitive to neurotoxins, implying all of SNAP-25 or syntaxin or synaptobrevin must be assembling/disassembling during the action of neurotoxins.[11,21] Consistent with this, *Clostridium* neurotoxin completely abolished the Ca^{2+}-dependent exocytosis and elicited virtually no Ca^{2+} current rundown (Fig. 1B, inset). This suggested that the fusion of secretory vesicles in mouse melanotrophs fully depends on the availability of plasma membrane SNARE proteins.[11]

We found that MgATP is not required for Ca^{2+}-dependent exocytosis *per se*, confirming previous findings.[8,9] Membrane capacitance changes that occurred in the presence of either Mg^{2+} or $PI(4,5)P_2$ mimicked the control secretory responses (Fig. 4Ac). This indicated that melanotrophs from tissue slices maintained a stable pool of release-ready vesicles, as previously found in cultured rat melanotrophs.[9] However, further stimulation reduced Ca^{2+}-dependent exocytosis likely because of a depletion of the pool of readily releasable vesicles (Fig. 4Ac). The inclusion of neomycin and MgATP in the pipette solution halved the secretory response (Fig. 4Ad), meaning that a subset of vesicles likely undergoes unpriming. In summary, these data indicate that MgATP and $PI(4,5)P_2$ are necessary for the activity-dependent mobilization of release-ready vesicles and that MgATP is indispensable for the $PI(4,5)P_2$-dependent regulation of releasable vesicle pool size.

Acknowledgments

We thank Marion Niebeling, Heiko Röhse, and Rudi Mlakar for excellent technical support. The work was supported by the Center for Molecular Physiology of the Brain (DFG) to SS and MR. The European Neuroscience Institute Göttingen (ENI-G) is jointly funded by the Göttingen University Medical School, the Max-Planck Society, and Schering AG. MR is supported from the Slovenian Research Agency (J3-7186-2334).

Conflicts of Interest

The authors declare no conflicts of interest.

References

1. Martin, T.F. 2001. PI(4,5)P(2) regulation of surface membrane traffic. *Curr. Opin. Cell Biol.* **13:** 493–499.
2. Eberhard, D.A. *et al.* 1990. Evidence that the inositol phospholipids are necessary for exocytosis. Loss of inositol phospholipids and inhibition of secretion in permeabilized cells caused by a bacterial phospholipase C and removal of ATP. *Biochem. J.* **268:** 15–25.
3. Hay, J.C. *et al.* 1995. ATP-dependent inositide phosphorylation required for Ca(2+)-activated secretion. *Nature* **374:** 173–177.
4. Hay, J.C. & T.F. Martin. 1993. Phosphatidylinositol transfer protein required for ATP-dependent priming of Ca(2+)-activated secretion. *Nature* **366:** 572–575.
5. Mains, R.E. & B.A. Eipper. 1979. Synthesis and secretion of corticotropins, melanotropins, and endorphins by rat intermediate pituitary cells. *J. Biol. Chem.* **254:** 7885–7894.

6. Sedej, S. *et al.* 2005. cAMP increases Ca2+-dependent exocytosis through both PKA and Epac 2 in mouse pituitary tissue. slices. *J. Physiol.* **567**(3): 799–813.
7. Hay, J.C. & T.F. Martin. 1992. Resolution of regulated secretion into sequential MgATP-dependent and calcium-dependent stages mediated by distinct cytosolic proteins. *J. Cell Biol.* **119:** 139–151.
8. Holz, R.W. *et al.* 1989. MgATP-independent and MgATP-dependent exocytosis. Evidence that MgATP primes adrenal chromaffin cells to undergo exocytosis. *J. Biol. Chem.* **264:** 5412–5419.
9. Parsons, T.D. *et al.* 1995. Docked granules, the exocytic burst, and the need for ATP hydrolysis in endocrine cells. *Neuron* **15:** 1085–1096.
10. Thomas, P., J.G. Wong & W. Almers. 1993. Millisecond studies of secretion in single rat pituitary cells stimulated by flash photolysis of caged Ca2+. *EMBO J.* **12:** 303–306.
11. Xu, T. *et al.* 1998. Multiple kinetic components of exocytosis distinguished by neurotoxin sensitivity. *Nat. Neurosci.* **1:** 192–200.
12. Otto, H., P.I. Hanson & R. Jahn. 1997. Assembly and disassembly of a ternary complex of synaptobrevin, syntaxin, and SNAP-25 in the membrane of synaptic vesicles. *Proc. Natl. Acad. Sci. USA* **94:** 6197–6201.
13. Morgan, A. & R.D. Burgoyne. 1995. A role for soluble NSF attachment proteins (SNAPs) in regulated exocytosis in adrenal chromaffin cells. *EMBO J.* **14:** 232–239.
14. Martin, T.F. & J.A. Kowalchyk. 1997. Docked secretory vesicles undergo Ca2+-activated exocytosis in a cell-free system. *J. Biol. Chem.* **272:** 14447–14453.
15. Eliasson, L. *et al.* 2003. SUR1 regulates PKA-independent cAMP-induced granule priming in mouse pancreatic B-cells. *J. Gen. Physiol.* **121:** 181–197.
16. Hatakeyama, H. *et al.* 2007. Two cAMP-dependent pathways differentially regulate exocytosis of large dense-core and small vesicles in mouse beta-cells. *J. Physiol.* **582:** 1087–1098.
17. Ozaki, N. *et al.* 2000. cAMP-GEFII is a direct target of cAMP in regulated exocytosis. *Nat. Cell Biol.* **2:** 805–811.
18. Seino, S. & T. Shibasaki. 2005. PKA-dependent and PKA-independent pathways for cAMP-regulated exocytosis. *Physiol. Rev.* **85:** 1303–1342.
19. Sikdar, S.K., M. Kreft & R. Zorec. 1998. Modulation of the unitary exocytic event amplitude by cAMP in rat melanotrophs. *J. Physiol.* **511**(Pt 3): 851–859.
20. Smith, P.A., M.R. Duchen & F.M. Ashcroft. 1995. A fluorimetric and amperometric study of calcium and secretion in isolated mouse pancreatic beta-cells. *Pflugers Arch.* **430:** 808–818.
21. Rupnik, M. *et al.* 2000. Rapid regulated dense-core vesicle exocytosis requires the CAPS protein. *Proc. Natl. Acad. Sci. USA* **97:** 5627–5632.
22. Sedej, S. *et al.* 2004. Voltage-activated Ca(2+) channels and their role in the endocrine function of the pituitary gland in newborn and adult mice. *J. Physiol.* **555:** 769–782.
23. Neher, E. & A. Marty. 1982. Discrete changes of cell membrane capacitance observed under conditions of enhanced secretion in bovine adrenal chromaffin cells. *Proc. Natl. Acad. Sci. USA* **79:** 6712–6716.
24. Olsen, H.L. *et al.* 2003. Phosphatidylinositol 4-kinase serves as a metabolic sensor and regulates priming of secretory granules in pancreatic beta cells. *Proc. Natl. Acad. Sci. USA* **100:** 5187–5192.
25. Zorec, R., S.K. Sikdar & W.T. Mason. 1991. Increased cytosolic calcium stimulates exocytosis in bovine lactotrophs. Direct evidence from changes in membrane capacitance. *J. Gen. Physiol.* **97:** 473–497.
26. Horrigan, F.T. & R.J. Bookman. 1994. Releasable pools and the kinetics of exocytosis in adrenal chromaffin cells. *Neuron* **13:** 1119–1129.
27. Horrigan, F.T. & R.J. Bookman. 1993. Na channel gating charge movement is responsible for the transient capacitance increase evoked by depolarization in rat adrenal chromaffin cells. *Biophys. J.* **64:** A101–A101.
28. Speier, S. *et al.* 2005. KATP-channels in beta-cells in tissue slices are directly modulated by millimolar ATP. *Mol. Cell Endocrinol.* **230:** 51–58.
29. Montecucco, C. & G. Schiavo. 1994. Mechanism of action of tetanus and botulinum neurotoxins. *Mol. Microbiol.* **13:** 1–8.
30. Rettig, J.R. & A. Folch. 2005. Large-scale single-cell trapping and imaging using microwell arrays. *Anal. Chem.* **77:** 5628–5634.
31. Gillis, K.D., R. Mossner & E. Neher. 1996. Protein kinase C enhances exocytosis from chromaffin cells by increasing the size of the readily releasable pool of secretory granules. *Neuron* **16:** 1209–1220.
32. Wu, L. *et al.* 2002. Dual regulation of voltage-gated calcium channels by PtdIns(4,5)P2. *Nature* **419:** 947–952.
33. Milosevic, I. *et al.* 2005. Plasmalemmal phosphatidylinositol-4,5-bisphosphate level regulates the releasable vesicle pool size in chromaffin cells. *J. Neurosci.* **25:** 2557–2565.
34. Bai, J., W.C. Tucker & E.R. Chapman. 2004. PIP2 increases the speed of response of synaptotagmin and steers its membrane-penetration activity toward the plasma membrane. *Nat. Struct. Mol. Biol.* **11:** 36–44.
35. Cremona, O. & P. De Camilli. 2001. Phosphoinositides in membrane traffic at the synapse. *J. Cell Sci.* **114:** 1041–1052.

36. Bittner, M.A. & R.W. Holz. 2005. Phosphatidylinositol-4,5-bisphosphate: actin dynamics and the regulation of ATP-dependent and -independent secretion. *Mol. Pharmacol.* **67:** 1089–1098.
37. Malacombe, M., M.F. Bader & S. Gasman. 2006. Exocytosis in neuroendocrine cells: new tasks for actin. *Biochim. Biophys. Acta* **1763:** 1175–1183.
38. Turner, J.E., S. Sedej & M. Rupnik. 2005. Cytosolic Cl- ions in the regulation of secretory and endocytotic activity in melanotrophs from mouse pituitary tissue slices. *J. Physiol.* **566:** 443–453.

Multiple Roles of $G_{i/o}$ Protein-coupled Receptors in Control of Action Potential Secretion Coupling in Pituitary Lactotrophs

Stanko S. Stojilkovic, Takayo Murano, Arturo E. Gonzalez-Iglesias, Silvana A. Andric, Marko A. Popovic, Fredrick Van Goor, and Melanija Tomić

Section on Cellular Signaling, Program in Developmental Neuroscience, National Institute of Child Health and Human Development, National Institutes of Health, Bethesda, Maryland, USA

$G_{i/o}$ protein-coupled receptors, signaling through G protein-dependent and protein-independent pathways, have prominent effects on secretion by modulating calcium signaling and regulating the size of the releasable secretory pool, the rates of exocytosis and endocytosis, and *de novo* synthesis. Pituitary cells fire action potentials spontaneously, and the associated calcium influx is sufficient to maintain prolactin (PRL) release but not gonadotropin release at high and steady levels for many hours. Such secretion, termed *intrinsic, spontaneous*, or *basal*, reflects fusion of secretory vesicles triggered by the cell type-specific pattern of action potentials. In lactotrophs, activation of endothelin ET_A and dopamine D_2 receptors causes inhibition of spontaneous electrical activity and basal adenylyl cyclase activity accompanied with inhibition of basal PRL release. Agonist-induced inhibition of cAMP production and firing of action potentials is abolished in cells with blocked pertussis toxin (PTX)-sensitive $G_{i/o}$ signaling pathway. However, agonist-induced inhibition of PRL release is only partially relieved in such treated cells, indicating that both receptors also inhibit exocytosis downstream of cAMP/calcium signaling. The PTX-insensitive step in agonist-induced inhibition of PRL release is not affected by inhibition of phosphoinositide 3-kinase and glycogen synthase kinase-3 but is partially rescued by downregulation of the $G_z\alpha$ expression. Thus, ET_A and D_2 receptors inhibit basal PRL release not only by blocking electrical activity but also by desensitizing calcium-secretion coupling.

Key words: **action potential; calcium; cAMP; GPCR; lactotrophs; gonadotrophs; dopamine D_2 receptors; endothelin ET_A receptors; PI3-kinase, and GSK-3**

Introduction

The anterior pituitary functions are carried out by five cell types defined by the hormone they produce and secrete: corticotrophs secreting adrenocorticotrophic hormone (ACTH), thyrotrophs secreting thyroid-stimulating hormone (TSH), somatotrophs secreting growth hormone (GH), lactotrophs secreting prolactin (PRL), and gonadotrophs secreting luteinizing hormone (LH) and follicle-stimulating hormone (FSH). Although all these cells arise from progenitors in Rathke's pouch, the embryonic primordial of the pituitary gland, they differ with respect to their secretory patterns *in vivo* and *in vitro*.[1] *In vivo,* removal of hypothalamopituitary connections leads to increased release of PRL but not gonadotropins. Animals bearing ectopic pituitary grafts also release high levels of PRL and low levels of LH for a prolonged period, leading to pseudopregnancy.

Address for correspondence: Dr. Stanko Stojilkovic, NICHD; Bldg. 49, Room 6A-36; 49 Convent Drive, Bethesda, MD 20892-4510. Voice: +301-496-1638; fax: +301-594-7031. stankos@helix.nih.gov

Mechanisms of Exocytosis: Ann. N.Y. Acad. Sci. 1152: 174–186 (2009).
doi: 10.1111/j.1749-6632.2008.03994.x

In vitro, basal (in the absence of agonists) PRL and GH secretion from pituitary fragments, dispersed pituitary cells, and immortalized lactosomatotrophs is high, occurs in an extracellular Ca^{2+}-dependent manner, and is also termed intrinsic or spontaneous secretory activity.[2] Slow spontaneous secretion from single lactotrophs could be monitored with confocal microscopy using FM4-64 loaded cells.[3,4] By contrast, basal LH secretion *in vitro* is low and is not affected by removal of extracellular calcium.[5] Consistent with the high basal levels of PRL and GH secretion, lactotrophs and somatotrophs are under negative hypothalamic control by the $G_{i/o}$-coupled dopamine and somatostatin receptors, respectively, in addition to positive regulation by G_q and G_s-coupled receptors.[6,7] Conversely, LH secretion from gonadotrophs is under positive hypothalamic control by Ca^{2+}-mobilizing gonadotropin-releasing hormone (GnRH) receptors[8] and only in neonatal gonadotrophs is GnRH-stimulated gonadotropin secretion inhibited by the pineal hormone melatonin.[9] Here we review roles of the $G_{i/o}$-coupled dopamine D_2 and endothelin $(ET)_A$ receptors in control of signaling and secretion in pituitary cells.

In the late 1970s, the dopamine D_2 receptor subtype was shown to mediate the tonic inhibitory control of hypothalamic dopamine on PRL release in lactotrophs.[10] Later investigations showed that two subtypes of D_2 receptors, termed D_{2S} and D_{2L}, are generated by alternative splicing in lactotrophs and other cell types.[11,12] Lactotrophs may express a different ratio of these two receptor subtypes, depending on the level of gonadal steroids.[13] In addition to lactotrophs, dopamine D_2 receptors have been identified in the intermediate lobe of the pituitary gland.[10] Consistent with the expression of these receptors in two pituitary cells types, the knockout D_2 mice showed chronic hyperprolactinemia, pituitary hyperplasia, and a moderate decrease in MSH content.[14] The pituitary dopamine receptors are functionally associated with pertussis toxin (PTX)-sensitive G proteins.[15] Two intracellular messengers affected by activation of PTX-sensitive pathways in pituitary cells, Ca^{2+} and cAMP, play major roles in controlling the fusion of secretory vesicles with the plasma membrane to release hormones in endocrine cells.[16,17] Dopamine D_{2S} and D_{2L} receptors also couple to the same extent to the PTX-sensitive $G_{i/o}$ protein and to the PTX-insensitive G_z proteins *in vitro*[18] and *in vivo*,[19] but the relevance of this coupling in pituitary hormone release has not been evaluated. The D_2 type of dopamine receptors can also exert their actions independently of G proteins by promoting the formation of a signaling protein complex composed of β-arrestin, protein kinase B (Akt), and protein phosphatase-2A.[20]

The actions of ETs in pituitary cells are mediated by ET_A receptors (ET_ARs), which are selective for ET-1 and ET-2 compared to ET-3.[21,22] Functional ET_ARs are expressed in all five secretory cell types.[23,24] In gonadotrophs, stimulation of these receptors leads to activation of the $G_{q/11}$ signaling pathway accompanied with the oscillatory Ca^{2+} release from intracellular pools and gonadotropin secretion.[25] The stimulatory action of these receptors on Ca^{2+} signaling and secretion in gonadotrophs is transient because of their rapid desensitization and internalization.[21] In somatotrophs and lactotrophs, ETs also activate the Ca^{2+} mobilization pathway and transiently stimulate hormone release.[26,27] In contrast to gonadotrophs, the stimulatory effect of ET is followed by inhibition of PRL and GH release below the basal levels.[26,28] This inhibitory phase lasts for several hours,[24,29] arguing against rapid desensitization of these receptors in lactotrophs and somatotrophs. ETRs also signal through G_s,[30] $G_{i/o}$,[31] G_z,[32] and $G_{12/13}$[33] pathways. Like dopamine D_2 receptors[12] but in contrast to over 90% of all other G protein-coupled receptors (GPCRs),[34] the ET_AR is encoded by an intron-containing gene.[35] Our recent studies addressed the role of novel spliced forms of these receptors in G protein-specific coupling.[36]

Spontaneous Electrical Activity and Basal Hormone Secretion

The membrane potential (V_m) of cultured anterior pituitary cells is not stable but oscillates from a baseline potential of about −60 mV. When V_m oscillations reach the threshold level, pituitary cells fire action potentials (APs). This is a common feature of all secretory anterior pituitary cells. However, cells differ with respect to the pattern of electrical activity and AP-driven calcium signaling and secretion *in vivo* and *in vitro*.[37–39] Cultured lactotrophs and somatotrophs frequently exhibit larger V_m oscillations, on top of which the depolarizing plateau and bursts of APs are generated, with spikes that do not usually reach the reverse potential. Such complex but organized superimposition of APs is termed *plateau bursting activity*. In contrast, V_m oscillations in cultured gonadotrophs are small and an uncharacterized pacemaking generates single APs that are short in duration and reverse polarity of the plasma membrane.[40,41] Consequently, they are termed *single axonal type APs* (Fig. 1A). Immortalized GH pituitary cells can exhibit both single AP spikes and the plateau-bursting pattern of firing.[42–44]

The differences in the patterns of spontaneous firing of APs among secretory pituitary cells are reflected in their respective pattern of calcium signaling, as demonstrated in experiments with simultaneous measurements of V_m, performed by perforated patch clamp, and intracellular calcium concentration ($[Ca^{2+}]_i$), performed by loading cells with the calcium indicator Indo-1. In somatotrophs and lactotrophs, slow baseline V_m oscillations with superimposed bursts of APs, with an average duration of seconds, are accompanied with high amplitude calcium transients that range from 0.3 μmol/L to 1.2 μmol/L (Fig. 1A). These signals are termed *global* Ca^{2+} *signals* because they reflect increases in intracellular calcium concentration ($[Ca^{2+}]_i$) in all regions of cytosolic compartments and are accompanied with high basal GH and PRL release (Fig. 1B). Conversely, the high-amplitude, millisecond-lived, single APs

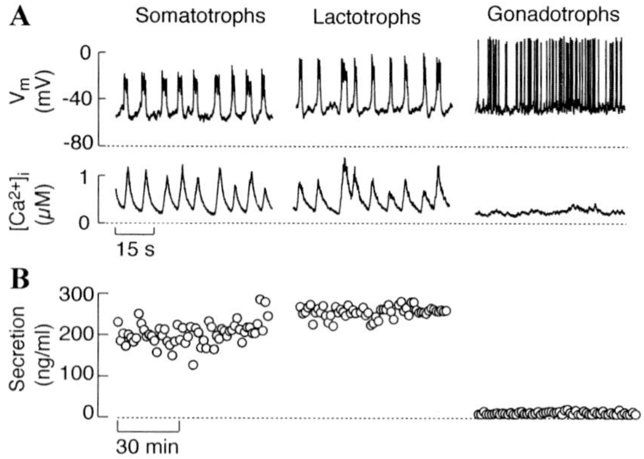

Figure 1. Characterization of spontaneous firing of action potentials (APs) and basal hormone release in pituitary cells. **(A)** Simultaneous measurements of membrane potential (V_m) and $[Ca^{2+}]_i$ in pituitary cells. **(B)** Basal hormone release in perifused pituitary cells. Basal secretion was normalized to account for the difference in the number of somatotrophs, lactotrophs, and gonadotrophs in mixed population of cells.

in spontaneously active gonadotrophs either cause no obvious changes in $[Ca^{2+}]_i$ or generate low-amplitude signals ranging from 20 to 70 nmol/L (Fig. 1A). Such APs generate localized Ca^{2+} signals below the plasma membrane, known also as *calcium domains*.[45,46] In parallel, basal LH release is low, at the level of detection by radioimmunoassay (Fig. 1B).

Pituitary gonadotrophs can also generate plateau-bursting type of electrical activity but only during the activation of Ca^{2+}-mobilizing GnRH receptors.[47,48] GnRH is associated with marked changes in the electrical activity in single gonadotrophs, including a transitory cessation of spontaneous AP firing, followed by a complex pattern of V_m oscillations in current-clamped cells. Oscillations occur over a frequency range (5–20/min) that is dependent on agonist concentration. Each cycle of V_m oscillations is initiated by rapid hyperpolarization followed by slow depolarization that often leads to the firing of multiple APs. Such a pattern of electrical activity resembles the plateau-bursting type of V_m activity in spontaneously active somatotrophs and lactotrophs.[39] At high nanomolar GnRH concentrations, a

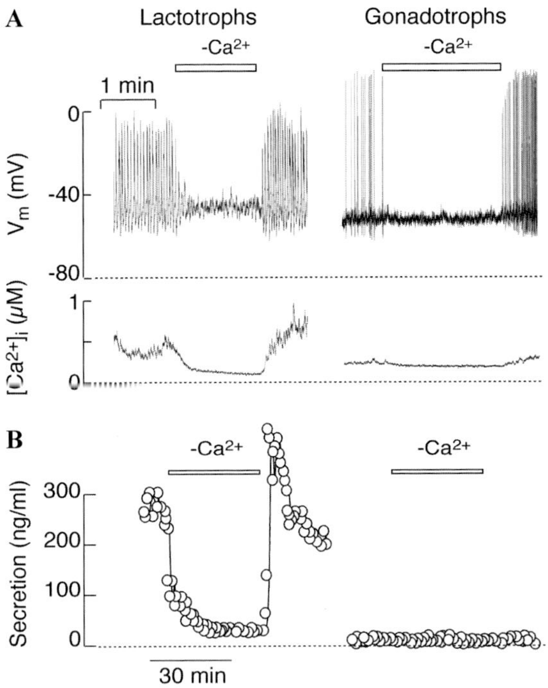

Figure 2. Extracellular calcium dependence of spontaneous electrical activity, calcium signaling, and secretion in pituitary lactotrophs (left) and gonadotrophs (right). (**A**) Effects of removal of extracellular calcium on firing of APs (top) and $[Ca^{2+}]_i$ (bottom). (**B**) Dependence of basal lactotrophs secreting prolactin (PRL) release (left) but not basal luteinizing hormone (LH) release (right) on spontaneous electrical activity. In this and the next figure, horizontal bars indicate duration of removal of extracellular calcium.

rapid and prolonged hyperpolarization of the plasma membrane is observed and is followed by sustained oscillations in V_m.[48]

Firing of spontaneous APs in all secretory cell types depends on the presence of calcium in extracellular medium.[40] Figure 2A illustrates that a transient removal of calcium leads to abolition of spiking in both lactotrophs and gonadotrophs. In lactotrophs but not in gonadotrophs this was accompanied by abolition of calcium transients and basal hormone release (Fig. 2A and B). These results indicate that the prolonged duration of the APs in lactotrophs accounts for the high amplitude $[Ca^{2+}]_i$ signals and the high level of basal PRL release. Increase in the duration of the APs in spontaneously active gonadotrophs induced by BayK 8644, an L-type calcium channel agonist, also generated global calcium signal in gonadotrophs and initiated LH secretion comparable to that observed in spontaneously firing lactotrophs.[40] Thus, basal pituitary hormone secretion is dependent on the duration of the AP wave form, which determines their capacity to drive calcium entry through voltage-gated calcium (Ca_v) channels.

Dependence of Dopamine and ET-1 Secretory Actions on Prestored PRL

In our experiments, hormone secretion was monitored using a cell column perifusion system, with 1.2–1.5 × 10^7 cells per column. Prior to any treatment, cells were perifused at a flow rate of 0.8 mL/min at 37°C for 2 h to establish stable basal secretion, followed by appropriate treatments of variable durations. Under such experimental conditions, perifusion of cells with cycloheximide, a protein synthesis inhibitor, was ineffective during the first 60 min of application, indicating that the prestored PRL predominantly accounts for AP-driven secretion during short-term experiments, whereas inhibition of *de novo* synthesis of hormone plays a major role in long-term experiments (Fig. 3A). To avoid contribution of *de novo* synthesis in secretory studies, in further experiments we focused on the short-term effects of two agonists, dopamine and ET-1, on PRL release.

Dopamine-induced inhibition of basal PRL release in perifused pituitary cells occurred rapidly, within 6–7 min of application, and the washout of agonist was accompanied with the full recovery of basal secretion (Fig. 3B). Dopamine was fully effective in 1 and 0.5 μmol/L concentrations, whereas at lower concentrations only partial inhibition was observed.[49] Perifusion experiments also revealed the presence of two phases in the action of ET-1 on PRL release: a rapid and transient increase in PRL release and a sustained decrease in secretion below the basal levels (Fig. 3C, central). The level of inhibition of PRL release during

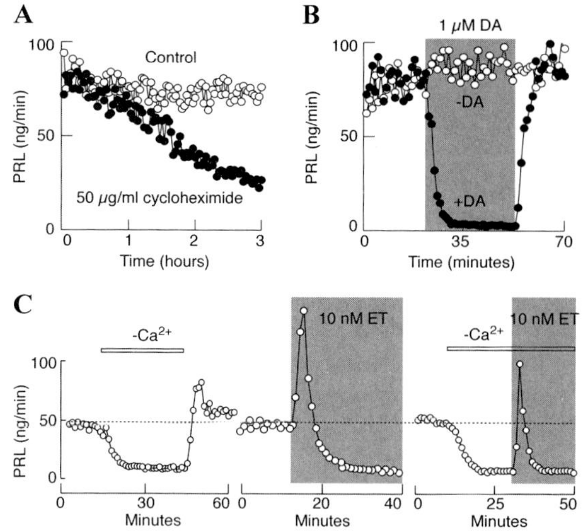

Figure 3. Dependence of dopamine and endothelin (ET)-1 actions on PRL release on prestored hormone. (**A**) Time course of cycloheximide effects on basal PRL release. (**B** and **C**) Time course of dopamine- (**B**) and ET-1- (**C**) induced inhibition of basal PRL release. Dotted line illustrates basal PRL release.

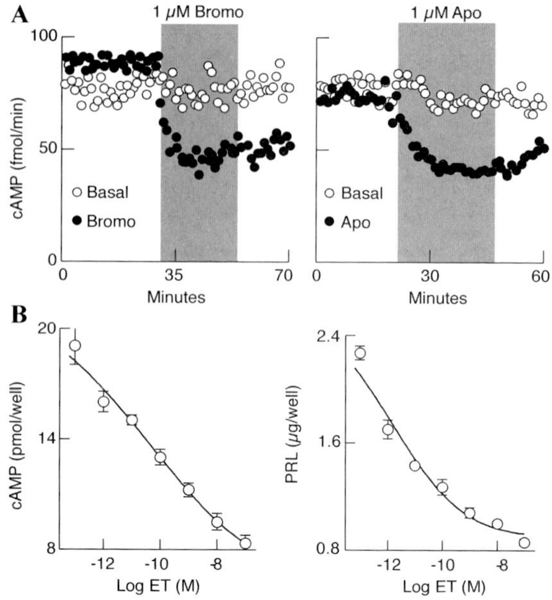

Figure 4. Inhibition of basal adenylyl cyclase activity by activation of pituitary dopamine D_2 and endothelin ET_A receptors. (**A**) Time course of bromocriptine (Bromo)- and apomorphine (Apo)-induced inhibition of cAMP release in perifused pituitary cells. (**B**) Concentration-dependent effects of ET-1 on cAMP levels (released + cell content) in pituitary cells in static culture.

the sustained agonist application was comparable to that observed in cells perifused with calcium-deficient medium (Fig. 3C, left and central). Consistent with this observation, ET-1 induced only monophasic response in cells perifused in calcium-deficient medium (Fig. 3C, right). Several experiments revealed that the extracellular calcium-independent stimulatory action of ET_A receptors on PRL release reflects coupling of these receptors to the $G_{q/11}$ signaling pathway, leading to $InsP_3$-mediated intracellular calcium mobilization and hormone secretion.[32] Like dopamine and extracellular calcium removal, ET-1-induced inhibition of PRL release required about 5–8 min to reach the steady-state level (Fig. 3B and C).

Inhibition of PRL Release in Cells with Elevated Cyclic AMP

In order to study the dependence of agonist-induced inhibition of PRL release on cAMP production, we determined both PRL and cyclic nucleotide contents in samples obtained from perifusion experiments, using the appropriate radioimmunoassay techniques. Figure 4 illustrates such an experiment. In parallel to changes in the rate of PRL release, dopamine agonists bromocriptine and apomorphine inhibited cAMP release, which was sustained after their removal. This finding is consistent with the coupling of dopamine D_2 receptors to $G_{i/o}$ and inhibition of adenylyl cyclase (AC) activity by alpha subunits.[50,51] ET also inhibited AC activity in a dose-dependent manner, with the EC_{50} value in a picomolar concentration range (Fig. 4B, left). Similar dose dependence of ET-1 was observed in studies on PRL release (Fig. 4B, right). Our recent studies also revealed that inhibition of basal cAMP production in pituitary lactotrophs does not abolish spontaneous firing of APs but partially inhibits basal PRL release,[52] a finding consistent with previous work by Zorec's group showing stimulatory effects of cAMP downstream of voltage-gated

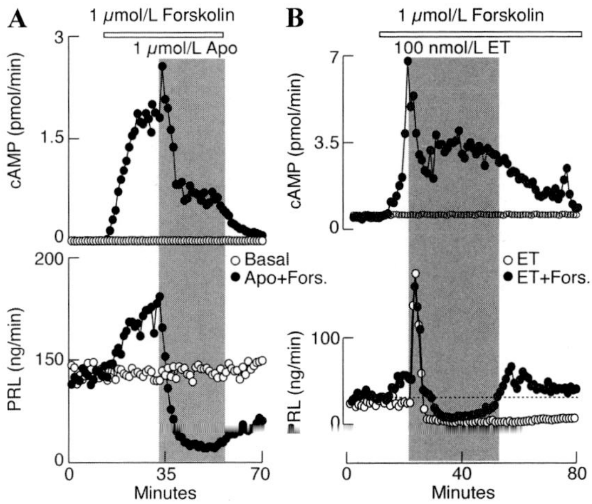

Figure 5. Agonist-mediated inhibition of basal PRL release in pituitary cells with forskolin-induced elevation of cAMP production. (**A**) Effects of forskolin and apomorphine on cAMP (top) and PRL (bottom) release. (**B**) Effects of forskolin and ET-1 on cAMP (top) and PRL (bottom) release. Horizontal bars indicate duration of forskolin treatment and gray areas represent duration of G protein-coupled receptor (GPCR) agonist application. Open circles, cells treated with agonists only.

calcium influx (VGCI).[16,17] Thus, parallelism in the actions of dopamine and ET-1 on cAMP production and PRL release raised the possibility that agonist-induced inhibition of PRL release could reflect downregulation of cAMP production.

To test this hypothesis, AC activity was elevated by challenging anterior pituitary cells with forskolin in perfusion. Figure 5A and B, top, shows that such a treatment elevates cAMP production within a few minutes of application. Measurements of PRL content in the same samples revealed that the increase in cAMP production was accompanied with stimulation of PRL release (Fig. 5A and B, bottom). Application of apomorphine and ET-1 attenuated AC activity in the presence of forskolin, but the residual cAMP production was several fold higher than in untreated cells. At such elevated cAMP levels, both agonists inhibited PRL release in a manner highly comparable to that observed in controls. Forskolin treatment also elevated $[Ca^{2+}]_i$ in a fraction of lactotrophs,[52] and both apomorphine and ET-1 abolished stimulatory effect of forskolin on Ca^{2+} transients in a majority of lactotrophs; in the absence of agonists, there was a sustained elevation in cAMP production, VGCI, and PRL release in forskolin-treated cells.[32,52] These results do not argue against the relevance of AC signaling pathway in dopamine and ET-1-regulated PRL release but demonstrate that both agonists can stop secretion through other signaling pathways as well.

Inhibition of Spontaneous VGCI by Dopamine and ET-1

Simultaneous measurements of electrical activity and intracellular calcium in perforated cells revealed that apomorphine and ET-1 hyperpolarized the membrane, leading to the abolition of AP firing and a decrease in $[Ca^{2+}]_i$ (Fig. 6A and B). Apomorphine and bromocriptine also inhibit spontaneous VGCI in intact cells loaded with Fura-2, indicating that inhibition of electrical activity in perforated cells does not reflect the side effects of the intrapipete chloride concentration.[52] In contrast to apomorphine action on $[Ca^{2+}]_i$, ET-1-induced inhibition of VGCI was preceded by a transient elevation in $[Ca^{2+}]_i$ (Fig. 6C), which was also preserved in cells bathed in calcium-deficient medium and was mediated by $InsP_3$-induced calcium release.[32] There was a strong parallelism in the actions of ET on spontaneous calcium transients and basal PRL release (Fig. 6C), consistent with the role of VGCI in basal PRL release.[40]

The role of dopamine in activation of inward rectifier potassium (K_{ir}) channels in pituitary lactotrophs is well established[53,54] and was also observed in our experimental conditions (Fig. 7A). ET-1 also increased the magnitude of the K_{ir} current in lactotrophs, which probably contributes to the sustained hyperpolarization of the cells (Fig. 7A). Although extracellular cesium in 2–5 mmol/L concentrations fully blocks K_{ir} channels in these cells,[31]

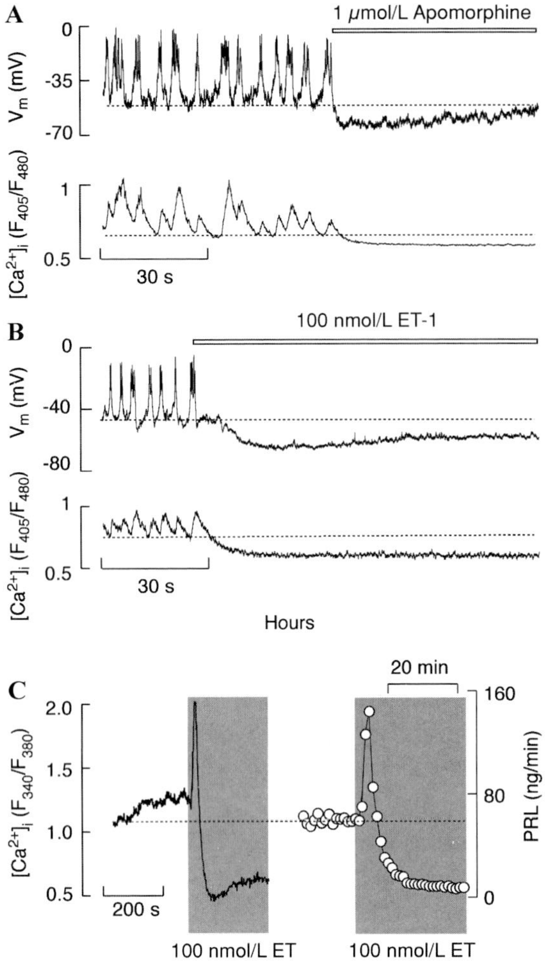

Figure 6. Agonist-induced inhibition of spontaneous electrical activity, calcium signaling, and secretion in pituitary lactotrophs. (**A** and **B**) Effects of apomorphine (**A**) and ET-1 (**B**) on spontaneous firing of APs (top) and calcium signaling (bottom) in identified lactotrophs. Horizontal bars indicate duration of agonist treatment. (**C**) Bidirectional effects of ET-1 on calcium signaling (left) and PRL release (right). Gray areas represent duration of ET-1 application.

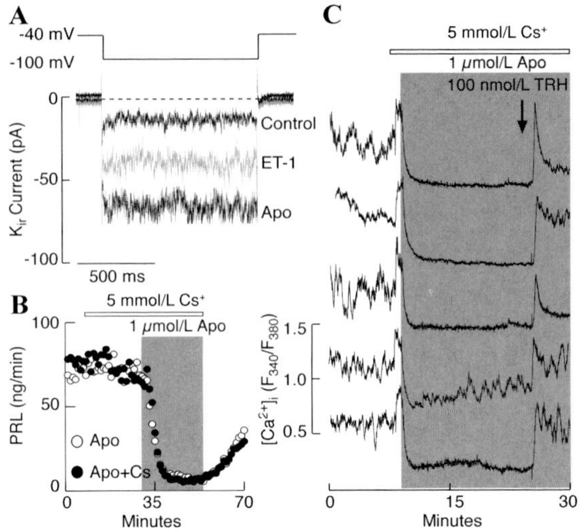

Figure 7. Role of inwardly rectifying potassium channels (K_{ir}) in spontaneous and agonist-controlled calcium signaling and secretion. (**A**) Activation of K_{ir} channels by ET-1 and apomorphine. (**B** and **C**) Apomorphine-induced inhibition of basal PRL release (**B**) and calcium transients (**C**) in cells with inhibited K_{ir} channels with cesium. Note increase in $[Ca^{2+}]_i$ after application of cesium. Horizontal bars indicate duration of application of 5 mmol/L cesium and gray areas represent duration of apomorphine treatment. Arrow indicates time of thyrotropin-releasing hormone (TRH) addition.

the addition of 5 mmol/L cesium did not affect apomorphine-induced inhibition of PRL release (Fig. 7B). In the presence of cesium, agonist-induced inhibition of AC activity is also preserved.[49] In parallel with previously published data in immortalized lactosomatotrophs,[55] cesium added extracellularly elevated $[Ca^{2+}]_i$ in single lactotrophs. Furthermore, in the presence of cesium, apomorphine inhibited spontaneous Ca^{2+} transients in all lactotrophs (Fig. 7C). The same effects were also observed in the presence of 1 mmol/L barium, another blocker of K_{ir} channels.[55] These observations are consistent with the finding that both dopamine[56,57] and ET-1[27,31] also inhibit Ca_v channels independently of the status of K_{ir} channels. Taken together, these experiments do not argue against the relevance of K_{ir} channels on dopamine and ET-1 action but show that agonist-induced inhibition of Ca_v channel gating alone is sufficient to block spontaneous Ca^{2+} influx.

Agonist-induced Inhibition of Secretion Downstream of VGCI

Earlier experiments have established that dopamine-induced inhibition of Ca_v channels[56,57] and activation of K_{ir} channels and consequent inhibition of VGCI[53] occurs

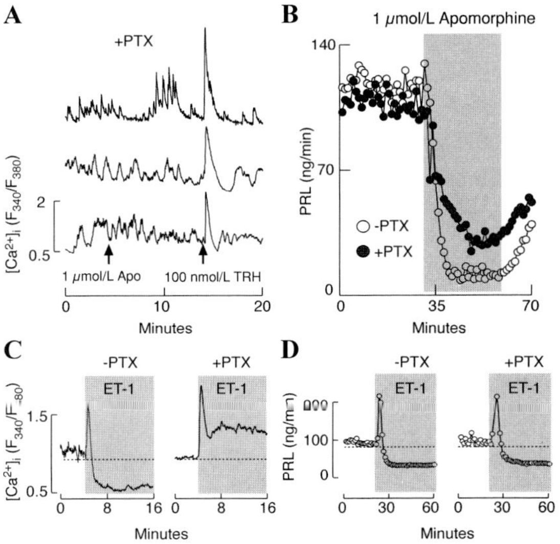

Figure 8. Dependence of agonist-induced inhibition of spontaneous calcium transients and basal PRL release on pertussis toxin (PTX)-sensitive $G_{i/o}$ signaling. (**A**) Lack of effects of apomorphine on calcium transients in anterior pituitary cells treated with **250** ng/mL PTX overnight. Arrows indicate time of apomorphine and TRH applications. (**B**) PTX-sensitive and PTX-insensitive components of apomorphine-induced inhibition of basal PRL release. (**C** and **D**) Patterns of ET-1-induced calcium signaling (**C**) and PRL secretion (**D**) in controls (left) and PTX-treated cells (right). Gray areas indicate duration of GPCR agonist treatment.

through coupling of D_2 receptors to the $G_{i/o}$ signaling pathway. Consistent with this, in cells treated with 250 ng/mL PTX overnight, the inhibitory effect of apomorphine on spontaneous Ca^{2+} transients was lost (Fig. 8A). However, the same PTX treatment could only exert a partial loss of inhibitory effects of apomorphine on PRL release (Fig. 8B). The dose-dependent studies with PTX revealed that 250 ng/mL overnight is a sufficient concentration to fully block the $G_{i/o}$ signaling pathway.[49] We also reported recently that depolarization-induced Ca^{2+} influx was reduced by apomorphine in controls but not in PTX-treated thyrotropin-releasing hormone (TRH)-responsive cells; although $[Ca^{2+}]_i$ levels were elevated in PTX-treated cells during the sustained depolarization, apomorphine effectively decreased PRL release below the basal levels.[49]

Experiments with PTX further revealed that the coupling of ET_A receptors to $G_{i/o}$ proteins provides an effective mechanism for the inhibition of spontaneous VGCI.[27,31] As shown in Figure 8C, the bidirectional response of $[Ca^{2+}]_i$ typically observed in control cells during the application of ET-1 (left) was replaced with a biphasic response in cells treated overnight with PTX, which was composed of an early spike phase and a sustained plateau phase of elevated $[Ca^{2+}]_i$ (right). In contrast to Ca^{2+} signaling, the bidirectional pattern of ET-induced PRL secretion was not obviously affected by PTX treatment (Fig. 8D, left vs. right). ET-1 also inhibited PRL release in PTX-treated static cultures of pituitary cells.[32] In parallel with the experiments described above using dopamine as an agonist, ET-1 was unable to decrease $[Ca^{2+}]_i$ in high (25 mmol/L) potassium-depolarized PTX-treated cells but inhibited PRL release to the levels observed in control cells.[32] Thus, both agonists can inhibit basal PRL release downstream of VGCI.

G Protein-independent Signaling by GPCRs and PRL Secretion

In general, dopamine and ET-1 could inhibit PRL release through G protein-dependent and independent pathways, the later being discovered recently by Caron's group.[20] The emerging components of this pathway include Akt, protein phosphatase-2A, and glycogen synthase kinase (GSK).[58] Via its pleckstrin homology domain, Akt is regulated by phosphoinositide 3 (PI3)-kinase,[59] which is effectively blocked by wortmannin. In our hands, however, wortmannin did not influence dopamine-induced inhibition of PRL release.[49] In the mouse striatum, dopamine-stimulated dephosphorylation of Akt leads to a reduction in kinase activity and activation of GSK-3, and lithium antagonizes dopamine-dependent effects mediated by Akt/GSK-3 signaling cascade.[60] In perifused pituitary cells with and without

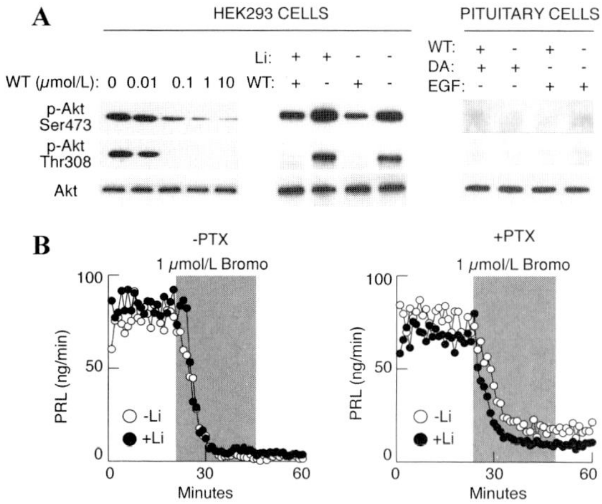

Figure 9. Independence of dopamine D_2 receptor action on PRL release of phosphoinositide 3 (PI3)-kinase and protein kinase B (Akt)/β-arrestin signaling pathways. (**A**) Western blot analysis showing that wortmanin (WT) and lithium respectively inhibit and stimulate Akt phosphorylation in HEK293 cells but not in pituitary cells. (**B**) Lack of effects of lithium, an inhibitor of glycogen synthase kinase (GSK)-3, on bromocriptine-induced inhibition of basal PRL release in anterior pituitary cells without (left) and with (right) blocked $G_{i/o}$ signaling pathway. Gray areas indicate duration of bromocriptine application.

blocked $G_{i/o}$ signaling pathway, we observed no effects of extracellular lithium on dopamine- and bromocriptine-induced inhibition of PRL release (Fig. 9B). To test the efficacy of wortmannin and lithium treatments to inhibit Akt phosphorylation, we used HEK293 cells that show constitutively high levels of Akt phosphorylation on Ser473 and Thr308. In these cells, wortmannin inhibited phosphorylation of Akt in a dose-dependent manner (Fig. 9A, left). On the other hand, lithium stimulated Akt phosphorylation (Fig. 9A, central), which is consistent with observations in other cell types.[60] Akt is also present in pituitary cells, but we could not observe the presence of phosphorylated forms of this protein, suggesting that in our experimental conditions dopamine cannot signal through the β-arrestin signaling complex by dephosphorylating Akt on Thr308.

G_z Signaling Pathway and PRL Secretion

Dopamine D_{2S} and D_{2L} receptors couple to the same extent to the PTX-sensitive $G_{i/o}$ protein and to the PTX-insensitive sister protein G_z *in vitro*[18] and *in vivo*.[19] Other subtypes of dopamine receptors also couple to G_z proteins.[18,61] Anterior pituitary cells also express $G_z\alpha$ protein,[32] and the functional coupling of endothelin ET_A receptor to this protein is suggested by the finding that ET-1-induced inhibition of AC activity is not abolished by PTX treatment.[32] To prevent the coupling of ET receptors to G_z-dependent signaling pathways, we used two experimental approaches. In the first series of experiments, cells were stimulated with the phorbol ester phorbol 12-myristate 13-acetate (PMA), which should silence the G_z signaling pathway through the protein kinase C-dependent phosphorylation of its α subunit.[62,63] Consistent with the coupling of ET_A receptors to G_z trimeric proteins, in PMA-treated cells ET-1-induced inhibition of AC was abolished as was the agonist-induced inhibition of sustained PRL release.[32] Likewise, the inhibitory effect of dopamine on PRL secretion was dramatically reduced in PMA-treated cells with operative $G_{i/o}$ signaling pathway, and agonist-induced inhibition of secretion was further reduced in PTX + PMA treated cells.[49]

To downregulate $G_z\alpha$ expression in lactotrophs, we continuously treated cells for 72 h with specific antisense phosphorothio-modified oligodeoxynucleotides, whereas control cells were treated with equivalent missense phosphorothio-modified oligodeoxynucleotides. Treatment with $G_z\alpha$-specific antisense oligodeoxynucleotides but not missense oligodeoxynucleotides substantially decreased the integrated optical density value of the $G_z\alpha$ protein band (Fig. 10A). The negative coupling of ET receptors to the AC signaling pathway was practically abolished in $G_z\alpha$-specific antisense oligodeoxynucleotide-treated cells (Fig. 10B). In the same samples, ET-induced inhibition of PRL release was substantially

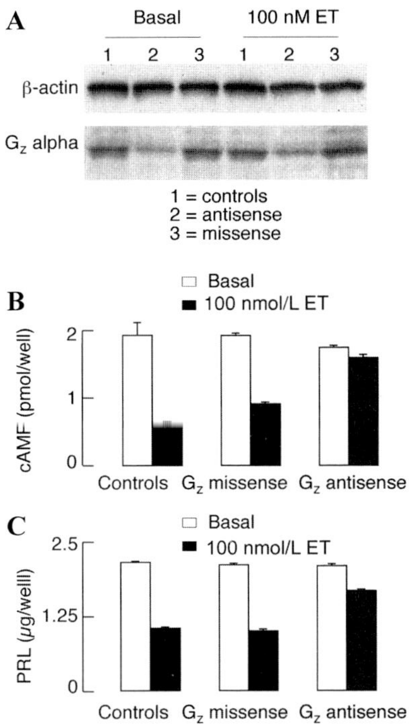

Figure 10. (**A**) Effects of downregulation of G_z-alpha protein expression in pituitary cells on (**B**) ET-1-induced cAMP signaling and (**C**) PRL release. Primary cultures of anterior pituitary cells were treated overnight with $G_z\alpha$-specific antisense and control missense phosphorothio-modified oligodeoxynucleotides prior to stimulation with 100 nmol/L ET-1. cAMP content, PRL release, and $G_z\alpha$ expression were later assayed (see text for further details).

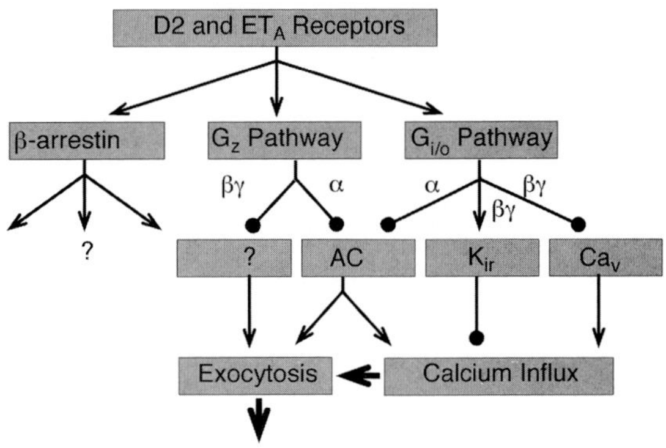

Figure 11. Activation of multiple signaling pathways by dopamine D_2 and endothelin ET_A receptors in anterior pituitary lactotrophs. Arrows indicate stimulatory actions and circles indicate inhibitory actions. AC, adenylyl cyclase; K_{ir}, inward rectifier potassium channels; Ca_v, voltage-gated calcium channels.

reduced (Fig. 10C), confirming the involvement of a G_z signaling pathway in the inhibitory actions of ET receptors on AC activity and Ca^{2+}-triggered exocytosis. Various experiments with AC activators in $G_{i/o}$-silenced cells further revealed that both arms of G_z proteins, α and βγ, participate in the ET-induced inhibition of PRL release downstream of VGCI.[32]

The contribution of $G_z\alpha$-mediated attenuation of AC activity in pituitary cells appeared not to be sufficient in dopamine-induced rapid inhibition of PRL release because such inhibition was preserved in forskolin-treated cells.[32,49] Thus, liberation of $G_z\beta\gamma$ may represent the primary factor in dopamine-induced inhibition of calcium-secretion coupling. Such a conclusion is in accordance with the finding in other cell types about the role of Gβγ subunits in control of secretion independently of their actions on AC, phospholipase C-β2, and several tyrosine kinases.[64] In fact, recent experiments by Martin's group in permeabilized PC12 cells also demonstrated that Gβγ directly regulates soluble *N*-ethylmaleimide-sensitive factor attachment protein receptor (SNARE) protein fusion machinery to modulate secretory granule exocytosis. This action occurs rapidly and affects ATP-primed secretory vesicles.[65]

Conclusions

Both dopamine D_2 and endothelin ET_A receptors activate multiple signaling pathways in lactotrophs, and such redundancy reinforces the blockade of basal PRL release (Fig. 11). Our results indicate that dopamine rapidly reduced basal cAMP production, a second messenger involved in both synthesis and release of PRL as well as in control of VGCI. Independently of the status of AC activity, however, these receptors inhibited spontaneous firing of APs and the associated VGCI in lactotrophs. Such inhibition was also observed in cells with blocked K_{ir} channels, suggesting that inhibition of Ca_v channels *per se* by these agonists is enough to stop spontaneous calcium influx. Dopamine-induced inhibition of AC and electrical activity required the coupling of receptors to PTX-sensitive $G_{i/o}$ proteins. In contrast, the coupling of dopamine receptors to both PTX-sensitive and PTX-insensitive class of G proteins and/or to other signaling pathways was required for full inhibition of AC activity and PRL release. The presence of the PTX-insensitive step in dopamine action on PRL release does not argue against the relevance of AC, K_{ir} channels, and Ca_v channels in this process *in vivo* but demonstrates that pharmacological exclusion of a single pathway does not provide a valid experimental model for a system simultaneously controlled by multiple pathways. *In vitro*, dopamine-mediated control of VGCI is more than sufficient to block PRL release, as was shown in experiments with removal of extracellular calcium.[40,66] The triple control of VGCI by inhibiting AC and Ca_v and activating K_{ir} channels combined with the mechanism for desensitization of calcium-secretion coupling not only secures inhibition of PRL release but also provides the possibility for maintaining such inhibition when cells are exposed to multiple agonists. For example, the G_z signaling pathway probably contributes to the control of PRL release by dopamine when the electrical activity in lactotrophs is stimulated by other factors. Based on the experimental observations outlined above with ET-1, an elegant theoretical model was developed to describe the complex signaling pathway in these cells, with the main idea that one receptor is coupled to many G proteins.[67] More recent findings further suggested the existence of functional splice forms of ET_A receptors, which could add to the remarkable complexity of G protein-specific coupling.[36]

Acknowledgments

This research was supported by the Intramural Research Program of the National Institute of Child Health and Human Development, National Institutes of Health.

Conflicts of Interest

The authors declare no conflicts of interest.

References

1. Zhu, X. *et al.* 2007. Signaling and epigenetic regulation of pituitary development. *Curr. Opin. Cell. Biol.* **19:** 605–611.
2. Ben-Jonathan, N. & R. Hnasko. 2001. Dopamine as a prolactin (PRL) inhibitor. *Endocr. Rev.* **22:** 724–763.
3. Stenovec, M. *et al.* 2004. Slow spontaneous secretion from single large dense-core vesicles monitored in neuroendocrine cells. *FASEB J.* **18:** 1270–1272.
4. Vardjan, N. *et al.* 2007. Subnanometer fusion pores in spontaneous exocytosis of peptidergic vesicles. *J. Neurosci.* **27:** 4737–4746.
5. Stojilkovic, S.S., S. Izumi & K.J. Catt. 1988. Participation of voltage-sensitive calcium channels in pituitary hormone release. *J. Biol. Chem.* **263:** 13054–13061.
6. Mayo, K.E. *et al.* 1995. Growth hormone-releasing hormone: synthesis and signaling. *Recent Prog. Horm. Res.* **50:** 35–73.
7. Freeman, M.E. *et al.* 2000. Prolactin: structure, function, and regulation of secretion. *Physiol. Rev.* **80:** 1523–1631.
8. Stojilkovic, S.S., J. Reinhart & K.J. Catt. 1994. Gonadotropin-releasing hormone receptors: structure and signal transduction pathways. *Endocr. Rev.* **15:** 462–499.
9. Balik, A. *et al.* 2004. Melatonin action in neonatal gonadotrophs. *Physiol. Res.* **53**(Suppl 1): S153–166.
10. Missale, C. *et al.* 1998. Dopamine receptors: from structure to function. *Physiol. Rev.* **78:** 189–225.
11. Giros, B. *et al.* 1989. Alternative splicing directs the expression of two D2 dopamine receptor isoforms. *Nature* **342:** 923–926.
12. Monsma, F.J., Jr. *et al.* 1989. Multiple D2 dopamine receptors produced by alternative RNA splicing. *Nature* **342:** 926–929.
13. Kukstas, L.A. *et al.* 1991. Different expression of the two dopaminergic D2 receptors, D2415 and D2444, in two types of lactotroph each characterised by their response to dopamine, and modification of expression by sex steroids. *Endocrinology* **129:** 1101–1103.
14. Kelly, M.A. *et al.* 1997. Pituitary lactotroph hyperplasia and chronic hyperprolactinemia in dopamine D2 receptor-deficient mice. *Neuron* **19:** 103–113.
15. Senogles, S.E. *et al.* 1987. The D2-dopamine receptor of anterior pituitary is functionally associated with a pertussis toxin-sensitive guanine nucleotide binding protein. *J. Biol. Chem.* **262:** 4860–4867.

16. Zorec, R., S.K. Sikdar & W.T. Mason. 1991. Increased cytosolic calcium stimulates exocytosis in bovine lactotrophs. Direct evidence from changes in membrane capacitance. *J. Gen. Physiol.* **97:** 473–497.
17. Sikdar, S.K., M. Kreft & R. Zorec. 1998. Modulation of the unitary exocytic event amplitude by cAMP in rat melanotrophs. *J. Physiol.* **511**(Pt 3): 851–859.
18. Obadiah, J. *et al.* 1999. Adenylyl cyclase interaction with the D2 dopamine receptor family; differential coupling to Gi, Gz, and Gs. *Cell. Mol. Neurobiol.* **19:** 653–664.
19. Leck, K.J. *et al.* 2006. Gz proteins are functionally coupled to dopamine D2-like receptors in vivo. *Neuropharmacology* **51:** 597–605.
20. Beaulieu, J.M. *et al.* 2005. An Akt/beta-arrestin 2/PP2A signaling complex mediates dopaminergic neurotransmission and behavior. *Cell* **122:** 261–273.
21. Stojilkovic, S.S. *et al.* 1992. Endothelin ETA receptors mediate the signaling and secretory actions of endothelins in pituitary gonadotrophs. *Endocrinology* **130:** 465–474.
22. Samson, W.K. 1992. The endothelin-A receptor subtype transduces the effects of the endothelins in the anterior pituitary gland. *Biochem. Biophys. Res. Commun.* **187:** 590–595.
23. Stojilkovic, S.S. *et al.* 1991. Calcium signaling and secretory responses in endothelin-stimulated anterior pituitary cells. *Mol. Pharmacol.* **39:** 762–770.
24. Kanyicska, B., T.P. Burris & M.E. Freeman. 1991. Endothelin-3 inhibits prolactin and stimulates LH, FSH and TSH secretion from pituitary cell culture. *Biochem. Biophys. Res. Commun.* **174:** 338–343.
25. Stojilkovic, S.S. *et al.* 1990. Endothelin stimulation of cytosolic calcium and gonadotropin secretion in anterior pituitary cells. *Science* **248:** 1663–1666.
26. Lachowicz, A. *et al.* 1997. Uncoupling of calcium mobilization and entry pathways in endothelin-stimulated pituitary lactotrophs. *J. Biol. Chem.* **272:** 28308–28314.
27. Tomic, M. *et al.* 1999. Expression of Ca(2+)-mobilizing endothelin(A) receptors and their role in the control of Ca(2^+) influx and growth hormone secretion in pituitary somatotrophs. *J. Neurosci.* **19:** 7721–7731.
28. Kanyicska, B., J.D. Livingstone & M.E. Freeman. 1995. Long term exposure to dopamine reverses the inhibitory effect of endothelin-1 on prolactin secretion. *Endocrinology* **136:** 990–994.
29. Samson, W.K. *et al.* 1990. Pituitary site of action of endothelin: selective inhibition of prolactin release in vitro. *Biochem. Biophys. Res. Commun.* **169:** 737–743.
30. Rebsamen, M.C. *et al.* 1997. Role of cAMP and calcium influx in endothelin-1-induced ANP release in rat cardiomyocytes. *Am. J. Physiol.* **273:** E922–931.
31. Tomic, M. *et al.* 2002. Ca(2^+)-mobilizing endothelin-A receptors inhibit voltage-gated Ca(2^+) influx through G(i/o) signaling pathway in pituitary lactotrophs. *Mol. Pharmacol.* **61:** 1329–1339.
32. Andric, S.A. *et al.* 2005. Endothelin-induced, long lasting, and $Ca2^+$ influx-independent blockade of intrinsic secretion in pituitary cells by Gz subunits. *J. Biol. Chem.* **280:** 26896–26903.
33. Gohla, A., G. Schultz & S. Offermanns. 2000. Role for G(12)/G(13) in agonist-induced vascular smooth muscle cell contraction. *Circ. Res.* **87:** 221–227.
34. Gentles, A.J. & S. Karlin. 1999. Why are human G-protein-coupled receptors predominantly intronless? *Trends Genet.* **15:** 47–49.
35. Miyamoto, Y. *et al.* 1996. Alternative RNA splicing of the human endothelin-A receptor generates multiple transcripts. *Biochem. J.* **313**(Pt 3): 795–801.
36. Hatae, N. *et al.* 2007. Cloning and functional identification of novel endothelin receptor type A isoforms in pituitary. *Mol. Endocrinol.* **21:** 1192–1204.
37. Kwiecien, R. & C. Hammond. 1998. Differential management of $Ca2^+$oscillations by anterior pituitary cells: a comparative overview. *Neuroendocrinology* **68:** 135–151.
38. Stojilkovic, S.S., H. Zemkova & F. Van Goor. 2005. Biophysical basis of pituitary cell type-specific $Ca2^+$ signaling-secretion coupling. *Trends Endocrinol. Metab.* **16:** 152–159.
39. Stojilkovic, S.S. 2006. Pituitary cell type-specific electrical activity, calcium signaling and secretion. *Biol. Res.* **39:** 403–423.
40. Van Goor, F. *et al.* 2001. Dependence of pituitary hormone secretion on the pattern of spontaneous voltage-gated calcium influx. Cell type-specific action potential secretion coupling. *J. Biol. Chem.* **276:** 33840–33846.
41. Van Goor, F., Y.X. Li & S.S. Stojilkovic. 2001. Paradoxical role of large-conductance calcium-activated K+ (BK) channels in controlling action potential-driven $Ca2^+$ entry in anterior pituitary cells. *J. Neurosci.* **21:** 5902–5915.
42. Schlegel, W. *et al.* 1987. Oscillations of cytosolic Ca2+ in pituitary cells due to action potentials. *Nature* **329:** 719–721.
43. Kwiecien, R. *et al.* 1998. Endogenous pacemaker activity of rat tumour somatotrophs. *J. Physiol.* **508**(Pt 3): 883–905.
44. Giraldez, T. *et al.* 2002. Correlation between electrical activity and intracellular $Ca2^+$ oscillations in GH3 rat anterior pituitary cells. *Cell Calcium* **31:** 65–78.
45. Li, Y.X. *et al.* 1995. Spontaneous electrical and calcium oscillations in unstimulated pituitary gonadotrophs. *Biophys. J.* **69:** 785–795.

46. Li, Y.X. *et al.* 1997. Sensing and refilling calcium stores in an excitable cell. *Biophys. J.* **72:** 1080–1091.
47. Tse, A. & B. Hille. 1992. GnRH-induced $Ca2^+$ oscillations and rhythmic hyperpolarizations of pituitary gonadotropes. *Science* **255:** 462–464.
48. Kukuljan, M. *et al.* 1992. Apamin-sensitive potassium channels mediate agonist-induced oscillations of membrane potential in pituitary gonadotrophs. *FEBS Lett.* **301:** 19–22.
49. Gonzalez-Iglesias, A.E. *et al.* 2008. Dopamine inhibits basal prolactin release in pituitary lactotrophs through pertussis toxin-sensitive and -insensitive signaling pathways. *Endocrinology* **149:** 1470–1479.
50. Enjalbert, A. & J. Bockaert. 1983. Pharmacological characterization of the D2 dopamine receptor negatively coupled with adenylate cyclase in rat anterior pituitary. *Mol. Pharmacol.* **23:** 576–584.
51. McDonald, W.M. *et al.* 1984. Dopaminergic inhibition of adenylate cyclase correlates with high affinity agonist binding to anterior pituitary D2 dopamine receptors. *Mol. Cell. Endocrinol.* **36:** 201–209.
52. Gonzalez-Iglesias, A.E. *et al.* 2006. Dependence of electrical activity and calcium influx-controlled prolactin release on adenylyl cyclase signaling pathway in pituitary lactotrophs. *Mol. Endocrinol.* **20:** 2231–2246.
53. Einhorn, L.C., K.A. Gregerson & G.S. Oxford. 1991. D2 dopamine receptor activation of potassium channels in identified rat lactotrophs: whole-cell and single-channel recording. *J. Neurosci.* **11:** 3727–3737.
54. Gregerson, K.A. *et al.* 2001. Identification of G protein-coupled, inward rectifier potassium channel gene products from the rat anterior pituitary gland. *Endocrinology* **142:** 2820–2832.
55. Charles, A.C. *et al.* 1999. L-type $Ca2^+$ channels and K+ channels specifically modulate the frequency and amplitude of spontaneous $Ca2^+$ oscillations and have distinct roles in prolactin release in GH3 cells. *J. Biol. Chem.* **274:** 7508–7515.
56. Lledo, P.M. *et al.* 1990. Dopamine inhibits two characterized voltage-dependent calcium currents in identified rat lactotroph cells. *Endocrinology* **127:** 990–1001.
57. Malgaroli, A. *et al.* 1987. Dopamine inhibits cytosolic $Ca2^+$ increases in rat lactotroph cells. Evidence of a dual mechanism of action. *J. Biol. Chem.* **262:** 13920–13927.
58. Bibb, J.A. 2005. Decoding dopamine signaling. *Cell* **122:** 153–155.
59. Burgering, B.M. & P.J. Coffer. 1995. Protein kinase B (c-Akt) in phosphatidylinositol-3-OH kinase signal transduction. *Nature* **376:** 599–602.
60. Beaulieu, J.M. *et al.* 2004. Lithium antagonizes dopamine-dependent behaviors mediated by an AKT/glycogen synthase kinase 3 signaling cascade. *Proc. Natl. Acad. Sci. USA* **101:** 5099–5104.
61. Sidhu, A. *et al.* 1998. Multiple coupling of human D5 dopamine receptors to guanine nucleotide binding proteins Gs and Gz. *J. Neurochem.* **70:** 2459–2467.
62. Fields, T.A. & P.J. Casey. 1995. Phosphorylation of Gz alpha by protein kinase C blocks interaction with the beta gamma complex. *J. Biol. Chem.* **270:** 23119–23125.
63. Wang, J. *et al.* 1999. Reciprocal signaling between heterotrimeric G proteins and the p21-stimulated protein kinase. *J. Biol. Chem.* **274:** 31641–31647.
64. Blackmer, T. *et al.* 2001. G protein betagamma subunit-mediated presynaptic inhibition: regulation of exocytotic fusion downstream of $Ca2^+$ entry. *Science* **292:** 293–297.
65. Blackmer, T. *et al.* 2005. G protein betagamma directly regulates SNARE protein fusion machinery for secretory granule exocytosis. *Nat. Neurosci.* **8:** 421–425.
66. Lamberts, S.W. & R.M. Macleod. 1990. Regulation of prolactin secretion at the level of the lactotroph. *Physiol. Rev.* **70:** 279–318.
67. Bertram, R. *et al.* 2006. Endothelin action on pituitary lactotrophs: one receptor, many GTP-binding proteins. *Sci STKE* **2006:** er2.

Exocytotic Properties of Human Pancreatic β-cells

Matthias Braun,[a] Reshma Ramracheya,[a] Paul R. Johnson,[b] and Patrik Rorsman[a]

[a] *Oxford Centre for Diabetes Endocrinology and Metabolism, University of Oxford, Churchill Hospital, Oxford, United Kingdom*

[b] *Nuffield Department of Surgery, John Radcliffe Hospital, Oxford, United Kingdom*

Pancreatic β-cells secrete insulin in response to elevated blood glucose via Ca^{2+}-dependent fusion of secretory granules with the plasma membrane (regulated exocytosis). While exocytosis has been extensively investigated in rodent β-cells, studies on human β-cells are scarce. We have characterized the exocytotic properties of human β-cells by insulin release measurements, carbon fiber amperometry, and capacitance measurements using the patch-clamp technique. Voltage-clamp depolarizations evoked capacitance increases in single β-cells in a time- and voltage-dependent manner. The capacitance responses as well as insulin release from intact islets were strongly amplified by elevation of intracellular cAMP levels. Exocytosis was more dependent on Ca^{2+} influx through P/Q-type than L-type Ca^{2+} channels, reflecting the relative contribution of these channels to the total Ca^{2+} current. Exocytosis (as monitored by capacitance or amperometric measurements) decreased during repetitive stimulation as a result of inactivation of Ca^{2+} channels as well as depletion of a readily releasable pool of granules. These results reveal both similarities and differences between human and rodent β-cells.

Key words: **β-cells; capacitance; exocytosis; human; insulin**

Introduction

Insulin, the body's only blood-glucose-lowering hormone, is produced by the β-cells in pancreatic islets of Langerhans. β-cells are of considerable clinical importance as the failure of these cells to secrete adequate amounts of insulin leads to type 2 diabetes.[1] The importance of insufficient insulin secretion (as opposed to insulin resistance of insulin-sensitive tissues) for the development of the disease is underscored by the fact that almost all type 2 diabetes-related genes discovered so far are expressed in the pancreas.[2]

Studies on rodent islets have shown that glucose stimulates insulin secretion from β-cells by raising the intracellular ATP concentration at the expense of ADP with subsequent closure of ATP-sensitive K^+ channels in the plasma membrane. The resulting membrane depolarization leads to electrical activity and Ca^{2+} influx through voltage-gated Ca^{2+} channels, which in turn trigger the exocytosis of insulin-containing, large, dense-core vesicles (LDCVs).[3] Synaptotagmins 7 and 9 have been identified as the putative Ca^{2+} sensors of insulin exocytosis.[4,5] Glucose-induced secretion is modulated by a variety of hormones (e.g., glucagon-like peptide [GLP]-1, glucagon, somatostatin) and neurotransmitters (e.g., acetylcholine). Many of these factors act at a late stage of the stimulus-secretion coupling via protein kinase A (PKA)- and PKC-dependent pathways, although the molecular targets of these pathways are still largely unknown. There is some evidence, however, indicating that phosphorylation/dephosphorylation of SNAP25 (as

Address for correspondence: Matthias Braun, Oxford Centre for Diabetes Endocrinology and Metabolism, University of Oxford, Churchill Hospital, Oxford OX3 7LJ, UK. Voice: +44-1865-857072; fax: +44-1865-857074. matthias.braun@drl.ox.ac.uk

Mechanisms of Exocytosis: Ann. N.Y. Acad. Sci. 1152: 187–193 (2009).
doi: 10.1111/j.1749-6632.2008.03992.x

previously documented for chromaffin cells) may be of significance.

In isolated single β-cells, insulin exocytosis has been extensively characterized by membrane capacitance measurements using the patch-clamp technique. This method reports changes in cell surface area from fusion / fission of secretory granules. While β-cells also contain small synaptic-like microvesicles capable of undergoing Ca^{2+}-dependent exocytosis, these appear to contribute little to the capacitance signal under most experimental conditions.[4] Similar to neurons and other endocrine cell types, insulin-containing secretory granules in β-cells can be functionally subdivided into distinct pools with different release probabilities. A limited pool of readily releasable granules (RRP) undergoes exocytosis rapidly upon stimulation and is more slowly replenished from a large reserve pool of granules. Transition from the reserve pool to the RRP is ATP dependent and may involve docking of granules to the plasma membrane.[6,7] It has been proposed that exocytosis of the RRP of insulin granules underlies the first phase of glucose-induced insulin secretion observed both *in vivo* and *in vitro* in isolated islets.[8,9] As first phase insulin secretion is often selectively disturbed in type-2 diabetes, the characterization of its cellular basis is important.

Because of the limited availability of human tissue, β-cell exocytosis has so far been studied almost exclusively in rodent preparations or clonal cells. Here we summarize some recently published data on exocytosis in human β-cells and present new results relating to the modulation of depolarization-evoked exocytosis and insulin secretion.

Materials and Methods

Islet Isolation and Cell Culture

Human islets were, with ethical consent, isolated from the pancreas of nondiabetic heart-beating donors in the Oxford Diabetes Research and Wellness Foundation Human Islet Isolation Facility. Islets were dispersed into single cells by incubation in Ca^{2+}-free medium upon arrival and cultured in RPMI medium containing 10 mmol/l glucose for 1–5 days prior to the experiments.

Hormone Secretion

Batches of 10–20 islets were pre-incubated in 1 mL Krebs–Ringer buffer (KRB) containing 1 mmol/l glucose for 1 h followed by 1-h incubation in 1 mL of test KRB supplemented as indicated.

Electrophysiology

Patch-clamp experiments were performed using an EPC-9 amplifier and Pulse software (HEKA Electronics, Lambrecht, Germany). All experiments were performed in the standard whole-cell configuration at approximately 32°C. Exocytosis was detected as changes in cell capacitance, which was estimated by the Lindau–Neher technique, implementing the "Sine + dc" feature of the Pulse lock-in module. The bath solution was composed of (mmol/L) 128 NaCl, 10 TEACl, 5.6 KCl, 2.6 $CaCl_2$, 1.2 $MgCl_2$, 5 HEPES, and 5 glucose (pH 7.4, adjusted with NaOH). The intracellular solution contained (mmol/L) 125 Cs-glutamate, 10 CsCl, 10 NaCl, 1 $MgCl_2$, 5 HEPES, 0.05 EGTA, 3 MgATP, 0.5 mg/mL biocytin, and (unless indicated otherwise) 0.1 cAMP (pH 7.2, adjusted with CsOH). The identity of the β-cells was confirmed after the experiments by immunocytochemistry.[10]

Amperometry

β-cells were cultured with 0.5 mmol/L serotonin and 0.5 mmol/L 5-hydroxytryptophan in the medium for 6–20 h prior to the experiments. Serotonin release was detected using a carbon fiber electrode (ProCFE, Dagan, Minneapolis, MN) held at 650 mV and positioned within approximately 1-(micro-)m distance of the cell.

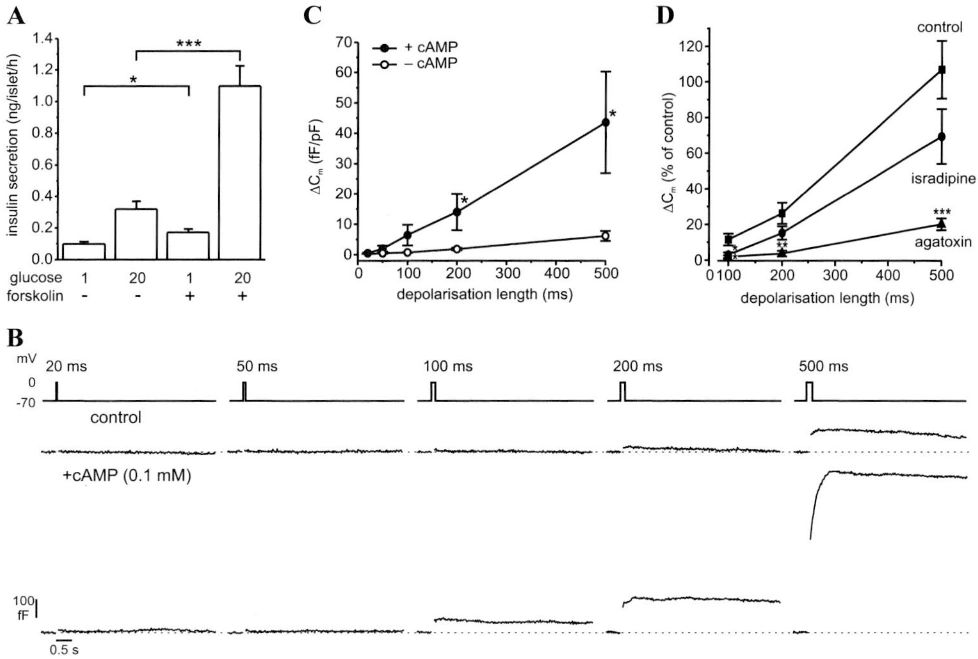

Figure 1. cAMP- and Ca^{2+}-channel dependence of insulin secretion and exocytosis from human β-cells. (**A**) Insulin secretion was measured from human islets at 1 or 20 (micro-)mol/L glucose and in the absence or presence of 10 mmol/L forskolin as indicated ($^{*}P < 0.05$, $^{***}P < 0.001$, $n = 12$). (**B**) Capacitance traces recorded in individual human β-cells stimulated by voltage-clamp depolarizations of increasing duration (20–500 ms) at 20-s intervals (top) in the absence (middle) or presence (bottom) of 0.1 mmol/L cAMP in the intracellular solution. (**C**) Relationship between depolarization length and capacitance increase (ΔC_m, normalized to the cell size) in the absence (open circles) and presence (filled circles) of 0.1 mmol/L cAMP, summary of 8–17 experiments performed as in (B) ($^{*}P < 0.05$). (**D**) ΔC_m plotted against depolarization length under control conditions (squares; $n = 8$) and in the presence of 10 (micro-)mol/L isradipine (circles; $n = 6$) or 200 nmol/L ω-agatoxin IVA (triangles; $n = 5$). Four series of depolarizations of increasing length were applied to the same cells at 70-s intervals. Where applicable, inhibitors were added after completion of the 2nd series. The responses to the 4th depolarization series, expressed in percent (%) of the response to the 500-ms depolarization in the 2nd series are displayed. Only cells with a capacitance increase of ≥100 fF to control 500-ms depolarizations were included in the analysis ($^{*}P < 0.05$; $^{**}P < 0.01$; $^{***}P < 0.001$).

Results and Discussion

It has been previously demonstrated that human β-cells respond to voltage-clamp depolarizations (triggering Ca^{2+} influx through voltage-gated Ca^{2+} channels) with increases in membrane capacitance.[11] This is accompanied by release of serotonin preloaded into insulin granules. We have recently shown that robust capacitance increases are elicited in single cells by depolarization ≥100 ms and that the capacitance response is proportional to the 4th order of Ca^{2+} entry.[10] Significant responses were evoked only by depolarizations positive to −20 mV and were maximal at 0 mV. This voltage dependence most closely resembled that of the P/Q-type Ca^{2+} channels. Unlike what is normally observed in mouse β-cells, membrane capacitance continued to increase after the end of the depolarization pulse in many cells (see also Fig. 1B). Such a slow component has also been recorded from somatostatin-secreting δ-cells in the mouse where it reflects exocytosis triggered by Ca^{2+}-induced Ca^{2+} release.[12] Indeed, ryanodine receptors have been found in human β-cells, and low (activating) concentrations of ryanodine stimulate insulin secretion from human islets.[13]

In rodent β-cells, hormones, such as glucagon and GLP-1, stimulate glucose-dependent insulin secretion by activating adenylate cyclase with subsequent elevation of intracellular cAMP levels. We studied the effect of the adenylate cyclase activator forskolin on insulin release from isolated human islets. At basal (1 mmol/L) glucose levels, forskolin (10 (micro-)mol/L) induced a small (less than twofold) but significant increase of insulin release (Fig. 1A). Insulin secretion was stimulated approximately threefold by raising the glucose concentration from 1 to 20 mmol/L, an effect potentiated greater than threefold by inclusion of forskolin in the incubation medium (Fig. 1A). This is in agreement with data obtained from perifused human islets.[14]

In mouse β-cells, elevation of intracellular cAMP potentiates secretion principally by a direct effect on the exocytotic machinery and to a lesser extent by increasing Ca^{2+} influx through voltage-gated Ca^{2+} channels.[15] To characterize the mechanisms by which cAMP elevation stimulates secretion in human islets, we monitored exocytosis in single human β-cells by capacitance measurements. In the absence of cAMP, even prolonged (500 ms) depolarizations evoked only small increases in membrane capacitance (Fig. 1B-C). When cAMP (0.1 mmol/L) was included in the intracellular solution, short (20–50-ms) depolarizations were still ineffective at triggering exocytosis whereas the responses to longer (200–500-ms) depolarizations were substantially amplified (Fig. 1B-C). On average, the capacitance response to 500-ms depolarizations increased from 5.4 ± 1.4 fF/pF in the absence to 36.2 ±7.1 fF/pF in the presence of cAMP ($P < 0.001$, $n = 23$–24). This was accompanied by only a 25% increment in the Ca^{2+} current (from 0.23 ± 0.03 to 0.29 ± 0.03 pC/pF, not significantly different; measured during 40-ms depolarizations to 0 mV). In mouse β-cells, cAMP stimulates Ca^{2+}-dependent exocytosis by PKA-dependent and -independent pathways, the latter involving the cAMP–GEF/Epac family of cAMP-binding proteins.[16] The significance of these pathways for insulin exocytosis from human β-cells remains to be characterized.

Human β-cells express three types of voltage-gated Ca^{2+} channels: L-type, P/Q-type, and T-type channels.[10] T-type channels inactivate rapidly and therefore contribute little to the total Ca^{2+} influx during depolarizations of >50-ms duration. We studied the relative contribution of Ca^{2+} influx through L- and P/Q-type channels to depolarization-evoked exocytosis using the specific blockers isradipine and ω-agatoxin IVA, respectively. Both compounds effectively blocked the capacitance response to shorter (100-ms) depolarizations (Fig. 1D). During longer depolarizations, isradipine had only a minor inhibitory effect while ω-agatoxin remained an efficient antagonist (see also Ref. 10). L- and P/Q-type channels carry 38% and 46%, respectively, of the Ca^{2+} influx during 100-ms depolarizations.[10] In contrast to L-type channels, P/Q-type channels show little inactivation during this time period and their relative contribution to the total Ca^{2+} influx will thus increase with depolarization length. We estimated that P/Q-type channels will be responsible to >65% of the current during 500-ms pulses (not shown). As exocytosis increases with the 4th order of Ca^{2+} influx (see Ref. 10 and Fig. 2D, below), a reduction of Ca^{2+} influx by 2/3 can be expected to decrease exocytosis by >80%. Therefore, in isolated β-cells P/Q- and L-type channels seem to contribute to depolarization-evoked exocytosis proportionally to their contribution to the total Ca^{2+} current. However, preliminary data from β-cells within intact human islets have revealed a prominent and ultrafast (maximal within 20–50-ms) component of exocytosis that remains sensitive to ω-agatoxin (not shown). As this is too short for Ca^{2+} to equilibrate in the cell, these findings argue that P/Q-type Ca^{2+} channels may assemble into functional complexes with the secretory granules in a way reminiscent of that previously documented in mouse β-cells for L-type channels.[17]

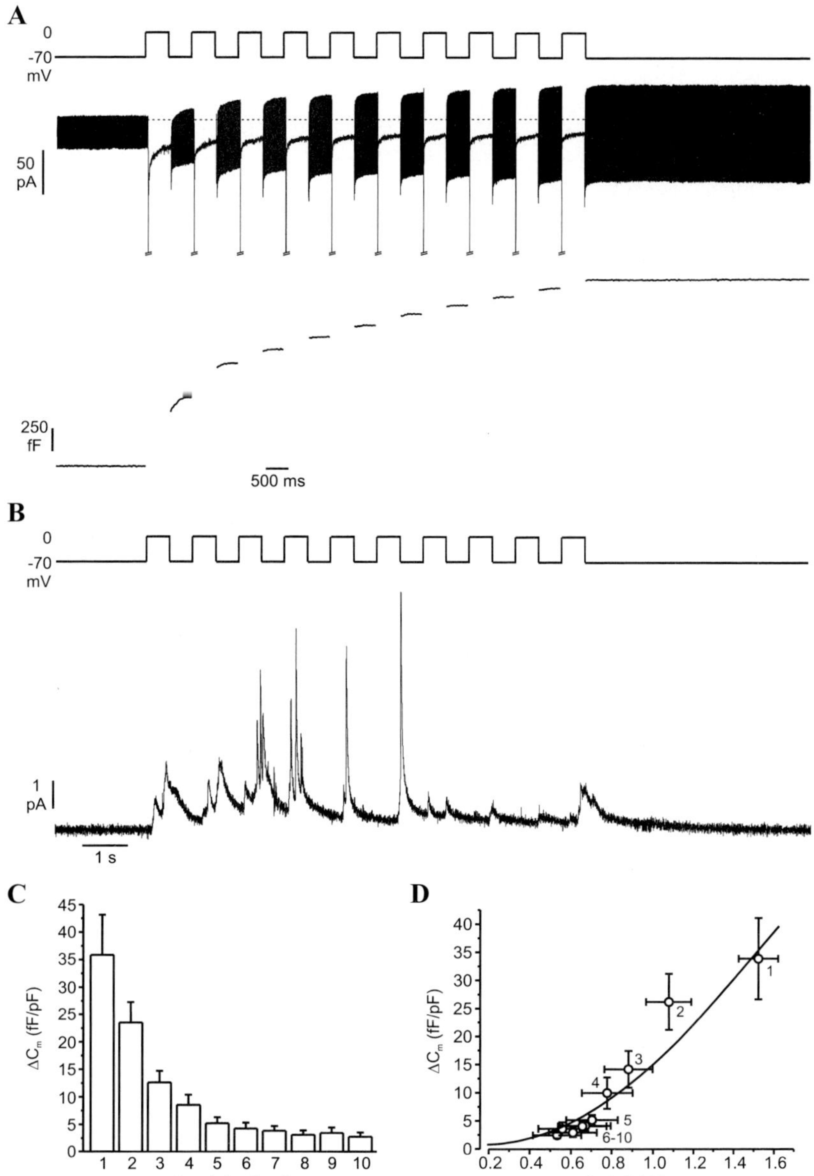

Figure 2. Exocytosis during repetitive stimulation. (**A**) Current response (middle) and capacitance increase (bottom) elicited by a train of ten 500-ms depolarizations (to 0 mV, 1 Hz) (top). Inward Na^+ currents have been truncated for display purposes. (**B**) Amperometric current recorded from a serotonin-loaded β-cell stimulated by a depolarization train as in (A). The recording shown is representative of five stimulations in three cells. (**C**) Capacitance increase (ΔC_m) per depolarization displayed against depolarization number during the train. Data are mean values ± SEM of 18 experiments. (**D**) Relationship between Ca^{2+} current (Q_{Ca}) and ΔC_m, both normalized to cell capacitance, for individual pulses during depolarization trains. Data were obtained from 11 different cells. The numbers next to the symbols indicate the number of the depolarization pulse. The curve was previously obtained by fitting data from experiments using depolarizations of increasing duration (20–500 ms) with a 4th order polynomial function.[10]

We also studied the capacitance response of human β-cells to repetitive stimulation. The cells were subjected to trains of ten 500-ms depolarizations to 0 mV (at 1 Hz, with cAMP in the pipette solution). A representative recording is shown in Figure 2A, and data from 18 cells are summarized in Figure 2C. The capacitance increase was maximal after the first depolarization and then gradually declined to a steady state during the 6th to 10th pulse. A similar pattern was observed when exocytosis was monitored by amperometry (Fig. 2B). β-cells take up serotonin from the medium and accumulate the compound in LDCVs. Quantal release of serotonin can then be detected using a carbon fiber electrode placed close to the cell. Similar to the capacitance increase, release of serotonin was triggered by the depolarizations and clearly decreased during the second half of the train. This suggests that the slowing in capacitance increase is not caused by simultaneous endocytosis.

From the whole-cell current trace (Fig. 2A, upper), it is apparent that the inward Ca^{2+} current elicited by each depolarization also decreased progressively during the train. As discussed above, depolarization-evoked exocytosis increases with the 4th order of Ca^{2+} influx (see Ref. 10). This is in agreement with the previously described relationship between $[Ca^{2+}]_i$ and exocytotic kinetics in chromaffin cells.[18] Small changes in the Ca^{2+} current can therefore be expected to have dramatic effects on exocytosis. Figure 2D shows the relationship between charge entry and capacitance increase for the individual pulses of the depolarization train. The fitted curve represents the previously described relationship for temporally separated depolarization pulses of 20–500-ms duration.[10] While the first pulse of the depolarization train corresponds well with the fitted curve, Ca^{2+} influx was more efficient in eliciting exocytosis during the 2nd to 4th pulse of the train. This may be due to cytosolic Ca^{2+} not returning to basal levels after the previous depolarizations (cf. residual calcium in the neuromuscular junction[19]). For the 5th to 10th depolarization of the train, Ca^{2+} influx was less effective in triggering exocytosis than observed for temporally separated depolarizations. The likely explanation for this is that human β-cells, like their rodent counterparts,[20] possess an RRP of granules that is exhausted during prolonged stimulation. Equating the cumulative response to the 1st to 4th depolarization to the RRP and using a conversion factor of 3 fF/granule, we estimate the RRP size to be approximately 320 granules. This is higher than the value previously obtained by KCl stimulation and total internal reflection fluorescent imaging (200 granules[21]), possibly from the presence of cAMP in the intracellular solutions used in the capacitance measurements, which may result in "overfilling" of RRP. In addition, the RRP can refill during prolonged stimulation.[22]

Summary

Human β-cells respond to voltage-clamp depolarizations with time- and voltage-dependent increases in membrane capacitance. The capacitance response, as well as insulin secretion, is strongly amplified by elevated intracellular cAMP. P/Q- and L-type Ca^{2+} channels contribute to the capacitance response in a manner reflecting their contribution to the total Ca^{2+} influx. Exocytotic rates decrease during repetitive stimulation due to both inactivation of voltage-gated Ca^{2+} channels and depletion of the RRP of granules.

Conflicts of Interest

The authors declare no conflicts of interest.

References

1. Kahn, S.E. 2003. The relative contributions of insulin resistance and beta-cell dysfunction to the pathophysiology of Type 2 diabetes. *Diabetologia* **46:** 3–19.
2. Zeggini, E. *et al*. 2008. Meta-analysis of genome-wide association data and large-scale replication identifies

additional susceptibility loci for type 2 diabetes. *Nat. Genet.* **40:** 638–645.

3. MacDonald, P.E., J.W. Joseph & P. Rorsman. 2005. Glucose-sensing mechanisms in pancreatic beta-cells. *Philos Trans. R. Soc. Lond. B. Biol. Sci.* **360:** 2211–2225.
4. Gustavsson, N. *et al.* 2008. Impaired insulin secretion and glucose intolerance in synaptotagmin-7 null mutant mice. *Proc. Natl. Acad. Sci. USA* **105:** 3992–3997.
5. Iezzi, M. *et al.* 2005. Adenovirus-mediated silencing of synaptotagmin 9 inhibits Ca2+-dependent insulin secretion in islets. *FEBS Lett.* **579:** 5241–346.
6. Ohara-Imaizumi, M. *et al.* 2007. Imaging analysis reveals mechanistic differences between first and second-phase insulin exocytosis. *J. Cell Biol.* **177:** 695–705.
7. Olofsson, C.S. *et al.* 2002. Fast insulin secretion reflects exocytosis of docked granules in mouse pancreatic B-cells. *Pflugers Arch.* **444:** 43–51.
8. Bratanova-Tochkova, T.K. *et al.* 2002. Triggering and augmentation mechanisms, granule pools, and biphasic insulin secretion. *Diabetes* **51**(Suppl 1): S83–S90.
9. Rorsman, P. & E. Renstrom. 2003. Insulin granule dynamics in pancreatic beta cells. *Diabetologia* **46:** 1029–1045.
10. Braun, M. *et al.* 2008. Voltage-gated ion channels in human pancreatic beta-cells: Electrophysiological characterization and role in insulin secretion. *Diabetes* **57**.
11. Misler, S., A. Dickey & D.W. Barnett. 2005. Maintenance of stimulus-secretion coupling and single beta-cell function in cryopreserved-thawed human islets of Langerhans. *Pflugers Arch.* **450:** 395–404.
12. Zhang, Q. *et al.* 2007. R-type Ca(2+)-channel-evoked CICR regulates glucose-induced somatostatin secretion. *Nat. Cell Biol.* **9:** 453–460.
13. Johnson, J.D. *et al.* 2004. Ryanodine receptors in human pancreatic beta cells: localization and effects on insulin secretion. *FASEB J.* **18:** 878–880.
14. Henquin, J.C., D. Dufrane & M. Nenquin. 2006. Nutrient control of insulin secretion in isolated normal human islets. *Diabetes* **55:** 3470–3477.
15. Ämmälä, C., F.M. Ashcroft & P. Rorsman. 1993. Calcium-independent potentiation of insulin release by cyclic AMP in single beta-cells. *Nature* **363:** 356–358.
16. Seino, S. & T. Shibasaki. 2005. PKA-dependent and PKA independent pathways for cAMP-regulated exocytosis. *Physiol. Rev.* **85:** 1303–1342.
17. Barg, S. *et al.* 2002. A subset of 50 secretory granules in close contact with L-type Ca2+ channels accounts for first-phase insulin secretion in mouse beta-cells. *Diabetes* **51**(Suppl 1): S74–S82.
18. Heinemann, C. *et al.* 1994. Kinetics of the secretory response in bovine chromaffin cells following flash photolysis of caged Ca2+. *Biophys. J.* **67:** 2546–2557.
19. Katz, B. 1966. *Nerve, muscle and synapse*. McGraw-Hill. New York.
20. Eliasson, L. *et al.* 1997. Rapid ATP-dependent priming of secretory granules precedes Ca(2+)-induced exocytosis in mouse pancreatic B-cells. *J. Physiol.* **503**(Pt 2): 399–412.
21. Michael, D.J. *et al.* 2007. Human Insulin Vesicle Dynamics During Pulsatile Secretion. *Diabetes* **56:** 1277–1288.
22. Voets, T., E. Neher & T. Moser. 1999. Mechanisms underlying phasic and sustained secretion in chromaffin cells from mouse adrenal slices. *Neuron* **23:** 607–615.

Expression of Dense-core Vesicles and of Their Exocytosis Are Governed by the Repressive Transcription Factor NRSF/REST

Rosalba D'Alessandro,[a] Andrijana Klajn,[a] and Jacopo Meldolesi[a,b,c,d]

[a] *Vita-Salute San Raffaele University, Milan, Italy*

[b] *Scientific Institute San Raffaele, Milan, Italy*

[c] *National Institute of Neurosciences, Milan, Italy*

[d] *Istituto Italiano di Tecnologia Network, Research Unit of Molecular Neuroscience, Milan, Italy*

The mechanism by which neurons and neurosecretory cells govern the expression and the exocytic discharge of their clear and dense-core vesicles had remained unclear until recently when studies in the neurosecretory cell model PC12 revealed these processes to be orchestrated by the transcriptional repressor neuron restrictive silencer factor (NRSF)/repressor element-1 silencing transcription factor (REST). In wild-type PC12 fully competent for neurosecretion, NRSF/REST is low. The genes of the proteins involved in neurosecretion [from the secretory to vesicle membrane and plasma membrane proteins, including the soluble *N*-ethylmaleimide-sensitive factor attachment protein receptors (SNAREs) of exocytosis] were all repressed by increases of NRSF/REST expression to various extents when the increase was only a fewfold but were completely or almost completely repressed when the increase was large, as in spontaneously defective PC12 clones. In the first case the dense-core vesicles were still competent for exocytosis but were smaller and less dense than in wild-type cells; in the second they were no longer visible but did reappear when the repression was attenuated by transfection of a dominant-negative construct of NRSF/REST combined with a secretory chromogranin or strengthened by treatment with a blocker of NRSF/REST-associated enzymes, the histone deacetylases.

Key words: **transcription; PC12; chromogranin; SNARE; histone deacetylase**

The expression of peculiar organelles, the clear and dense-core vesicles (CVs and DCVs), and their exocytic discharge in response to specific stimuli are common to the whole heterogeneous family of neurons and neurosecretory cells. During the last few decades, CVs and DCVs have attracted great attention, and many of their properties, in particular their molecular structure and the machinery of their exocytosis, have been dissected out and largely revealed (see, e.g., Refs. 1, 2). Until recently, however, the mechanisms that govern the appearance of these organelles during nerve cell differentiation and their persistence thereafter have remained largely undefined. Here we will summarize a number of results recently obtained in the field by us and others, together with a few perspectives with which we intend to develop our future work.

The Transcription Repressor Factor NRSF/REST Plays a Key Role in Nerve Cell Differentiation

NRSF/REST (neuron restrictive silencing factor/repressor element-1 silencing

Address for correspondence: Jacopo Meldolesi, Vita-Salute San Raffaele University, DIBIT, via Olgettina 58, 20132 Milan, Italy. Voice: +0039-02-26432770; fax: +0039-02-26434813. meldolesi.jacopo@hsr.it

Mechanisms of Exocytosis: Ann. N.Y. Acad. Sci. 1152: 194–200 (2009).
doi: 10.1111/j.1749-6632.2008.03988.x

transcription factor, from here on referred to as REST) was discovered in 1996 as a transcription silencer that restricts the expression of multiple neuron-specific genes.[3,4] Thereafter, a downregulation of the factor taking place at the transition of pluripotent precursors to postmitotic neurons was shown to orchestrate the acquisition of a large part of the neuronal phenotype. To explain the mechanisms of these events let us consider the structure of REST. Its DNA binding depends on a central domain, composed by eight zinc fingers addressed to a conserved 21–23 base sequence, the RE-1 sequence, included in the promoter (or other areas) of the silenced genes; the repression depends on two distinct domains, one distributed at the amino- the other at the carboxy-terminus of the protein, where specific corepressor complexes, including regulatory proteins and various chromatin-modifying enzymes (such as the histone deacetylases, HDACs), are recruited. As long as the REST levels are high, the chromatin structure precludes the access of the RNA polymerase II to the RE-1-containing genes harboring the REST system, thus silencing their transcription; the downregulation of REST relieves the repression and triggers transcription of those genes (for details see Refs. 5 and 6).

Throughout the years, a growing number of neuron-specific genes [encoding channels, receptors, neurotrophins, transcription factors, soluble *N*-ethylmaleimide-sensitive factor attachment protein receptor (SNAREs), and so on] were shown to be controlled by REST. More recently, bioinformatic and chromatin occupancy studies have identified in various genomes the RE-1-containing genes, that is, the potential targets of REST repression. Depending on the species and on the analyses, the number of these genes was shown to vary from over 500 to almost 2000.[7–10] Additional genes containing no RE-1 have been shown to be repressed indirectly, either because of being adjacent to or because of being controlled by RE-1-containing genes. The key role of REST in the regulation of a large network of genes essential for neuronal function is therefore undisputed. Until recently, however, it was unclear whether REST was enough to inhibit coordinate neuronal functions, such as neurosecretion. The recent studies carried out by us and others in the rat pheochromocytoma cell line PC12[11,12] demonstrate that this is indeed the case.

REST Governs Expression of Neurosecretion in PC12 Cells

PC12 cells are the most widely employed neurosecretory model, characterized by low levels of REST and equipped with both CVs containing acetylcholine and DCVs containing catecholamines together with ATP and secretory proteins, such as the chromogranins. When growing, the PC12 gross phenotype resembles that of chromaffin cells, however, with smaller DCVs (105 instead of 250 nm in diameter). Exposure to nerve growth factor (NGF) results, within several days, in the block of mitoses accompanied by the sprouting of fibers, which can reach hundreds of μm in length.[13]

In order to investigate the role of REST, both Bruce *et al.*[11] and D'Alessandro *et al.*[12] used wild-type (wt) neurosecretion-competent PC12 cells stably transfected with full-length REST constructs to study gene expression and cellular and functional activities. Bruce *et al.*[11] found that, in the transfected cells, over 300 out of 7000 sequences profiled with deoxyoligonucleotide microarrays exhibited more than twofold repression. Almost half of these 300 genes were linked to putative RE-1 sequences. Among these were five genes encoding proteins (chromogranin A and B, dopamine-β-hydroxylase, Mint1, and SNAP 25) known to participate in the structure and/or the exocytic discharge of CVs and/or DCVs. Moreover, in the transfected cells, the DCVs were fewer and less dense, discharge of catecholamines was reduced in proportion to the decreased uptake by the cells, and a few genes repressed by REST were derepressed by a dominant-negative construct of the same factor.[11]

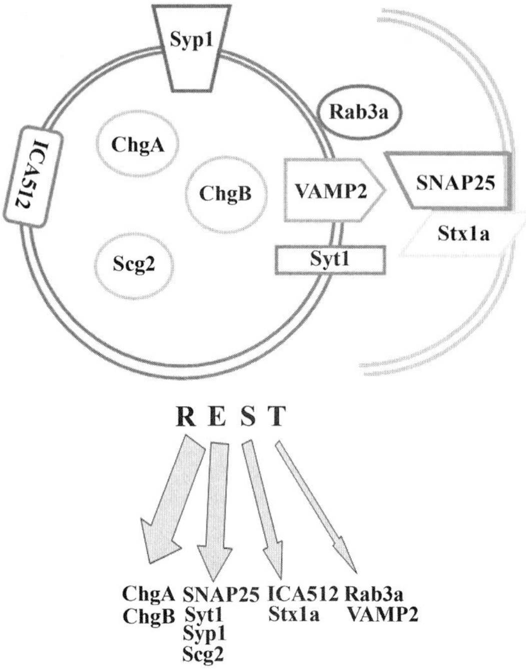

Figure 1. Differential sensitivity to neuron restrictive silencing factor/repressor element-1 silencing transcription factor (REST) repression of the various vesicle and exocytosis genes. The upper part illustrates the distribution of the various proteins in the dense-core vesicles (DCVs) and in the plasma membrane: the three secretory proteins chromogranin A and B (ChgA and ChgB) and secretogranin 2 (Scg2) in the lumen; the vesicle membrane proteins synaptotagmin 1 (Syt1), synaptophysin 1 (Syp1), ICA 512, and the R-soluble *N*-ethylmaleimide-sensitive factor attachment protein receptor (R-SNARE) VAMP 2; Rab 3a, associated with the vesicle surface; and the two Q-SNAREs [SNAP 25 and syntaxin 1a (Stx1a)] in the plasma membrane. In the lower part, the effect of REST on the expression of the corresponding genes, investigated at the mRNA and protein level (see D'Alessandro *et al.*[12]), is illustrated quantitatively by the thickness of the arrows: very high for the chromogranins; high for SNAP 25, secretogranin 2, synaptotagmin, and synaptophysin; moderate for ICA 512 and syntaxin 1a; small for VAMP 2 and Rab 3a.

The approach of D'Alessandro *et al.*[12] was focused exclusively on neurosecretion. Expression of 10 genes (eight containing and two lacking RE-1) encoding proteins of the vesicles and of their exocytic machinery were investigated in the REST-transfected cells at both the mRNA [by quantitative real time (qRT)-PCR] and protein (by Western blot) levels. Interestingly, the changes of the various genes appeared quite variable. In PC12 cells exhibiting a fourfold increase in their REST protein level, the two major secretory proteins of DCVs, the chromogranins, were decreased of more than 90% with respect to the levels in wt cells, while the vesicle membrane proteins synaptotagmin 1 and synaptophysin 1 together with the Q-SNARE SNAP 25 were decreased by 60–80%. An additional membrane protein, ICA 512, the two SNAREs, VAMP 2 and syntaxin 1a, and the small G protein Rab 3a also decreased in expression, however to minor extents (Fig. 1). At the single cell level, the decreases in the various proteins were found to vary depending on the REST construct expression, reaching at the highest levels almost 100% in the case of chromogranins and 70% in the case of synaptotagmin 1. When, on the other hand, the transfected cells were overtransfected with a dominant-negative construct of REST, even the expression of chromogranins reverted to the wt level. All neurosecretion genes, therefore, appear to be repressed by REST; however, they exhibit different sensitivities. In contrast, expression of a few other RE-1-containing genes (tyrosine hydroxylase and synapsin 1), also linked to neurosecretion but not directly, was little affected by the full-length REST construct.

From the functional point of view, the DCVs of the REST-transfected cells, still fully competent for regulated exocytosis, were greatly changed in structure, that is, they were not only less dense but also much smaller (only approximately 25% in volume) with respect to wt DCVs (Fig. 2). This change, which had never been reported in the literature (neurosecretory cells keep the size of their DCVs constant), was probably a consequence of the unbalanced decreases of the various components with relative excess of the membrane with respect to the core proteins.[12]

The paper by D'Alessandro *et al.*[12] also included experiments in a PC12 clone selectively defective of neurosecretion that had

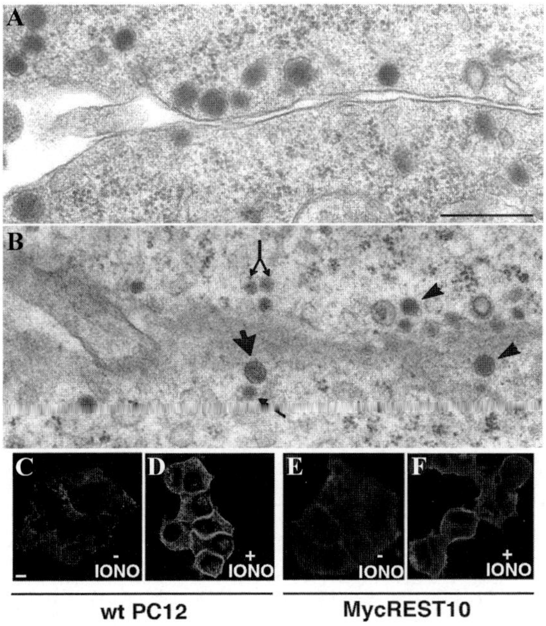

Figure 2. In the PC12 cells expressing a fourfold increase in REST, the dense-core vesicles are smaller but still competent for exocytosis. Panel **A** shows DCVs in the cytoplasm of wild-type (wt) PC12. Notice their size (on average 105 nm in diameter) and the considerable density of their core. Panel **B** shows the vesicles in the wt PC12 cells transfected with REST. Notice their variably smaller size and the variable density of their core, in some almost as high as in the nontransfected cells (arrowheads) while in others much less dense (arrows). The thick arrow points to a lysosome. Panels **C** and **D** show the surface immunofluorescence labeling for the secretory protein chromogranin B in resting and stimulated (2 μmol/L ionomycin, 1 min), nonpermeabilized wt PC12 cells; panels **E** and **F** show the same experiment in the cells transfected with REST. Notice that, although weaker with respect to the nontransfected cells, the increase in the surface chromogranin B induced by stimulation is clearly appreciable also in the transfected cells, documenting the competence for neurosecretion of their small DCVs. The bar in (**A**), valid also for (**B**), corresponds to 0.5 μm; that in (**C**), valid also for (**D–F**), corresponds to 5 μm.

been isolated in the laboratory and found to express high levels of REST.[14] Stable transfection by others of a dominant-negative truncated construct of REST in another defective clone had been found previously to induce increased expression of three neurosecretory genes, appreciable, however, only at the mRNA level, with no reappearance of the corresponding proteins.[15] In contrast, when the dominant-negative construct was transfected in our clone, only the chromogranins remained inappreciable. Expression of all other neurosecretory genes was recovered, at least in part, at the protein level, with VAMP 2 and syntaxin 1a reaching levels as high as in the wt PC12. In their cytoplasm, the defective cells transfected with the dominant-negative construct exhibited discrete puncta, positive for DCV membrane proteins, which, however, failed to show the typical dense ultrastructure of the organelles because of core protein defects. Expression in the transfected clone of a chromogranin and/or attenuation of their REST silencing by a blocker of the HDAC enzymes resulted in the reappearance of DCVs resembling, in their small size and low density, the DCVs already described in the wt PC12 transfected with the full-length REST (Fig. 3).

Taken together, the data obtained by Bruce *et al.*[11] and D'Alessandro *et al.*[12] document that gene expression, not only of various proteins but of the whole neurosecretory process, is indeed orchestrated by REST (Fig. 4) not only during nerve cell differentiation but even in the differentiated cells. This observation is of great interest because the levels of REST do not seem to remain constant throughout the life of nerve cells but may change depending on oxygenation and stimulation conditions.[6,16] In this respect, the differential sensitivity to REST of the various neurosecretion genes might be of importance. An increased repression could in fact change neurosecretion not only in terms of quantity but also of quality because of the stronger downregulation of chromogranins and SNAP 25 with respect to the other SNAREs, VAMP 2, and syntaxin 1a. Upon increase of their REST levels, the nerve cells may well maintain some neurosecretory activity characterized, however, by smaller exocytic quanta and slower discharge as a result of the decreased volume of the DCVs and the replacement of the SNAP 25 by its analogue, the ubiquitous

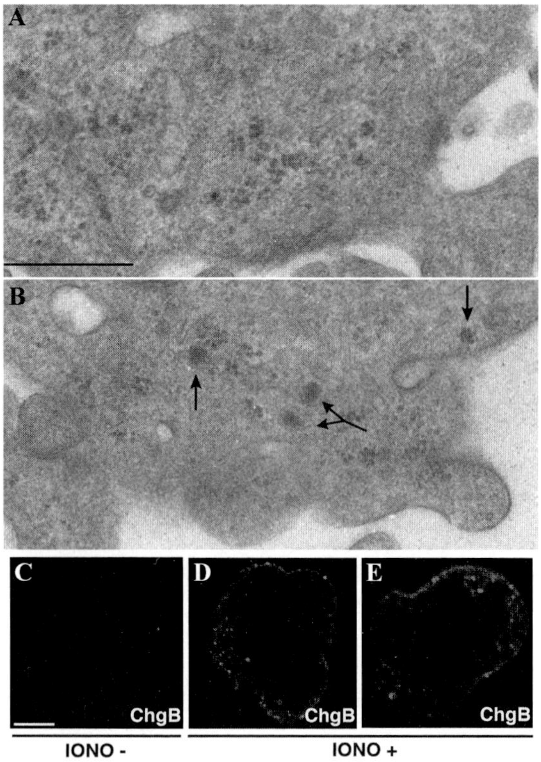

Figure 3. Reappearance of DCVs and of their exocytosis in defective PC12 transfected with the dominant-negative construct of REST, DNA-binding domain (DBD), and treated with trichostatin A (TSA). Panel **A** shows that the defective PC12 cells (PC12–27 clone) lack recognizable DCVs, which in contrast reappear, although small, in the cells transfected with DBD and treated with the blocker of histone deacetylase (HDACs), TSA (arrows in panel **B**). Panels **C–E** show the surface immunofluorescence labeling for chromogranin B in the defective nonpermeabilized cells pretransfected/treated with DBD/TSA, analyzed before (**C**) and after stimulation (**D**) and (**E**) with ionomycin (2 μmol/L, 1 min). Notice in (**D**) and (**E**) the discrete immunofluorescent puncta that most likely correspond to single discharged DCVs. The bar in (**A**), valid also for (**B**), corresponds to 0.5 μm; that in (**C**), valid also for (**D**) and (**E**), to 3 μm.

SNAP 23, which is encoded by a gene containing no RE-1.

Ongoing and Future Developments

The studies summarized in this presentation have introduced REST in the cell biological studies of neurosecretion. In most previous studies, in fact, the products of the various genes orchestrated by REST had been often considered as targets of gene expression or markers of cell differentiation and not for their functional role within the cell. The investigation of this research area remains, therefore, to be largely developed. DCVs and their exocytosis are not the only neuroscientific processes depending on proteins encoded by genes controlled, directly or indirectly, by REST. A role of the repressor in a variety of other cell functions can in fact be envisaged.

At present, our laboratory is active along new but adjacent lines, investigating in particular two specific problems. First, we wondered whether indeed the action of REST depends only on its intracellular levels. Recent studies have shown that, at least in non-neurosecretory cells, the REST half-life is very short ($t_{1/2} \sim 1$ h). Considerable intracellular fluctuations of the repressor, dependent on changes of its turnover, appear, therefore, to take place.[17,18] On the other hand, the action of the repressor is mediated by the multiprotein complexes associated with its N- and C-terminal domains, which seem to function depending on their precise and dynamic architecture. In particular, changes of such architecture, resulting from changes of the ratios between the various components of the complexes, might have a role in the tuning of the enzymes and thus of the REST molecules where the complexes are harbored. Our present evidence strongly suggests that a protein of the BRAF–HDAC complex, named BHC80, abundant in the brain and poor in the REST-rich cells, works indeed as a tuner of the REST repressor function (A. Klajn, unpublished data).

A further line of research deals with fiber sprouting in PC12 cells exposed to NGF. This process takes place at apparently similar rates in both low-REST wt and high-REST-defective clones. Sprouting, therefore, was believed to be completely independent of neurosecretion expression.[19] This conclusion, however, appears only partially true, inasmuch as the fibers

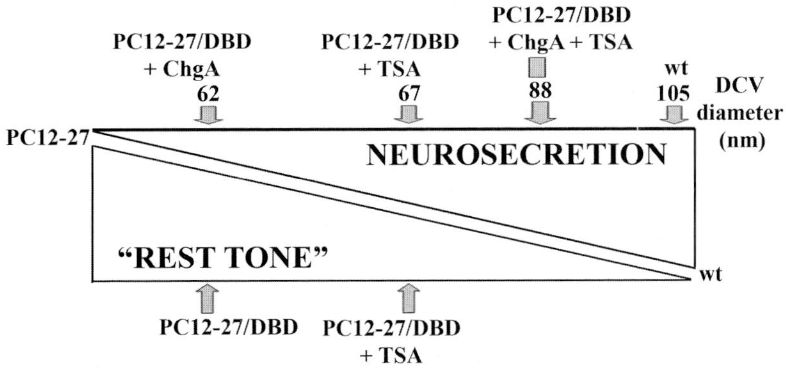

Figure 4. Relationship between the "REST tone" and neurosecretion in wt and defective PC12 cells. This is a model illustrating the reverse relationship between the expression of REST (the "REST tone") and that of DCVs ("neurosecretion") reported by D'Alessandro *et al.*[12] and summarized in this review. In the defective PC12 cells (PC12–27 clone, left), the REST tone is high (approximately 60-fold the wt) and DCVs are absent. The transfection of the dominant negative construct of REST, DBD, induces a decrease of the tone, which in the cells cotransfected with chromogranin A (ChgA) is accompanied by the reappearance of DCVs much smaller than the wt CVs (see the numbers on top of the model specifying the diameter of the DCVs). The combination of DBD with the blocker of HDAC, TSA, induces a further decrease of the REST tone and an increase in the vesicle size. The size increases further when DBD and TSA are combined with the transfection of ChgA, approaching the size of wt DCVs (right).

sprouted by the high-REST PC12 cells exhibit at least a few properties different from the wt fibers. These studies are expected to expand our understanding of the mechanisms governing nerve cell differentiation. In addition, they could provide an explanation of the defects of axonal pathfinding previously observed in REST-overexpressing spinal cord neurons of chicken embryos.[20]

Conclusion

Frontier studies in which knowledge and know-how most often employed by distinct research groups are combined and used concomitantly can be particularly fruitful. This has been the case with the combination of REST with exocytosis studies in neurosecretory cells. REST is a master gene that works both directly and indirectly through its orchestration of a large network of other regulatory genes. Studies on its role in cell biology could therefore be expanded, for example, by identifying the differential pathways of its regulatory cascade within distinct cell types.[6] Another aspect of the field is worth mentioning. The expression of many RE-1-containing genes appears to be controlled not by REST alone but by double-negative feedback loops that include REST (transcriptional level) and several microRNAs (posttranscriptional level). In addition, the activity of both arms of the loops may be regulated by other transcription factors, such as cAMP response element binding (CREB).[21–23] The complexity of these regulatory networks has begun to emerge only recently and has invaded areas of biology and biomedicine, such as cancer (see, e.g., Refs. 24, 25). It appears likely, therefore, that knowledge of these networks will further increase. The present time may therefore be important for strategic decisions by which the REST cell biology laboratories, based on realistic considerations of their funding opportunities, understanding of scientific problems, and experimental know-how, could choose the areas of their future studies.

Acknowledgments

We thank E. Chieregatti, C. Ferrai, and E. Cocucci for discussions; P. Podini, G. Racchetti, and I. Prada for participating in some of the experiments, and F. Floriani for assistance. The original work of this review was supported by grants from the European Community (APOPIS-LSHM-CT-2003-503330), Telethon Fondazione Onlus (GGGP030234), and the Italian Ministry of Research (FIRB 2004 and PRIN 2007 Programs).

Conflicts of Interest

The authors declare no conflicts of interest.

References

1. Takamori, S. *et al.* 2006. Molecular anatomy of a trafficking organelle. *Cell* **127:** 831–846.
2. Jahn, R. & R.H. Scheller. 2006. SNAREs–engines for membrane fusion. *Nat. Rev. Mol. Cell. Biol.* **7:** 631–643.
3. Schoenherr, C.J. & D.J. Anderson. 1995. The neuron-restrictive silencer factor (NRSF): a coordinate repressor of multiple neuron-specific genes. *Science* **267:** 1360–1363.
4. Chong, J.A. *et al.* 1995. REST: a mammalian silencer protein that restricts sodium channel gene expression to neurons. *Cell* **80:** 949–957.
5. Ballas, N. & G. Mandel. 2005. The many faces of REST oversee epigenetic programming of neuronal genes. *Curr. Opin. Neurobiol.* **15:** 500–506.
6. Ooi, L. & I.C. Wood. 2007. Chromatin crosstalk in development and disease: lessons from REST. *Nat. Rev. Genet.* **8:** 544–554.
7. Bruce, A.W. *et al.* 2004. Genome-wide analysis of repressor element 1 silencing transcription factor/neuron-restrictive silencing factor (REST/NRSF) target genes. *Proc. Natl. Acad. Sci. USA* **101:** 10458–10463.
8. Mortazavi, A. *et al.* 2006. Comparative genomics modeling of the NRSF/REST repressor network: from single conserved sites to genome-wide repertoire. *Genome. Res.* **16:** 1208–1221.
9. Otto, S.J. *et al.* 2007. A new binding motif for the transcriptional repressor REST uncovers large gene networks devoted to neuronal functions. *J. Neurosci.* **27:** 6729–6739.
10. Johnson, D.S. *et al.* 2007. Genome-wide mapping of in vivo protein-DNA interactions. *Science* **316:** 1497–1502.
11. Bruce, A.W. *et al.* 2006. The transcriptional repressor REST is a critical regulator of the neurosecretory phenotype. *J. Neurochem.* **98:** 1828–1840.
12. D'Alessandro, R. *et al.* 2008. Expression of the neurosecretory process in pc12 cells is governed by rest. *J. Neurochem.* **105:** 1369–1383.
13. Greene, L.A. & A.S. Tischler. 1976. Establishment of a noradrenergic clonal line of rat adrenal pheochromocytoma cells which respond to nerve growth factor. *Proc. Natl. Acad. Sci. USA* **73:** 2424–2428.
14. Malosio, M.L. *et al.* 1999. Neurosecretory cells without neurosecretion: evidence of an independently regulated trait of the cell phenotype. *J. Physiol.* **520**(Pt 1): 43–52.
15. Pance, A., F.J. Livesey & A.P. Jackson. 2006. A role for the transcriptional repressor REST in maintaining the phenotype of neurosecretory-deficient PC12 cells. *J. Neurochem.* **99:** 1435–1444.
16. Calderone, A. *et al.* 2003. Ischemic insults derepress the gene silencer REST in neurons destined to die. *J. Neurosci.* **23:** 2112–2121.
17. Guardavaccaro, D. *et al.* 2008. Control of chromosome stability by the beta-TrCP-REST-Mad2 axis. *Nature* **452:** 365–369.
18. Westbrook, T.F. *et al.* 2008. SCFbeta-TRCP controls oncogenic transformation and neural differentiation through REST degradation. *Nature* **452:** 370–374.
19. Leoni, C. *et al.* 1999. Neurite extension occurs in the absence of regulated exocytosis in PC12 subclones. *Mol. Biol. Cell.* **10:** 2919–2931.
20. Paquette, A.J., S.E. Perez & D.J. Anderson. 2000. Constitutive expression of the neuron-restrictive silencer factor (NRSF)/REST in differentiating neurons disrupts neuronal gene expression and causes axon pathfinding errors in vivo. *Proc. Natl. Acad. Sci. USA* **97:** 12318–12323.
21. Conaco, C. *et al.* 2006. Reciprocal actions of REST and a microRNA promote neuronal identity. *Proc. Natl. Acad. Sci. USA* **103:** 2422–2427.
22. Wu, J. & X. Xie. 2006. Comparative sequence analysis reveals an intricate network among REST, CREB and miRNA in mediating neuronal gene expression. *Genome Biol.* **7:** R85.
23. Visvanathan, J. *et al.* 2007. The microRNA miR-124 antagonizes the anti-neural REST/SCP1 pathway during embryonic CNS development. *Genes Dev.* **21:** 744–749.
24. Westbrook, T.F. *et al.* 2005. A genetic screen for candidate tumor suppressors identifies REST. *Cell* **121:** 837–848.
25. Coulson, J.M. 2005. Transcriptional regulation: cancer, neurons and the REST. *Curr. Biol.* **15:** R665–R658.

The Coffin–Lowry Syndrome-associated Protein RSK2 Controls Neuroendocrine Secretion through the Regulation of Phospholipase D1 at the Exocytotic Sites

Maria Zeniou-Meyer, Aurélie Béglé, Marie-France Bader, and Nicolas Vitale

Département Neurotransmission and Sécrétion Neuroendocrine, Institut des Neurosciences Cellulaires et Intégratives (UMR 7168), Centre National de la Recherche Scientifique and Université Louis Pasteur, Strasbourg, France

Together with the soluble *N*-ethylmaleimide-sensitive factor attachment protein receptor (SNARE) proteins, fusogenic cone-shaped lipids, such as phosphatidic acid (PA), have been recently shown to be important actors in membrane fusion during exocytosis. Phospholipase D (PLD) appears to be the main provider of PA at the exocytotic site in neuroendocrine cells. We show here that ribosomal S6 kinase 2 (RSK2) stimulates PLD activity through the phosphorylation of Thr147 in the PLD1 amino-terminal Phox-homology domain. In PC12 cells, depletion of RSK2 dramatically prevents PA synthesis at exocytotic sites and inhibits hormone release. Expression of PLD1 phosphomimetic mutants fully restores secretion in cells depleted of RSK2, suggesting that RSK2 is a critical upstream signaling element in the activation of PLD1 to produce the lipids required for exocytosis.

Key words: **RSK2; phospholipase D; PC12 cells; exocytosis**

Introduction

Release of neurotransmitters and hormones occurs through a highly similar process called regulated exocytosis that involves the fusion of specialized cargo-containing vesicles with the plasma membrane. This process is governed by a set of proteins called SNAREs (soluble *N*-ethylmaleimide-sensitive factor attachment protein receptors).[1] Overwhelming evidence accumulated over the last two decades indicates that the SNARE proteins syntaxin, synaptobrevin, and SNAP-25 are intimately involved in the final steps of membrane fusion. In fact, a physically plausible mechanism for fusion is one in which formation of a tight coiled-coil complex between these proteins pulls the participating membranes together.[1] In this model, the energy liberated by the formation of the SNARE complex is used to overcome the energy barrier of lipid rearrangements during fusion. The role of lipids in membrane fusion is less well understood, but recent findings indicate that in addition to the well-known signaling function of phosphoinositol derivatives, lipids might be important for the fusion process through their shape. *In vitro,* cone-shaped lipids, which spontaneously form negative membrane curvatures, favor the formation of hemifusion intermediates when present in the contacting leaflets of apposed membranes.[2–5] Moreover, the lipid-modifying enzyme phospholipase D (PLD), which produces the cone-shaped phosphatidic acid (PA), emerges as a major actor in a varied set of cellular processes that have in common membrane fusion.[6–12]

Address for correspondence: Nicolas Vitale, INCI, CNRS UMR-7168, 5 rue Blaise Pascal, 67084 Strasbourg, France. Voice: +33-388-45-67-12; fax: +33-388-60-16-64. vitalen@neurochem.u-strasbg.fr

Mechanisms of Exocytosis: Ann. N.Y. Acad. Sci. 1152: 201–208 (2009).
doi: 10.1111/j.1749-6632.2008.04001.x

In neuroendocrine cells, between the two PLD isoforms, PLD1 is activated in response to various secretagogues to produce PA at the granule-docking sites on the plasma membrane.[13] PLD1 is a multimodule protein that displays a very low basal activity and requires interactions with small GTPases of the ARF, Ral, and Rho families and with protein kinase C (PKC) to be fully activated.[14,15] We previously reported that ARF6 and RalA are essential activators for PLD1 in the exocytotic machinery from neuroendocrine cells.[16–18] Interestingly, PLD1 is also regulated by phosphorylation in many cellular processes,[14,15] which could be by itself sensitive to intracellular calcium level. This is especially relevant since calcium is the main trigger of neurotransmitter and hormone exocytosis. Threonine 147 in PLD1, known as being phosphorylated by PKC, lies also within a consensus phosphorylation site for another kinase family, namely the ribosomal S6 serine/threonine kinases or RSKs. RSK are activated through the extracellular signal-regulated kinase (ERK) pathway and are thus able to phosphorylate their targets. In the present study, we demonstrate that RSK2 is an essential modulator of PLD1 activity in calcium-regulated exocytosis.

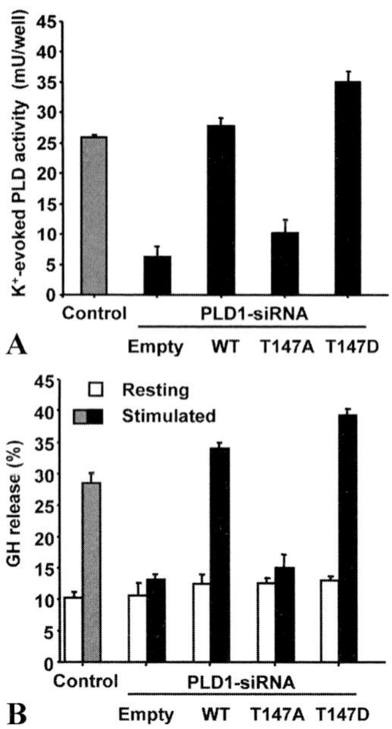

Figure 1. Phosphorylation on Thr147 is required for phospholipase D (PLD) activation and exocytosis in stimulated neuroendocrine cells. PC12 cells were cotransfected with pEGFP-N2 and pCGN (control, gray bars) or pEGFP-N2-PLD1 small interfering (si)RNA and either pCGN or pCGN-PLD1 rescue wild-type (WT), pCGN-PLD1 rescue T147A, pCGN-PLD1 rescue T147D. (**A**) Cells were stimulated for 10 min with 59 mmol/L K^+, and PLD activity was assayed as described previously.[23] K^+-evoked PLD activity was obtained by subtracting the PLD activity detected in resting cells from the PLD activity measured in the K^+-stimulated cells. Data are normalized to the transfection efficiency (55–65%). (**B**) Growth hormone (GH) release was measured in parallel experiments from resting (white bars) and 59 mmol/L K^+-stimulated (dark bars) cells as shown previously.[23] Similar results were obtained in at least four independent experiments performed with triplicate samples.

Results

To determine the relevance of PLD1 phosphorylation in the course exocytosis, we expressed, in PLD1-depleted PC12 cells, PLD1 proteins carrying a phosphorylation-deficient mutation (T147A) or a phosphomimetic mutation (T147D) and examined their effect on growth hormone (GH) secretion as a reporter assay for exocytosis. PLD activity and GH release were measured in parallel in resting and secretagogue-stimulated PC12 cells expressing PLD1 small interfering (si)RNAs to reduce endogenous PLD1 levels and the various PLD1 mutants additionally mutated at wobble codons within the siRNA target sequence for siRNA-resistant expression. Inhibition of endogenous PLD1 expression strongly decreased K^+-stimulated PLD activity and GH release from PC12 cells. This inhibition could be completely rescued by expression of a wild-type PLD1 protein insensitive to siRNA-mediated degradation (Fig. 1A and B, $P < 0.001$, $n = 4$). Expression of the Thr-147 phosphorylation-deficient mutant (T147A) was not able to restore normal levels of PLD

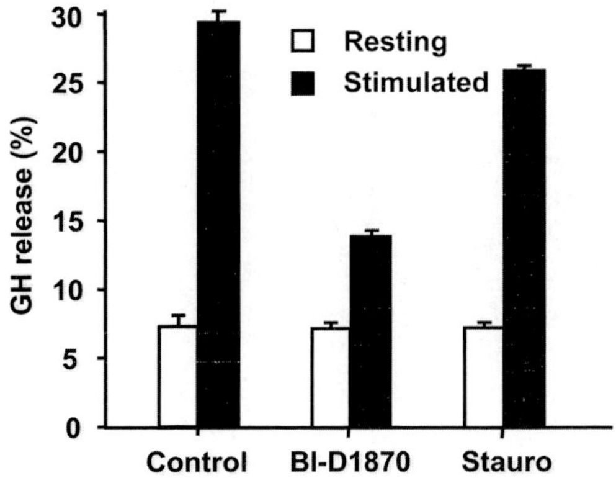

Figure 2. Inhibition of exocytosis by the ribosomal S6 kinase (RSK) inhibitor BI-D1870. PC12 cells expressing GH were incubated 30 min with 10 μmol/L of the RSK inhibitor BI-D1870 or the protein kinase C (PKC) inhibitor staurosporine (Stauro) and incubated 10 min in Locke's solution (Resting; white bars) or stimulated for 10 min with 59 mmol/L K^+ (Stimulated; dark bars). Similar results were obtained in three independent experiments performed with triplicate samples.

activity and secretion in K^+-stimulated cells expressing PLD1 siRNAs, whereas expression of the Thr-147 phosphomimetic mutant T147D completely restored PLD1 activation and GH release (Fig. 1A and B). These results suggest that phosphorylation of the Thr-147 residue is required for both secretagogue-evoked PLD1 activation at the plasma membrane and for the function that PLD1 fulfils in the exocytotic machinery.

To identify the kinase that might be able to phosphorylate and regulate PLD1 activation in stimulated neuroendocrine cells, we tested the effect on exocytosis of BI-D1870 and staurosporine, two specific inhibitors of RSK and PKC, respectively.[19,20] As illustrated in Figure 2, a 30-min preincubation with the RSK inhibitor BI-D1870 significantly inhibited GH secretion from PC12 cells stimulated with high K^+ ($P < 0.001$, $n = 3$), whereas staurosporine had little effect. This result indicates that a phosphorylation step by a member of the RSK family is important for exocytosis. Interestingly, loss of function mutations in the human gene encoding RSK2 leads to Coffin–Lowry syndrome (CLS), characterized by severe mental retardation, growth retardation, and progressive skeletal deformations,[21] all of which could be the consequence of defects in neurotransmitter and/or hormone release. Thus, RSK2 appears as an attractive candidate for being the kinase involved in regulated exocytosis. We first attempted to establish the role of the RSK2 isoform in exocytosis by examining the effect in PC12 cells of two dominant-negative mutants of RSK2 unable to be phosphorylated and activated.[22] Expression of both RSK mutants potently inhibited K^+-evoked GH secretion (Fig. 3A, $P < 0.001$, $n = 4$). The functional importance of RSK2 in exocytosis was then addressed directly by RNA interference. Selective knockdown of RSK2 did not modify basal GH release, but it resulted in a strong decrease in the amount of GH secreted in response to 59 mmol/L K^+ (Fig. 3B, $P < 0.001$, $n = 4$). A mutated siRNA sequence that did not affect endogenous RSK2 level had no effect on secretion (Fig. 3B). Furthermore, expression of human RSK2 (insensitive to the siRNA) in RSK2-depleted PC12 cells completely rescued K^+-evoked GH release, confirming the specificity of the silencing effect (Fig. 3B). All together, these results strongly argue for an important positive regulatory function of RSK2 in calcium-dependent exocytosis.

Using an antibody that specifically recognizes the Thr-147-phosphorylated form of PLD1, we were able to detect a significant increase in Thr-147 phosphorylation of PLD1 in cells stimulated with 59 mmol/L K^+ without detectable modification of the total amount of PLD1 (Fig. 4A). Depletion of endogenous RSK2 by siRNA strongly reduced the amount of Thr-147-phosphorylated PLD1 (Fig. 4A, $P < 0.001$, $n = 3$), in line with the idea that RSK2 is the upstream kinase that phosphorylates PLD1 in stimulated cells. Accordingly, a 30-min preincubation with the RSK inhibitor BI-D1870, but not with the PKC inhibitor staurosporine, inhibited Thr-147-phosphorylation of PLD1.[23] To probe the idea that PLD1 phosphorylation by RSK2 regulates its enzymatic

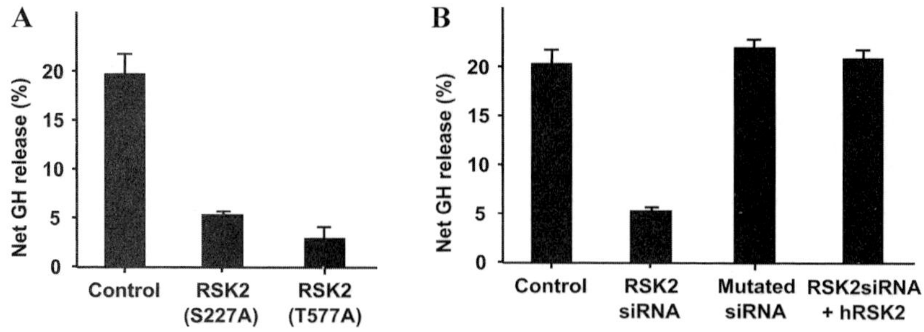

Figure 3. RSK2 is involved in calcium-regulated exocytosis. (**A**) PC12 cells were co-transfected with the pGH and pTL1 vectors (control) or the pGH vector and pTL1 vectors expressing dominant negative mutants of RSK2 (S227A and T577A). Forty-eight hours after transfection, cells were incubated 10 min in calcium-free Locke's solution or stimulated for 10 min with 59 mmol/L K^+. Basal GH release (approximately 7%) was unchanged and was subtracted from the GH release evoked by the high K^+ application to obtain the net GH release. (**B**) PC12 cells were transfected with pGHsuper (Control), pGHsuper-RSK2 siRNA, or pGHsuper-mutated RSK2 siRNA, with or without pTL1 or pTL1-human RSK2. Net GH release was measured 72 h after transfection as in (A). Similar results were obtained in at least four independent experiments performed with triplicate samples.

activity, we examined whether the reduction of endogenous RSK2 affected PLD1 activation in PC12 cells stimulated with a secretagogue. As shown in Figure 4B, reduction of endogenous RSK2 level by siRNA resulted in a strong inhibition of K^+-induced PLD stimulation compared to control cells ($P < 0.001$, $n = 3$). This inhibition could be completely rescued by expression of the siRNA-insensitive human RSK2.[23] In cells with reduced RSK2 levels, expression of wild-type PLD1 was unable to restore normal PLD activity and also had no effect on secretion (Fig. 4B, C). In contrast, the PLD1 phosphomimetic mutant was activated in stimulated RSK2-depleted cells and fully restored secretion despite the low level of endogenous RSK2 (Fig. 4B, C). These results support a sequence of events in which elevated K^+ and a rise in cytosolic calcium activates RSK2, which in turn phosphorylates and stimulates PLD1 to produce the PA required for the exocytotic reaction. In other words, RSK2 appears to be a critical upstream signaling element for PLD1 activation and the production of fusogenic lipids required for exocytosis.

We recently reported the use of green fluorescent protein (GFP)-Spo20p as a specific sensor for PA production in neuroendocrine cells.[13] GFP-Spo20p specifically binds to PA *in vitro* and is partially recruited to the plasma membrane after chromaffin cell stimulation.[13] Total internal reflection fluorescent (TIRF) microscopy and amperometry recording of secretion revealed a close correlation between the formation of PA at the plasma membrane and the exocytotic response.[13] A detailed morphometric analysis further revealed that PA synthesis occurred preferentially at the granule-docking sites on the plasma membrane.[13] To assess whether the PA produced at the plasma membrane upon cell stimulation required activation of PLD1 by phosphorylation by RSK2, we studied the distribution of the GFP-Spo20p-PA sensor in cells expressing the RSK2 siRNA. In these experiments, PC12 cells were first transfected with a plasmid driving the expression of GFP alone or GFP and the RSK2 siRNA and then 48 h later with the GFP-Spo20p plasmid. As shown in Fig. 5A, the PA sensor was recruited to the plasma membrane in stimulated cells expressing GH. However, the accumulation of GFP-Spo20p at the cell periphery was largely inhibited in cells expressing RSK2 siRNA (Fig. 5A). We quantified the percentage of colocalization of GFP-Spo20p with synaptosome-associated protein

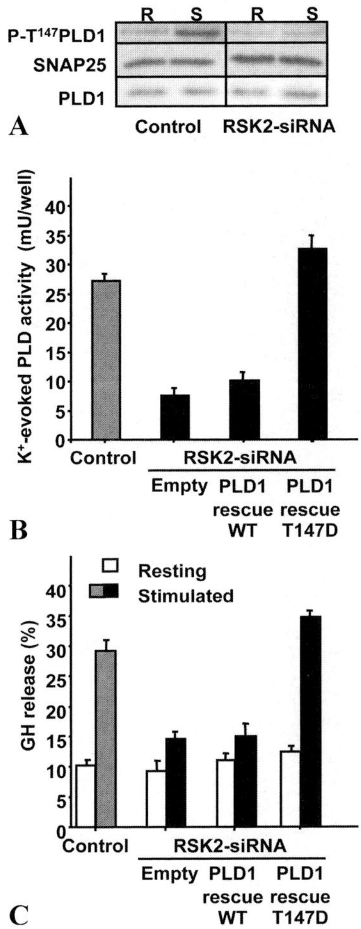

Figure 4. RSK2 phosphorylates PLD1 and regulates its activity in secretagogue-stimulated PC12 cells. (**A**) Resting or 59 mmol/L K^+-stimulated PC12 cells expressing GFP (Control) or GFP and RSK2 siRNA were scrapped and analyzed by Western blotting using anti-SNAP25 (synaptosome-associated protein of 25 kDa), anti-PLD1, and antiphospho-Threonine147-PLD1 antibodies. (**B–C**) Phosphorylation on Thr147 is required for PLD activation and exocytosis. PC12 cells were cotransfected with pEGFP-N2 and pCGN (Control) or pEGFP-N2-PLD1 siRNA and either pCGN, pCGN-PLD1 rescue WT, or pCGN-PLD1 rescue T147D. (**B**) Cells were stimulated for 10 min with 59 mmol/L K^+ and PLD activity was assayed. K^+-evoked PLD activity was obtained by subtracting the PLD activity detected in resting cells from the PLD activity measured in the K^+-stimulated cells. Data are normalized to the transfection efficiency (55–65%). (**C**) GH release measured in parallel experiments from resting (white bars) and 59 mmol/L K^+-stimulated (dark bars) cells. Similar results were obtained in at least three independent experiments performed with triplicate samples.

of 25 Kda (SNAP25) under resting or stimulated conditions (Fig. 5B). Expression of RSK2 siRNA dramatically reduced the recruitment of GFP-Spo20p to the plasma membrane in K^+-stimulated PC12 cells, indicating that PA was not formed at the plasma membrane in cells with reduced levels of endogenous RSK2. Taken together our results show that RSK2 is involved in the PA biosynthetic pathway activated in secretagogue-stimulated cells. Hence, RSK2 appears as a critical upstream signaling element for PLD1 activation and the production of plasma membrane-bound fusogenic lipids required for exocytosis.

Discussion

Using an RNA interference strategy, we have previously described[23] and show here that RSK2 plays a crucial role in the calcium-regulated exocytosis of large dense-core secretory granules. RSK2 coprecipitates with PLD1,[23] phosphorylates PLD1, and stimulates its enzymatic activity (PA production) in response to a secretagogue. Moreover, we demonstrate that PLD1 phosphorylation by RSK2 is essential for the function of PLD1 in exocytosis, i.e., the production of fusogenic PA at granule-docking sites with the plasma membrane.

RSK2 belongs to a kinase family (RSK1-4) that has been implicated in various cellular processes through phosphorylation of multiple substrates. Interestingly in humans, loss of function mutations in the gene encoding RSK2 leads to CLS.[21] The molecular bases of several major CLS features, such as growth and mental retardation, remain elusive. PLD1 has been implicated in the release process of a variety of hormones[6,8–11] and seems to be also essential for fast neurotransmitter release.[7] Thus, the functional link between RSK2 and PLD1 may bring some clues for understanding the molecular mechanisms underlying the clinical features of CLS through the idea that part of the growth and mental retardation phenotypes observed in CLS patients could be

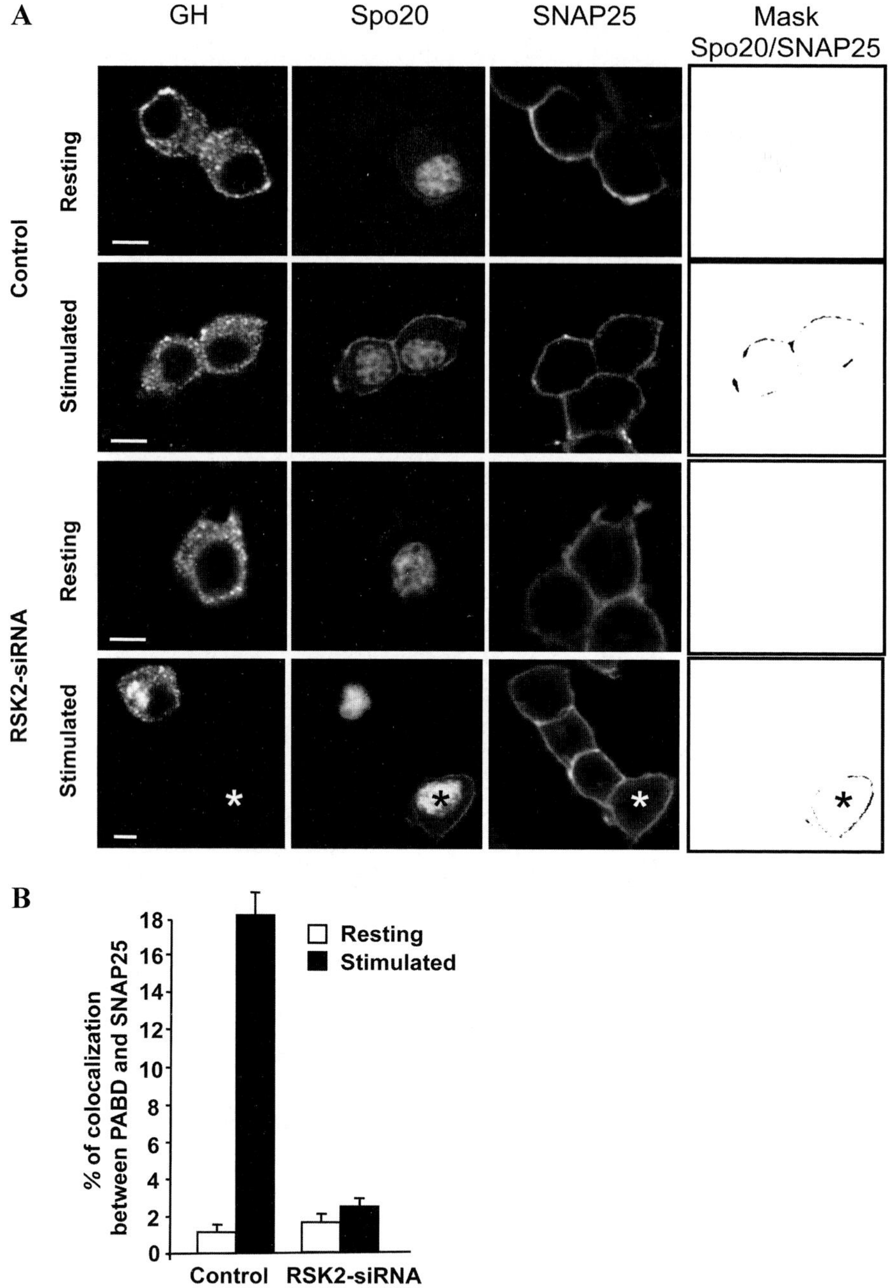

Figure 5. RSK2 is required for PLD1-mediated phosphatidic acid (PA) production at the plasma membrane. (**A**) PC12 cells were cotransfected with pEGFP-Spo20 and pGHsuper (Control) or pGHsuper-RSK2 siRNA. Cells were stimulated for 10 min with Locke's solution (Resting) or 59 mmol/L K^+ (Stimulated) and processed for immunostaining with anti-SNAP25 and anti-GH antibodies. Bar = 5 μm. In the mask images, the black staining indicates the areas of colocalization of Spo20-GFP and SNAP25 as illustrated by displaying the double-labeled pixels. (*) highlights a cell that is not expressing the RSK2 siRNA but expresses Spo20 in which the PA probe is recruited to the plasma membrane. (**B**) Histogram presenting a semiquantitative analysis of the percentage of Spo20-GFP that colocalizes with SNAP25 at the plasma membrane in resting and stimulated PC12 cells. Data are presented as mean values ± SEM ($n > 20$ cells for each experimental condition).

associated to defects in hormone or neurotransmitter release.

Finally, the upstream signal that links secretagogue-evoked cell stimulation to RSK2 activation remains an important open question. We have shown that membrane depolarization, which directly triggers a rise in cytosolic calcium through the activation of voltage-gated calcium channels, stimulates phosphorylation and activation of RSK2. ERKs appear to be good candidates to link the intracellular calcium levels to the activation of RSK2 since RSKs are classically activated by ERK-mediated phosphorylation events[24,25] and ERKs become phosphorylated and activated in cells stimulated with high potassium.[23] To what extent ERKs represent the upstream signal for RSK2-mediated PLD1 activation in the course of exocytosis is an interesting route for future investigations as ERKs have been implicated in secretion and in several forms of synaptic plasticity.

Acknowledgments

We thank T. Thahouly and V. Calco for technical assistance. This work was supported by the Agence Nationale de la Recherche (ANR-05-BLAN-0326-01 to NV), the French Ministry of Science (ACI-BCMS015 to MFB), the Fondation pour la Recherche Médicale (fellowship to MZM and grant to NV), the Association pour la Recherche sur le Cancer (Grant 4051 to NV). We also acknowledge the confocal microscopy facilities of Plateforme Imagerie In Vitro of IFR 37.

Conflicts of Interest

The authors declare no conflicts of interest.

References

1. Sollner, T. *et al.* 1993. A protein assemblydisassembly pathway in vitro that may correspond to sequential steps of synaptic vesicle docking,activation, and fusion. *Cell* **75:** 409–418.
2. Chernomordik, L., M.M. Kozlov & J. Zimmerberg. 1995. Lipids in biological membrane fusion. *J. Membr. Biol.* **146:** 1–14.
3. Kozlovsky, Y., L.V. Chernomordik, & M.M. Kozlov. 2002. Lipid intermediates in membrane fusion: formation, structure, and decay of hemifusion diaphragm. *Biophys. J.* **83:** 2634–2651.
4. Chernomordik, L.V. & M.M. Kozlov. 2003. Protein-lipid interplay in fusion and fission of biological membranes. *Annu. Rev. Biochem.* **72:** 175–207.
5. Vicogne, J. *et al.* 2006. Asymmetric phospholipid distribution drives in vitro reconstituted SNARE-dependent membrane fusion. *Proc. Natl. Acad. Sci. USA* **103:** 14761–14766.
6. Vitale, N. *et al.* 2001. Phospholipase D1: a key factor for the exocytotic machinery in neuroendocrine cells. *EMBO J.* **20:** 2424–2434.
7. Humeau, Y. *et al.* 2001. A role for phospholipase D1 in neurotransmitter release. *Proc. Natl. Acad. Sci. USA* **98:** 15300–15305.
8. Choi, W.S. *et al.* 2002. Regulation of phospholipase D and secretion in mast cells by protein kinase A and other protein kinases. *Ann. N. Y. Acad. Sci.* **968:** 198–212.
9. Hughes, W.E. *et al.* 2004. Phospholipase D1 regulates secretagogue-stimulated insulin release in pancreatic beta-cells. *J. Biol. Chem.* **279:** 27534–27541.
10. Waselle, L. *et al.* 2005. Role of phosphoinositide signaling in the control of insulin exocytosis. *Mol. Endocrinol.* **19:** 3097–3106.
11. Choi, S.Y. *et al.* 2006. A common lipid links Mfn-mediated mitochondrial fusion and SNARE-regulated exocytosis. *Nat. Cell Biol.* **8:** 1255–1262.
12. Corrotte, M. *et al.* 2006. Dynamics and function of phospholipase D and phosphatidic acid during phagocytosis. *Traffic* **7:** 365–377.
13. Zeniou-Meyer, M. *et al.* 2007. Phospholipase D1 production of phosphatidic acid at the plasma membrane promotes exocytosis of large dense-core granules at a late stage. *J. Biol. Chem.* **282:** 21746–21757.
14. McDermott, M., M.J. Wakelam & A.J. Morris. 2004. Phospholipase D. *Biochem. Cell Biol.* **82:** 225–253.
15. Jenkins, G.M. & M.A. Frohman. 2005. Phospholipase D: a lipid centric review. *Cell Mol. Life Sci.* **62:** 2305–2316.
16. Vitale, N. *et al.* 2002. Calcium-regulated exocytosis of dense-core vesicles requires the activation of ADP-ribosylation factor (ARF)6 by ARF nucleotide binding site opener at the plasma membrane. *J. Cell Biol.* **159:** 79–89.
17. Vitale, N. *et al.* 2005. The small GTPase RalA controls exocytosis of large dense core secretory granules by interacting with ARF6-dependent phospholipase D1. *J. Biol. Chem.* **280:** 29921–29928.

18. Caumont, A.S. *et al.* 1998. Regulated exocytosis in chromaffin cells. Translocation of ARF6 stimulates a plasma membrane-associated phospholipase D. *J. Biol. Chem.* **273:** 1373–1379.
19. Watson S.P. *et al.* 1988. The action of the protein kinase C inhibitor, staurosporine, on human platelets. Evidence against a regulatory role for protein kinase C in the formation of inositol trisphosphate by thrombin. *Biochem. J.* **249:** 345–350.
20. Sapkota G.P. *et al.* 2007. BI-D1870 is a specific inhibitor of the p90 RSK (ribosomal S6 kinase) isoforms in vitro and in vivo. *Biochem J.* **401:** 29–38.
21. Trivier, E. *et al.* 1996. Mutations in the kinase Rsk-2 associated with Coffin-Lowry syndrome. *Nature* **384:** 567–570.
22. Merienne, K. *et al.* 2000. Activation of RSK by UV-light: phosphorylation dynamics and involvement of the MAPK pathway. *Oncogene* **19:** 4221–4229.
23. Zeniou-Meyer M. *et al.* 2008. The Coffin-Lowey syndrom-associated protein RSK2 is implicated in calcium-regulated exocytosis through the regulation of PLD1. *Proc. Natl. Acad. Sci. USA* **105:** 8434–8439.
24. Sturgill, T.W. *et al.* 1988. Insulin-stimulated MAP-2 kinase phosphorylates and activates ribosomal protein S6 kinase II. *Nature* **334:** 715–718.
25. Erikson, E. & J.L. Maller. 1991. Purification and characterization of ribosomal protein S6 kinase I from Xenopus eggs. *J. Biol. Chem.* **266:** 5249–5255.

Calcium-regulated Exocytosis in Neuroendocrine Cells: Intersectin-1L Stimulates Actin Polymerization and Exocytosis by Activating Cdc42

Fanny Momboisse, Stéphane Ory, Valerie Calco, Magali Malacombe, Marie-France Bader, and Stéphane Gasman

Département Neurotransmission et Sécrétion Neuroendocrine, Institut des Neurosciences Cellulaires et Intégratives, Centre National de la Recherche Scientifique et Université Louis Pasteur, Strasbourg, France

Actin cytoskeleton remodeling is a critical step of regulated exocytosis in many secretory cell types, including neuroendocrine cells. While the classical model considers the cortical actin network as a physical barrier preventing the uncontrolled recruitment of secretory granules to the plasma membrane docking sites, recent evidence supports the idea that actin polymerization also plays a more active role in the late stages of exocytosis. However, the molecular machinery underlying this positive function of actin in the course of exocytosis remains largely unknown. Here, we propose that the neuronal guanine nucleotide exchange factor, intersectin-1L, activates the GTPase Cdc42, which in turn provides *de novo* actin filaments that are important for calcium-regulated exocytosis in PC12 cells.

Key words: **actin; Cdc42; exocytosis; intersectin; PC12 cells**

Introduction

Calcium-regulated exocytosis requires a fine and rapid remodeling of peripheral actin filaments. For a long time, the classical view of actin rearrangements in most secretory cells were based on the actin-physical-barrier model whereby local disassembly of the cortical actin network permitted secretory granules to gain access to exocytotic sites at the plasma membrane.[1] However, we and others have recently questioned this model by describing that actin polymerization may also play an important active role in the final stages of exocytosis.[1]

Over the past years, we have focused our efforts on elucidating the functional importance of Rho GTPases, which are well-known regulators of actin cytoskeleton dynamics in a wide range of membrane trafficking processes.[2] Using chromaffin and PC12 cells, we found that activation of Cdc42 stimulates secretion by inducing the formation of actin structures at the interface between granules and the plasma membrane.[3] In addition, we demonstrated that intersectin-1L, a Rho guanine nucleotide exchange factor (GEF) from the Dbl family, modulates hormone release in PC12 cells through the activation of Cdc42.[4] However, the participation of intersectin-1L in the actin remodeling events accompanying the exocytotic process remained to be investigated.

Intersectin-1L is mainly expressed in neuronal cells and contains, in addition to the Dbl homology domain (DH)-plekstrin homology

Address for correspondence: Stéphane Gasman, Département Neurotransmission et Sécrétion Neuroendocrine, Institut des Neurosciences Cellulaires et Intégratives (UMR 7168), Centre National de la Recherche Scientifique et Université Louis Pasteur, 5 rue Blaise Pascal, 67084 Strasbourg, France. Voice: +33 388 45 67 12; fax: +33 388 60 16 64. gasman@neurochem.u-strasbg.fr

Mechanisms of Exocytosis: Ann. N.Y. Acad. Sci. 1152: 209–214 (2009).
doi: 10.1111/j.1749-6632.2008.03998.x

domain (PH) region that catalyzes nucleotide exchange on Rho GTPases, two N-terminal Eps15 homology domains (EH1 and EH2), a central coiled-coil region, five Src homology 3 (SH3) domains, and a carboxy-terminal C2 domain.[5] Because of its capacity to promote actin polymerization[6,7] and to interact with various endocytic proteins, such as dynamin, Eps15, synaptojanin, and stonin,[8–10] intersectin has been essentially described as a scaffolding protein that couples the assembly of endocytic-protein complexes to the actin network reorganization at the sites of clathrin-mediated endocytosis. In the present study, we demonstrate the importance of intersectin-1L in the actin cytoskeletal rearrangements required for exocytosis in neuroendocrine cells. We also show that the C-terminal part of intersectin-1L, containing DH, PH, and the C2 domain, stimulates exocytosis by promoting actin polymerization in the cell periphery through the activation of Cdc42.

Results and Discussion

In view of their well-established function on actin organization, Rho GTPases are ideal candidates to promote actin structures required for calcium-induced exocytosis in neuroendocrine cells. Using an interference RNA strategy, we previously demonstrated that intersectin-1L is the GEF responsible for the activation of Cdc42 that occurs during hormone release in PC12 cells.[4] To directly probe the importance of the GEF domain of intersectin-1L for exocytosis, we compared the secretory response of PC12 cells expressing intersectin-1S (a shorter splice variant lacking DH-PH and C2 domains),[5] intersectin-1L, or the isolated tail region of intersectin1-L containing the DH, PH, and C2 domains (Fig. 1A). Using the growth hormone (GH) release assay,[4] we observed that expression of the tail DH-PH-C2 construct enhanced the K^+-evoked secretion by approximately 40%, whereas the full-length intersectin-1S had no effect (Fig. 1B). These

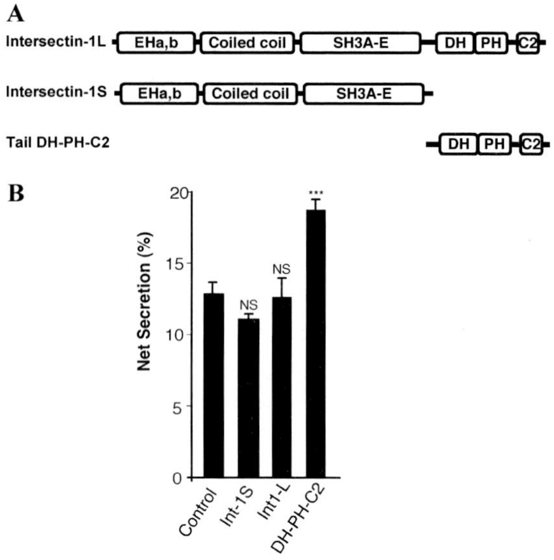

Figure 1. The DH-PH-C2 domain of intersectin-1L stimulates exocytosis in PC12 cells. (**A**) Schematic representation of the intersectin-1 constructs depicting the position of the various functional domains. EH, Eps15 homology domain; DH, Dbl homology domain; PH, pleckstrin homology domain. (**B**) PC12 cells were transfected with the indicated intersectin-1 constructs along with the plasmid encoding growth hormone (GH). At 48 h posttransfection, cells were maintained under resting conditions in Locke's solution for 10 min or stimulated for 10 min with 59 mM K^+ (Locke's containing 59 mmol/L KCl and 85 mmol/L NaCl) and GH release assay was performed as described.[4] GH release is expressed as the percentage of total GH present in the cells before the 10-min stimulation period. GH release detected in resting cells (approximately10%) was unchanged and was subtracted from the GH release evoked by 59 mmol/L K^+ to obtain the net secretory response. Data are given as the mean values ± SEM (n = 3), $^{***}P < 0.001$, NS (not significant) compared to control cells (ANOVA using Minitab® statistical software).

results are in line with a role of the guanine nucleotide exchange activity of intersectin-1L in its regulation of exocytosis. However, we also observed that expression of intersectin-1L full length did not affect exocytosis. One possible explanation is that intersectin-1L remained inactive because it is maintained in an auto-inhibited conformation by intramolecular bonds between the SH3 and the DH domain, which would block Cdc42 binding.[11] In

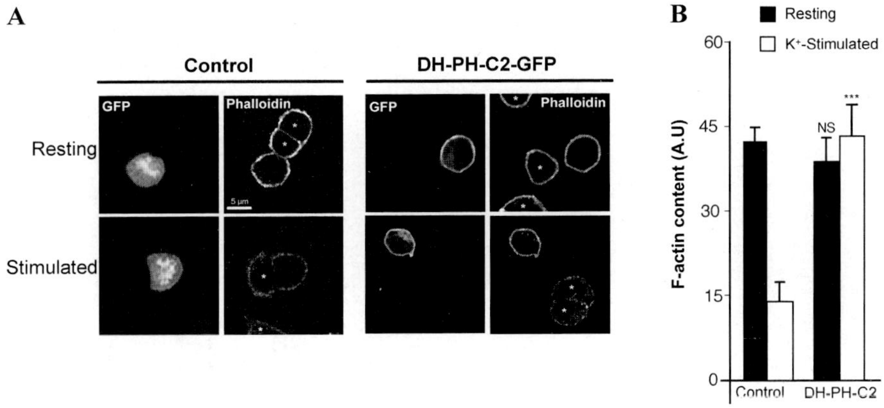

Figure 2. The tail DH-PH-C2 domain of intersectin-1L stimulates actin filament formation at the cell periphery. PC12 cells expressing green fluorescent protein (GFP) (control) or the GFP-tagged tail DH-PH-C2 were maintained in Locke's solution (resting) or stimulated for 10 min with 59 mmol/L K^+, fixed, and processed for actin-filament staining as previously described.[3] (**A**) Confocal images representing the distribution of GFP-DH-PH-C2 and actin filaments visualized with rhodamine-conjugated phalloidin. Asterisks indicate untransfected cells. (**B**) Semiquantitative analysis of the F-actin content in cells transfected with the indicated plasmid. The amount of F-actin present in PC12 cells was measured and expressed as average fluorescence intensity normalized to the corresponding surface area and divided by the total surface of each cell.[3] Data are given as the mean values ± SEM ($n = 25$ cells) and expressed in arbitrary units (AU). $^{***}P < 0.001$, NS (not significant) compared to control cells (ANOVA using Minitab® statistical software).

agreement with this possibility, the isolated tail region of intersectin-1L has been previously shown to be more efficient than the full-length protein at both activating Cdc42 and stimulating the formation of filopodia.[6]

Aside from Cdc42, intersectin directly interacts with various proteins that have been functionally linked to actin organization, such as Eps15, N-WASP, and dynamin.[6,8–10] N-WASP and the yeast homologue of Eps15, Pan1, can promote actin assembly through the activation of the Arp2/3 complex.[12,13] In addition to the established role of dynamin in fission of clathrin-coated pits, it has been shown to function in the formation of actin comets tails, which trigger endosomal compartment transport.[14,15] Moreover, dynamin interacts with cortactin, a protein that has been also shown to activate the nucleating activity of the Arp2/3 complex.[16] Interestingly, we have previously demonstrated the presence of both dynamin-2 and the Arp2/3 complex at the surface of secretory granules in chromaffin and PC12 cells.[3,17] Thus, given its capacity to promote actin polymerization in various cells and to mediate secretagogue-induced Cdc42 activation in PC12 cells, it is tempting to propose that intersectin-1L could be a component of the exocytotic machinery, which may be linked either directly or via Cdc42 to the actin rearrangements needed for exocytosis. To test this hypothesis, we expressed the green fluorescent protein (GFP)-tagged tail DH-PH-C2 domain of intersectin-1L and examined its effect on the cortical actin network organization in PC12 cells. In chromaffin and PC12 cells, the majority of actin filaments are concentrated in the subplasmalemmal region, forming a continuous cortical actin network that is partially disassembled upon activation of exocytosis.[18] Accordingly, phalloidin staining revealed intact and disrupted actin networks in resting cells and stimulated cells, respectively (Fig. 2A, B). In contrast, expression of the DH-PH-C2

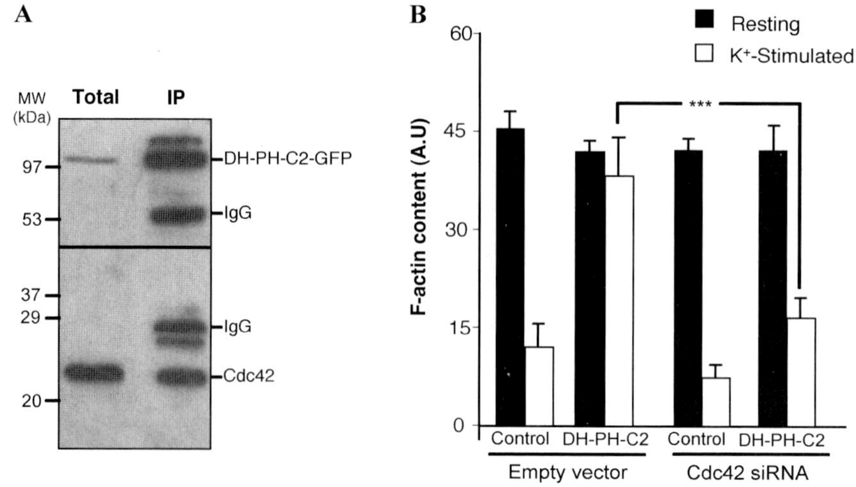

Figure 3. Intersectin 1L-mediated actin remodeling during exocytosis requires a functional interaction with Cdc42. (**A**) The tail DH-PH-C2 domain of intersectin-1L coprecipitates with Cdc42 in PC12 cells. Cell lysates from PC12 cells transfected with the pCB6-GFP-DH-PH-C2 plasmid were incubated with anti-GFP antibodies and then with A/G Sepharose beads. Immunoprecipited proteins (IP) or initial cell lysate extract (total) were separated by gel electrophoresis and transferred to nitrocellulose. Immunoblots were probed with anti-Cdc42 and anti-GFP antibodies. Bands at 50 and 25 KDa correspond to the heavy and light chains of immunoglobulin (IgG), respectively. (**B**) Cells were cotransfected with pCB6-GFP (control) or pCB6-GFP-DH-PH-C2 (DH-PH-C2) along with the pSuper vector either empty (empty vector) or encoding the small hairpin (sh)RNA for Cdc42 [Cdc42 small interfering (si)RNA]. Cells were then maintained in resting condition or stimulated for 10 min with 59 mmol/L K^+ (K^+ stimulated), fixed, and stained with rhodamine-conjugated phalloidin. Confocal image analysis of the F-actin content was then performed as previously described.[3] Data are given as the mean values ± SEM ($n = 25$ cells) and expressed as arbitrary units (AU). $^{***}P < 0.001$ (ANOVA using Minitab® statistical software).

domain enhanced the phalloidin fluorescence in the cell periphery in stimulated cells, indicating that this domain triggers actin polymerization in response to secretagogue stimulation (Fig. 2A, B). This effect is similar to what we observed in cells transfected with the GTPase-deficient Cdc42^{Q61L} mutant.[3]

To further probe the possibility that intersectin-1L regulates actin organization through the activation of Cdc42, we next examined whether the tail DH-PH-C2 region of intersectin-1L interacts with Cdc42. Endogenous Cdc42 could be precipitated with intersectin-1L DH-PH-C2 from PC12 cells expressing a GFP-tagged DH-PH-C2 construct (Fig. 3A), in line with a possible interaction between intersectin-1L and Cdc42. In an attempt to assess the role of Cdc42 in the stimulatory effect of intersectin-1L on actin filament polymerization, Cdc42 expression was silenced by interference RNA. PC12 cells were cotransfected with the plasmid containing the DH-PH-C2 region and either an empty vector or a plasmid coding for the small interfering (si)RNA against Cdc42. Interestingly, we found that expression of the tail region of intersectin-1L, which has nucleotide exchange activity, did not significantly modify the amount of peripheral F-actin in cells depleted of endogenous Cdc42 (Fig. 3B). These data taken together with our previous results[3,4] indicate a causal relationship between the intersectin-1L-mediated activation of Cdc42 and the *de novo* formation of actin filaments during the exocytotic process in neuroendocrine cells. The role of these newly formed actin filaments at the site of exocytosis remains an open question. One particularly interesting feature of intersectin-1L is its bifunctional aspect, which allows both the regulation of the actin cytoskeleton and recruitment of the endocytic machinery. Hence, the actin cytoskeleton plays a key role in various types of endocytic processes.[19] To maintain a constant cell surface area and to permit secretory granule recycling, regulated exocytosis is followed by compensatory endocytosis.[20] Whereas synaptic vesicle recycling has been extensively explored, little is known concerning the molecular basis for large dense-core granule endocytosis after exocytosis. The few studies performed on neuroendocrine cells suggest that there is a fast spatial and temporal coupling of compensatory endocytosis with exocytosis.[20–22] Intersectin-1L is associated with exocytotic sites in neuroendocrine cells where it activates Cdc42[4] and, as shown here, promotes actin polymerization upon cell stimulation. Therefore it is tempting to speculate that intersectin-1L couples the activation of Cdc42 and the subsequent actin remodeling necessary for hormone release to the recruitment of the endocytotic machinery for granule membrane retrieval. Further research is required to establish the validity of this rather attractive hypothesis.

Acknowledgments

We wish to thank Dr. Nancy Grant for critical reading of the manuscript and Dr. Peter S. McPherson (McGill University, Montreal, Canada) for his fruitful collaboration and for providing us the intersectin constructs. This work was supported by a Human Frontier Science Program (HFSP) grant (RGY40-2003C) and an Agence Nationale de la Recherche grant (ANR-07-JCJC-088-01) to SG. We acknowledge the confocal microscopy facilities of Plateforme Imagerie In Vitro of Institut Fédératif de Recherche 37.

Conflicts of Interest

The authors declare no conflicts of interest.

References

1. Malacombe, M., M.F. Bader & S. Gasman. 2006. Exocytosis in neuroendocrine cells: new tasks for actin. *Biochim. Biophys. Acta* **1763:** 1175–1183.
2. Ridley, A.J. 2006. Rho GTPases and actin dynamics in membrane protrusions and vesicle trafficking. *Trends Cell. Biol.* **16:** 522–529.
3. Gasman, S. *et al.* 2004. Regulated exocytosis in neuroendocrine cells: a role for subplasmalemmal Cdc42/N-WASP-induced actin filaments. *Mol. Biol. Cell.* **15:** 520–531.
4. Malacombe, M. *et al.* 2006. Intersectin-1L nucleotide exchange factor regulates secretory granule exocytosis by activating Cdc42. *EMBO J.* **25:** 3494–3503.
5. Guipponi, M. *et al.* 1998. Two isoforms of a human intersectin (ITSN) protein are produced by brain-specific alternative splicing in a stop codon. *Genomics* **53:** 369–376.
6. Hussain, N.K. *et al.* 2001. Endocytic protein intersectin-l regulates actin assembly via Cdc42 and N-WASP. *Nat. Cell Biol.* **3:** 927–932.
7. McGavin, M.K. *et al.* 2001. The intersectin 2 adaptor links Wiskott Aldrich Syndrome protein (WASp)-mediated actin polymerization to T cell antigen receptor endocytosis. *J. Exp. Med.* **194:** 1777–1787.
8. Sengar, A.S. *et al.* 1999. The EH and SH3 domain Ese proteins regulate endocytosis by linking to dynamin and Eps15. *EMBO J.* **18:** 1159–1171.
9. Martina, J.A. *et al.* 2001. Stonin 2: an adaptor-like protein that interacts with components of the endocytic machinery. *J. Cell Biol.* **153:** 1111–1120.
10. Yamabhai, M. *et al.* 1998. Intersectin, a novel adaptor protein with two Eps15 homology and five Src homology 3 domains. *J. Biol. Chem.* **273:** 31401–31407.
11. Zamanian, J.L. & R.B. Kelly. 2003. Intersectin 1L guanine nucleotide exchange activity is regulated by adjacent src homology 3 domains that are also involved in endocytosis. *Mol. Biol. Cell.* **14:** 1624–1637.
12. Duncan, M.C. *et al.* 2001. Yeast Eps15-like endocytic protein, Pan1p, activates the Arp2/3 complex. *Nat. Cell Biol.* **3:** 687–690.
13. Prehoda, K.E. *et al.* 2000. Integration of multiple signals through cooperative regulation of the N-WASP-Arp2/3 complex. *Science* **290:** 801–806.
14. Lee, E. & P. De Camilli. 2002. Dynamin at actin tails. *Proc. Natl. Acad. Sci. USA* **99:** 161–166.
15. Orth, J.D. *et al.* 2002. The large GTPase dynamin regulates actin comet formation and movement in living cells. *Proc. Natl. Acad. Sci. USA* **99:** 167–172.
16. Uruno, T. *et al.* 2001. Activation of Arp2/3 complex-mediated actin polymerization by cortactin. *Nat. Cell Biol.* **3:** 259–266.

17. Galas, M.C. *et al.* 2000. Presence of dynamin–syntaxin complexes associated with secretory granules in adrenal chromaffin cells. *J. Neurochem.* **75:** 1511–1519.
18. Trifaro, J.M., S. Gasman & L.M. Gutiérrez. 2008. Cytoskeletal control of vesicle transport and exocytosis in chromaffin cells. *Acta Physiol. (Oxf)* **192:** 165–172.
19. Smythe, E. & K.R. Ayscough. 2006. Actin regulation in endocytosis. *J. Cell Sci.* **119:** 4589–4598.
20. Smith, C. & E. Neher. 1997. Multiple forms of endocytosis in bovine adrenal chromaffin cells. *J. Cell Biol.* **139:** 885–894.
21. Artalejo, C.R. *et al.* 1995. Rapid endocytosis coupled to exocytosis in adrenal chromaffin cells involves Ca2+, GTP, and dynamin but not clathrin. *Proc. Natl. Acad. Sci. USA* **92:** 8328–8332.
22. Patzak, A. & H. Winkler. 1986. Exocytotic exposure and recycling of membrane antigens of chromaffin granules: ultrastructural evaluation after immunolabeling. *J. Cell Biol.* **102:** 510–515.

A Neurotoxic Secretory Phospholipase A_2 Induces Apoptosis in Motoneuron-like Cells

Zala Jenko Pražnikar, Toni Petan, and Jože Pungerčar

Department of Molecular and Biomedical Sciences, Jožef Stefan Institute, Ljubljana, Slovenia

Ammodytoxin A (AtxA) is a presynaptically neurotoxic secretory phospholipase A_2 from snake venom. The aim of this study was to investigate the mechanism of its cytotoxicity expressed against mouse motoneuronal NSC34 cells. AtxA displayed a potent dose- and time-dependent cytotoxicity that was associated with apoptosis and not necrosis, as revealed by a reduction of mitochondrial membrane potential, activation of caspase-3, and by the absence of propidium iodide staining. The cytotoxic- and apoptosis-inducing effects of AtxA were specific for the motoneuronal cells; human embryonic kidney (HEK293) and mouse myoblast (C2C12) cells were shown to be resistant to the toxin.

Key words: **motoneuronal cells; phospholipase A_2; ammodytoxin A; cytotoxicity; apoptosis**

Introduction

Phospholipases A_2 (PLA_2s) comprise a set of extracellular and intracellular enzymes that catalyze the hydrolysis of the *sn*-2 ester bond of glycerophospholipids. The various groups of PLA_2s are divided into three categories: secretory PLA_2s ($sPLA_2$s), cytosolic PLA_2s ($cPLA_2$s), and Ca^{2+}-independent PLA_2s ($iPLA_2$s). The $sPLA_2$ family consists of relatively small (13–18 kDa), disulfide-rich, Ca^{2+}-dependent enzymes that are expressed in many mammalian tissues and are major components of some animal venoms.[1] Mammalian $sPLA_2$s are involved in arachidonic acid release from cell membranes and play important roles in inflammation, atherosclerosis, host defense, and cancer, but their exact physiological functions remain to be fully elucidated.[2] Snake venom $sPLA_2$s, whose structure is very similar to that of their mammalian homologues, display a variety of pharmacological and toxic effects, including presynaptic neurotoxicity.[1]

Ammodytoxin A (AtxA) is a presynaptically neurotoxic $sPLA_2$ from venom of the long-nosed viper *Vipera ammodytes ammodytes*. AtxA causes a persistent blockade of acetylcholine release at vertebrate neuromuscular junctions, leading to respiratory failure and ultimately death of the envenomated animal.[3] The enzymatic activity of presynaptically toxic $sPLA_2$s is an absolute requirement for irreversible inhibition of neuromuscular transmission, but the exact molecular mechanism underlying the neurotoxicity of snake venom $sPLA_2$s is still not known.[1]

In a recent study,[4] we provided the first experimental evidence that an $sPLA_2$ neurotoxin is internalized and rapidly translocated into the cytosol of motoneuronal cells where it interacts with its binding proteins (calmodulin and 14-3-3 proteins). These results strongly support a molecular mechanism of presynaptic neurotoxicity in which the intracellular action of $sPLA_2$s plays a central role.[5] The mouse neuroblastoma x spinal cord hybrid cell line NSC34[6] used in the study, which has been proven to be a useful tool in a number of studies

Address for correspondence: Jože Pungerčar, Department of Molecular and Biomedical Sciences, Jožef Stefan Institute, Jamova cesta 39, SI-1000 Ljubljana, Slovenia. Voice: +386-1-477-3713; fax: +386-1-477-3984. joze.pungercar@ijs.si

Mechanisms of Exocytosis: Ann. N.Y. Acad. Sci. 1152: 215–224 (2009).
doi: 10.1111/j.1749-6632.2008.03999.x

on dysfunction, degeneration, and disease of motoneurons,[7–9] is very sensitive to exposure to AtxA.[4] Electron microscopic examination of NSC34 cells exposed to AtxA exhibited clear features of neurotoxicity and cellular damage. In addition to multiple, electron-dense, Ω-shaped invaginations of the plasma membrane, extensive vacuolization in the cytoplasm as well as round-shaped mitochondria with abnormal internal structure were also observed after prolonged treatment with AtxA.[4] Because some of the cellular injuries observed indicated induction of cell-death processes, the aim of the present work was to analyze this cytotoxic action of AtxA on NSC34 cells. Our results provide further support for the view that NSC34 cells are a pertinent model for studying the molecular mechanism(s) of action of presynaptically neurotoxic, as well as of endogenous nontoxic, $sPLA_2$s on motoneuronal cells and open the way for further studies aimed at elucidating the death signaling pathways in motoneurons during cytotoxic insults.

Materials and Methods

Materials

Recombinant AtxA was prepared in *Escherichia coli* similarly as described.[10] The mouse myoblast C2C12 cell line and the NSC34 cell line were kindly provided by Professor Angelo Poletti (University of Milan, Milan, Italy). The HEK293 cells (ATTC, CRC-1573) were provided by Ana Petelin (Jožef Stefan Institute, Ljubljana, Slovenia). Dulbecco's modified Eagle's medium (DMEM) and heat-inactivated fetal bovine serum (FBS) were obtained from Invitrogen (Carlsbad, CA, USA). Defined FBS was from HyClone (Logan, UT, USA) and Invitrogen. The chemicals 5,5′,6,6′-tetrachloro-1,1′,3,3′-tetraethyl-benzimidazolocarbocyanine iodide (JC-1), propidium iodide (PI), acetyl-*L*-Asp-*L*-Glu-*L*-Val-*L*-Asp-α-(4-methyl-coumaryl-7-amide) (Ac-DEVD-MCA), and MitoTracker Red CMXRos were from Invitrogen. The MTS [3-(4,5-dimethylthiazol-2-yl)-5-(3-carboxymethoxyphenyl)-2-(4-sulfophenyl)-2H-tetrazolium salt] assay kit was obtained from Promega Biosciences (San Luis Obispo, CA, USA). All other chemicals were of at least analytical grade and were from either Sigma-Aldrich (St. Louis, MO, USA) or Serva (Heidelberg, Germany).

Analytical Methods

Electrospray ionization mass spectrometry (ESI-MS) analysis of AtxA was performed using a high-resolution magnetic-sensor AutospecQ mass spectrometer (Micromass, Manchester, UK). N-terminal sequencing was performed on an Applied Biosystems (Foster City, CA, USA) Procise 492A protein sequencing system. SDS-PAGE was performed on a Mini Protean III (Bio-Rad, Hercules, CA, USA) in the presence of 150 mmol/L dithiothreitol on 15% (w/v) polyacrylamide gels with Coomassie Brilliant Blue R250 staining. The enzymatic activity of AtxA was assayed using a sensitive fluorometric $sPLA_2$ assay[11] with small, unilamellar, phospholipid vesicles composed of 1-palmitoyl-2-pyrenedecanoyl-*sn*-glycero-3-phosphoglycerol (Invitrogen, Carlsbad, CA, USA) on a Safire2 microplate fluorescence detection system (Tecan, Männedorf, Switzerland).

Cell Cultures

NSC34 cells were maintained in DMEM supplemented with 5% (v/v) defined FBS, 1 mmol/L glutamine, and antibiotics (100 units/mL penicillin, 100 μg/mL streptomycin) and grown at 37°C in a humidified atmosphere of 5% CO_2 in 25 cm^2 flasks. The growth medium was changed every 3 days. Cells were mechanically dissociated and seeded into new culture flasks on a weekly basis at a density of 5×10^5 cells/flask. Before each experiment, cytosine-β-D-arabinofuranoside hydrochloride (Ara-C) was added and the cells were incubated at 37°C for an additional

24 h. Ara-C incorporates into DNA and inhibits DNA replication but does not inhibit RNA synthesis. NSC34 cultures contain two populations of cells: small undifferentiated cells that have the capacity to undergo cell division and larger, differentiated, multinucleate cells. Therefore, Ara-C was used to increase the number of differentiated cells and thus attain a well-defined and controlled cell population. C2C12 and HEK293 cells were maintained in DMEM with 20% (v/v) FBS, 1 mmol/L glutamine, and antibiotics (as above) and grown at 37°C in a humidified atmosphere of 5% CO_2 in 25 cm^2 flasks. The medium was changed every 2–3 days. Cells were subcultured weekly using enzymatic dissociation (trypsin) and seeded into new flasks, typically at a density of 1×10^5 cells/flask.

Cell Viability (MTS) Assay

Cells were seeded in a volume of 100 μL of the appropriate supplemented medium at a density of 1×10^4 cells per well in 96-well cell-culture plates and incubated at 37°C for 24 h. Cell viability was assessed using the MTS assay, which is based on the ability of mitochondrial dehydrogenases of viable cells to selectively reduce the tetrazolium ring of MTS to yield soluble formazan. At specified times following continuous exposure of cells to a range of concentrations of AtxA, 10 μL of MTS solution was added to each well and incubated at 37°C for 2 h, after which absorbance readings at 490 nm were obtained, using a Safire2 microplate reader. Parallel controls using equal volumes of 1mol/L KCl were used to normalize the viability data for each treatment. Data were corrected for blanks, normalized, and expressed as a percentage.

Caspase-3 Activity Assay

Apoptotic cell death was assessed by detection of caspase-3 activation in cells treated with AtxA, using the fluorogenic substrate Ac-DEVD-MCA. Cells were treated with increasing concentrations of AtxA, and at specified times cell extracts were prepared in RIPA buffer (50 mM Tris, pH 8.0, 100 mM NaCl, 0.1% (w/v) SDS, 1% (v/v) Nonidet P-40, 0.5% (w/v) deoxycholic acid, 1 mmol/L EDTA) containing a protease inhibitor cocktail. Following 10 min incubation on ice, insoluble material was removed by centrifugation at 8,000 g for 10 min. Total protein concentration was determined using the Bradford assay (Bio-Rad). Cell lysates (50 μg of total protein) were transferred into a 96-well plate and caspase buffer [100 mM HEPES, 200 mM NaCl, 0.2% (w/v) CHAPS, 20% (w/v) sucrose, 2 mM EDTA, 20 mmol/L dithiothreitol, pH 7.0] was added to a final volume of 90 μmol/L per well. Following a 30-min incubation at 37°C, Ac-DEVD-MCA was added to a final concentration of 100 μmol/L and caspase-3 activity was measured continuously using a Safire2 microplate reader at excitation and emission wavelengths of 400 and 505 nm, respectively. The activity of caspase-3 is expressed in arbitrary units proportional to the amount of Ac-DEVD-MCA cleaved.

Measurement of Mitochondrial Membrane Potential

Changes in the transmembrane potential of mitochondria in AtxA-treated cells were monitored using the lipophilic cationic probe JC-1.[12] This molecule penetrates the cytosol of healthy cells and exhibits potential-dependent accumulation in mitochondria, forming reversible dimers (J-aggregates) with a fluorescence emission maximum at 585 nm. Upon collapse of the membrane electrochemical gradient, it dissociates into monomers, which accumulate in the cytosol and display a fluorescence emission maximum of 522 nm. Cells were seeded in a volume of 100 μL, in the appropriate supplemented medium, at a density of 1×10^4 cells per well in 96-well cell-culture plates and incubated at 37°C for 24 h. Following 10 min incubation in the presence of 10 μg/mL JC-1 at 37°C, the wells were washed and incubated with a range of concentrations (10–1000 nmol/L) of AtxA for specified time periods. Fluorescence emission scans were performed by varying the excitation wavelength from 400 to 650 nm with a differential of 15 nm, using a Safire2 microplate reader. The ratio of J-aggregate to monomer fluorescence intensity (580 nm/522 nm) was calculated and the data normalized to positive control responses obtained after incubating the cells with a 1.5-μmol/L concentration of the mitochondrial depolarizing agent CCCP (carbonyl cyanide 3-chlorophenylhydrazone, Sigma).

Propidium Iodide and MitoTracker Red Staining

NSC34 cells were plated onto sterile cover slips in 12-well cell-culture plates at a density of 5×10^4 cells per well in 500 μL of the standard supplemented medium and incubated at 37°C. After 24 h, the medium was replaced with DMEM supplemented with 2 mM $CaCl_2$ and the cells incubated with 100 nmol/L AtxA for different time periods. Cells were washed twice with cold PBS and fixed at room temperature with 3.7% paraformaldehyde in PBS for 10 min. After fixing, the cells were washed three times with PBS and incubated in PBS containing 200 μmol/L PI for 20 min. After three 10-min washes in PBS, the slides were mounted using Mowiol (Calbiochem, Darmstadt, Germany). Slides were examined with an inverted research microscope (Olympus IX71, equipped with a digital charge-coupled device camera DP50), using a fluorescent light illuminator. To selectively stain active mitochondria, we used a fixable cell-permeant mitochondrion-selective dye, MitoTracker Red CMXRos. The dye freely diffuses across the plasma membrane into the cytosol of living cells, passes through porin channels or the outer mitochondrial membrane, and enters the matrix of metabolically active mitochondria, driven by the proton motive force of the inner mitochondrial membrane. Once in the matrix, it is metabolized (oxidized) and accumulates within the mitochondria. Thirty minutes before fixing, MitoTracker Red CMXRos was added to cell preparations at a final concentration of 500 nmol/L.

Statistical Analysis

All data are shown as means ± SEM, and Student's t-test was used to determine significant differences between the means of two groups ($P < 0.05$).

Results and Discussion

The application of nanomolar concentrations of AtxA to NSC34 motoneuronal cells induces typical neurotoxic lesions closely resembling those observed at the rat neuromuscular junction intoxicated by the similar $sPLA_2$ neurotoxins β-bungarotoxin, notexin, and taipoxin.[4,13,14] After binding to NSC34 cells, AtxA released fatty acids from the plasmalemma and, within minutes, was internalized in the cell where it induced characteristic neurotoxic effects, such as irreversible block of synaptic vesicle cycling and reorganization of F-actin.[4] Upon longer treatment with AtxA (i.e., 3 h), however, the NSC34 cells displayed extensive cellular damage and underwent widespread cell death. The aim of this study was to investigate the mechanism(s) of this cell death triggered by AtxA.

AtxA Is Cytotoxic to NSC34 Cells

AtxA displayed a very potent cytotoxic effect on NSC34 cells. Indeed, their viability, examined using the MTS assay, was significantly affected by treatment with nanomolar concentrations of the toxin. The cytotoxic effect of AtxA displayed sigmoidal concentration and time dependences (Fig. 1). Approximately 50% reduction in the number of viable cells was observed when NSC34 cells were treated with 50 nmol/L AtxA for 3 h. However, the viability of the NSC34 cells differed by only 14% over the range of concentrations of AtxA from 10 to 100 nmol/L (Fig. 1A), indicating that the cytotoxic effects of AtxA had reached

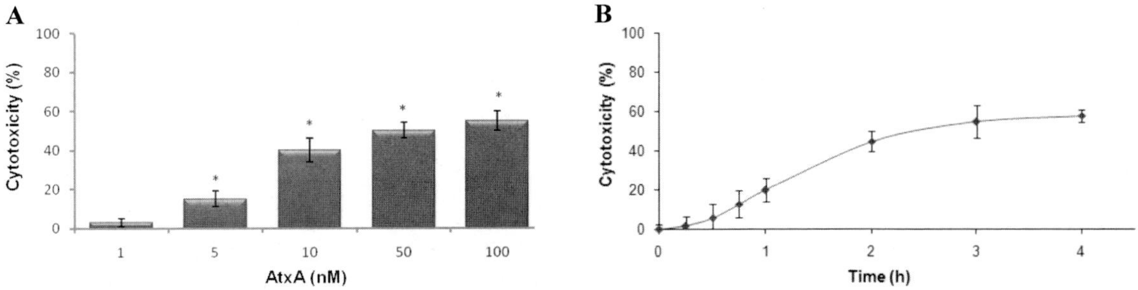

Figure 1. Cytotoxicity induced by Ammodytoxin A (AtxA) in NSC34 cells in a (**A**) concentration and (**B**) time-dependent manner. The NSC34 cells were treated with (**A**) various concentrations of AtxA for 3 h or (**B**) with 100 nmol/L AtxA for different time periods. Cytotoxicity was determined using the MTS [3-(4,5-dimethylthiazol-2-yl)-5-(3-carboxymethoxyphenyl)-2-(4-sulfophenyl)-2H-tetrazolium salt] assay. Data represent the mean ± SEM of triplicate samples from five separate experiments. *$P < 0.05$, for comparison with the control group.

a saturation level. After 2 h, 100 nmol/L AtxA generated a very significant level of cytotoxicity, which remained relatively constant up to 4 h (Fig. 1B) and even up to 12 h (data not shown). On the other hand, the lowest concentrations of AtxA used, 1 and 5 nmol/L, caused only 3 ± 1% and 16 ± 8% reduction, respectively, in NSC34 cell viability after 3 h, although after 12 h the level of cytotoxicity was in the range of that determined with the highest doses of AtxA [35 ± 43% and 47 ± 6% (data not shown)]. Thus, AtxA displays a potent cytotoxic activity on NSC34 cells, and the sigmoidal cytotoxicity curves suggest that the toxin causes cell injury via a saturable mechanism of action, for example, cells express a constant number of molecules that can be bound by AtxA.

The dimeric $sPLA_2$ β-bungarotoxin has been shown to be cytotoxic to mature cerebellar granule neurons but not to the associated glial cells or the nondifferentiated human neuroblastoma cell lines IMR-32 and SK-N-MC.[15,16] This apparent selectivity for cell type and differentiation stage may be a common characteristic of presynaptically neurotoxic $sPLA_2$s. We therefore performed experiments on NSC34 (neuronal), C2C12 (muscle), and HEK293 (kidney) cells to determine whether the cytotoxicity of AtxA is specific to neuronal cells. As shown in Table 1, while AtxA was cytotoxic to NSC34 cells, as described in detail above, the HEK293 and C2C12 cell lines were essentially unaffected as they survived and divided for at least 2 days in the presence of 1000 nmol/L AtxA with no apparent toxic effects (data not shown). These results indicate that AtxA, like other presynaptically neurotoxic $sPLA_2$s, is cytotoxic specifically to neuronal cells.

TABLE 1. Cytotoxicity Induced by Ammodytoxin A (AtxA) in Cultured Cells

	Cytotoxicity (%)		
Cell line	AtxA (100 nmol/L)	AtxA (500 nmol/L)	AtxA (1000 nmol/L)
NSC34	56 ± 5*	58 ± 4*	63 ± 5*
C2C12	1 ± 1	3 ± 1	1 ± 2
HEK293	1 ± 1	1 ± 1	2 ± 1

NSC34, C2C12, and HEK293 cells were treated for 3 h with AtxA at 10, 100, and 1000 nmol/L. Cytotoxicity was determined using the MTS [3-(4,5-dimethylthiazol-2-yl)-5-(3-carboxymethoxyphenyl)-2-(4-sulfophenyl)-2H-tetrazolium salt] assay. Data represent the mean ± SEM of triplicate samples from five separate experiments. *$P < 0.05$ for comparison with the control group.

AtxA Induces Apoptosis in NSC34 Motoneuronal Cells

Apoptosis contributes to the pathogenesis of a variety of diseases, including neurodegenerative disorders and immune deficiency.[17,18] The morphological and biological events of the apoptotic process include mitochondrial membrane potential reduction, activation of

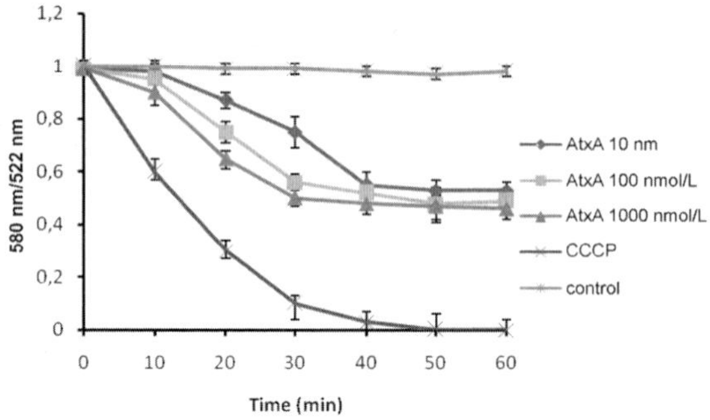

Figure 2. AtxA induces loss of mitochondrial membrane potential in NSC34 cells. Cells were cultured in serum-free medium without or with AtxA (10, 100, 1000 nmol/L) for different time periods. Changes in mitochondrial membrane potential were determined using the mitochondrion-selective fluorescent dye JC-1. The mitochondrial depolarizing agent carbonyl cyanide 3-chlorophenylhydrazone (CCCP) was used as a positive control (see text for details). The results, expressed relative to CCCP, are means ± SEM of six replicate samples from four separate experiments.

TABLE 2. Determination of Mitochondrial Membrane Potential Changes Induced by AtxA

	Reduction of mitochondrial potential (%)		
Cell line	AtxA (10 nmol/L)	AtxA (100 nmol/L)	AtxA (1000 nmol/L)
NSC34	42 ± 6*	45 ± 4*	49 ± 6*
C2C12	3 ± 2	2 ± 1	4 ± 1
HEK293	3 ± 1	4 ± 2	4 ± 2

NSC34, C2C12, and HEK293 cells were treated with AtxA (10, 100, and 1000 nmol/L) for 1 h. Mitochondrial membrane potential was determined using the JC-1 fluorescent probe and carbonyl cyanide 3-chlorophenylhydrazone (CCCP) was used as a positive control for complete membrane depolarization (see text for details). The results, expressed relative to CCCP, are means ± SEM of six replicate samples from three separate experiments. $^*P < 0.05$ for comparison with the control group.

caspases, cell shrinkage, and DNA fragmentation.[19–21] To evaluate whether the cytotoxic potency of AtxA is associated with induction of either necrotic or apoptotic cell death, we examined the mitochondrial membrane potential, caspase-3 activity, and susceptibility to PI staining of cells exposed to AtxA.

To examine whether a collapse of mitochondrial membrane potential, which is an early event in apoptotic cell death, is involved in AtxA-induced cytotoxicity, we measured the changes in mitochondrial membrane potential of NSC34 cells following treatment with AtxA, using the fluorescent probe JC-1. Treatment with AtxA (100 nmol/L) was shown to result in a gradual reduction of mitochondrial membrane potential (Fig. 2) from an approximately 25% loss after 20 min to 52 ± 2% of control after exposure for 1 h. Interestingly, the reduction after 20-min treatment with AtxA concentrations ranging from 10 to 1000 nmol/L differed by only 20% (Fig. 2). These data are in agreement with the results obtained using the MTS assay, confirming that mitochondrial membrane depolarization and cytotoxicity resulting from exposure to AtxA are the consequence of a saturable mechanism of action. Thus, AtxA causes a dissipation of mitochondrial membrane potential, indicating the onset of apoptosis in NSC34 cells.

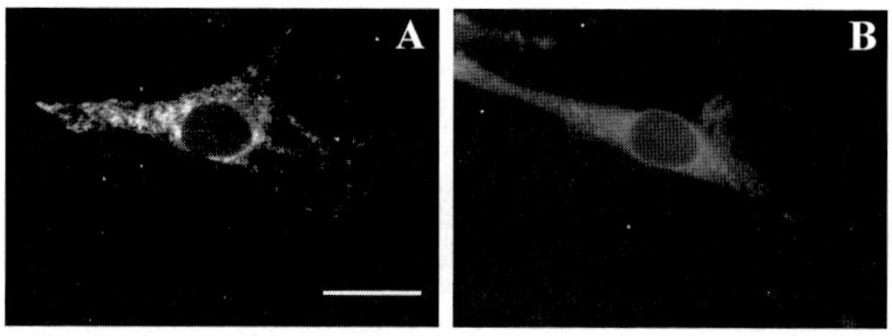

Figure 3. Fluorescence microscopy analysis of mitochondrial integrity in NSC34 cells treated with AtxA. (**A**) Control cells and (**B**) cells treated with 100 nmol/L AtxA for 1 h. Before fixing, cells were incubated with MitoTracker Red CMXRos for 30 min and the subcellular distribution of the dye was analyzed using a fluorescence microscope. The scale bar represents 20 μm.

The loss of mitochondrial membrane potential induced by AtxA was observed only in NSC34 cells, while the HEK293 and C2C12 cells were insensitive to AtxA (Table 2). These results strengthen the suggestion that AtxA targets the neuronal NSC34 cells specifically and that its cytotoxic activity is most probably associated with induction of neuronal apoptosis.

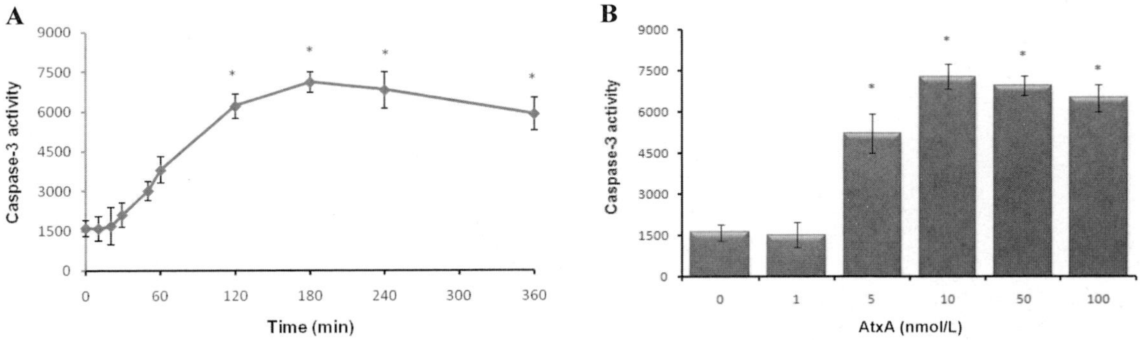

Figure 4. Caspase-3 activity induced by AtxA in NSC34 cells as a function of time and concentration. Cells were treated with (**A**) 100 nmol/L AtxA for various time intervals or (**B**) with different concentrations of AtxA for 3 h. Caspase-3 activity was determined using the acetyl-*L*-Asp-*L*-Glu-*L*-Val-*L*-Asp-α-(4-methyl-coumaryl-7-amide) (Ac-DEVD-MCA) substrate. Data represent the mean ± SEM of triplicate samples from three separate experiments. $^*P < 0.05$, for comparison with the control group not treated with AtxA.

In order to confirm that the above mitochondrial changes are characteristic of early apoptosis, we used a fixable cell-permeant dye, MitoTracker Red CMXRos, to visualize the mitochondria. The dye accumulates in active mitochondria, depending on their membrane potential, and strong fluorescence was observed in nontreated NSC34 cells just within intact mitochondria (Fig. 3A). In contrast, NSC34 cells treated with 100 nmol/L AtxA for 60 min exhibited no selective mitochondrial staining but only a diffuse cytosolic fluorescence (Fig. 3B). These results clearly show that AtxA induces dissipation of mitochondrial membrane potential in NSC34 cells, indicating initiation of the cellular responses that lead inevitably to apoptotic cell death.

The interpretation of the above results suggests the expectation that caspase-3 could be activated in NSC34 cells after exposure to AtxA. Caspase-3 is one of the key enzymes involved in the execution of the apoptotic process in many tissues, and several lines of evidence suggest that it plays an important role in neuronal apoptosis.[16] We therefore treated NSC34 cells with AtxA for various time periods and determined the changes in caspase-3 activity, using a specific tetrapeptide substrate, Ac-DEVD-AMC. After 3 h of exposure to 100 nmol/L toxin, caspase-3 activity was fourfold higher than that in nontreated controls (Fig. 4A). Additionally, the increased activity displayed a saturable concentration dependence, (Fig. 4B) evidenced by the fact that very little variation of caspase-3 activation in NSC34 cells treated with concentrations of AtxA in the range from 10 to 100 nmol/L was observed (Fig. 4B). These data indicate that AtxA-induced cell death is caspase-3 dependent and, in agreement with the results presented above, further suggest that AtxA affects NSC34 cells through a saturable mechanism.

It is notable that AtxA induced caspase-3 activation only in NSC34 cells; HEK293 and C2C12 cells were insensitive to the toxin (Table 3), in agreement with the cytotoxic and mitochondrial permeability-inducing effects of AtxA, which were also specific for the neuronal cells.

To confirm that the cytotoxicity induced by AtxA in NSC34 cells was apoptotic and not necrotic, we stained cells with the cell-impermeable DNA-binding dye PI. In contrast to apoptosis, the cell membrane has been shown to become permeable very early in the process of necrotic cell death.[22] In the absence of AtxA, the NSC34 cells were viable, as evident from the low level of PI staining from weak labeling of RNA in the cytoplasm (Fig. 5A). Treatment with AtxA caused no significant change in PI staining (less then 5% of cells in respect to the control) in the first 30 min of treatment

TABLE 3. Caspase-3 Activity Induced by AtxA in Cultured Cells

	Caspase-3 activity (U)			
Cell line	Control	AtxA (10 nmol/L)	AtxA (100 nmol/L)	AtxA (1000 nmol/L)
NSC34	1800 ± 200	7200 ± 500*	6800 ± 350*	7150 ± 450*
C2C12	1560 ± 340	1650 ± 360	1700 ± 430	1730 ± 350
HEK293	1460 ± 250	1470 ± 200	1650 ± 280	1580 ± 250

Cells were treated with AtxA (10, 100, and 1000 nmol/L) for 3 h. Caspase-3 activity was determined using the acetyl-*L*-Asp-*L*-Glu-*L*-Val-*L*-Asp-α-(4-methyl-coumaryl-7-amide) (Ac-DEVD-MCA) substrate. Data represent the mean ± SEM of triplicate samples from three separate experiments. $^*P < 0.05$ for comparison with the control group.

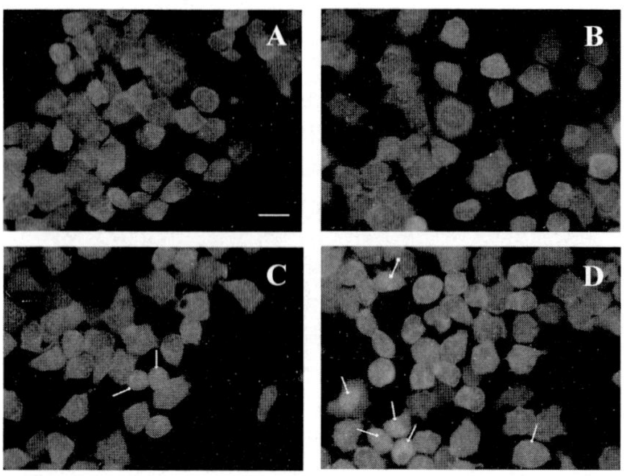

Figure 5. Susceptibility of NSC34 cells to propidium iodide (PI) staining in the presence of AtxA. Cells were treated with different concentrations of AtxA for different time periods. (**A**) Control cells, (**B**) cells treated with 100 nmol/L AtxA for 30 min, and (**C**) cells treated with 100 nmol/L for 1 h and (**D**) 3 h are shown. After fixation, cells were incubated with PI and observed by fluorescence microscopy. Arrows denote PI labeling of NSC34 cells. Scale bar represents 40 μm.

(Fig. 5B), strongly indicating that the cytotoxic activity of AtxA is primarily apoptotic and not necrotic. During this time, the plasma membrane of NSC34 cells remained intact, as revealed by the almost complete absence of PI labeling. In the later phase, after 3 h, we observed stronger PI staining, suggesting that the cells are in late apoptosis[22] (Fig. 5D). These results provide further confirmation that apoptosis, and not necrosis, is the primary cause of cell death of NSC34 cells treated with AtxA.

Mammalian $sPLA_2$s of groups IB and IIA have been reported to induce apoptosis in neuronal cultures,[23,24] and induction of apoptosis in primary cortical neurons was also shown in the case of group III $sPLA_2$.[25] The concentrations of the mammalian $sPLA_2$s required to induce neuronal apoptosis ranged from nanomolar to micromolar,[23–26] but, in all cases, the apoptotic effects were observed only after 24 h or longer treatment. Our results indicate that AtxA, a group IIA $sPLA_2$, is significantly more potent than mammalian $sPLA_2$s in inducing apoptosis in neuronal cultures. Indeed, AtxA caused a rapid induction of apoptosis in NSC34 motoneuron-like cells even at a concentration of 5 to 10 nmol/L in the time frame of minutes to hours. The differences observed between various types of $sPLA_2$s could be explained by their different affinities for a binding site and/or different outcomes upon binding. Hernandez *et al.* have suggested that the engagement of the M-type receptor by $sPLA_2$ activates the mitogen-activated protein kinase cascade[27] and upregulates the cell surface expression of Fas ligand.[28] The human group IIA $sPLA_2$ has been shown to bind to apoptotic cell surface heparan sulfate proteoglycans with high affinity, which may enhance its ability to generate inflammatory signals.[29] It has also been shown that $sPLA_2$s induce apoptosis through the tumor necrosis factor (TNF)-α and cytochrome *c*-mediated caspase cascade in macrophage cells.[30] Our results demonstrate that caspase-3 activation is involved in AtxA-induced motoneuronal cell death. This is in accordance with reports on another presynaptically neurotoxic $sPLA_2$, β-bungarotoxin, which induces apoptosis on cerebellar granule neurons.[15,16] The fact that the mitochondria of NSC34 cells treated with AtxA undergo a collapse of membrane potential suggests that the mitochondrion-mediated pathway is involved in AtxA-induced apoptosis. The question whether caspase-3 activation is a consequence of mitochondrial disruption is currently under investigation in our laboratory as part of our ongoing study of the mechanism(s) by which AtxA induces neurotoxicity and apoptosis in motoneuronal cells.

In conclusion, AtxA, a snake venom presynaptically neurotoxic $sPLA_2$, acts as a potent inducer of apoptosis in cultured motoneuronal cells via a pathway that includes caspase-3 activation and mitochondrial membrane depolarization.

Acknowledgments

We are grateful to Professor Neil R. Cashman, University of Toronto, Canada, for giving us permission to work on the NSC34 cells and Dr. Roger H. Pain for critical reading of the manuscript. This work was supported by grant P1-0207 from the Slovenian Research Agency.

Conflicts of Interest

The authors declare no conflicts of interest.

References

1. Kini, R.M. 2003. Excitement ahead: structure, function and mechanism of snake venom phospholipase A2 enzymes. *Toxicon.* **42:** 827–840.
2. Kudo, I. & M. Murakami. 2002. Phospholipase A2 enzymes. *Prostaglandins Other Lipid Mediat.* **68–69:** 3–58.
3. Lee, C.Y. *et al.* 1984. Mode of neuromuscular blocking action of toxic phospholipases A2 from *Vipera ammodytes venom. Arch. Int. Pharmacodyn. Ther.* **268:** 313–324.
4. Jenko Pražnikar, Z. *et al.* (2008. A presynaptically toxic secreted phospholipase A2 is internalized into motoneuron-like cells where it is rapidly translocated into the cytosol. *Biochim. Biophys. Acta* **1783:** 1129–1139.
5. Pungerčar, J. & I. Križaj. 2007. Understanding the molecular mechanism underlying the presynaptic toxicity of secreted phospholipases A2. *Toxicon.* **50:** 871–892.
6. Cashman, N.R. *et al.* 1992. Neuroblastoma x spinal cord (NSC) hybrid cell lines resemble developing motor neurons. *Dev. Dyn.* **194:** 209–221.
7. Piccioni, F. *et al.* 2002. Androgen receptor with elongated polyglutamine tract forms aggregates that alter axonal trafficking and mitochondrial distribution in motoneuronal processes. *FASEB J.* **16:** 1418–1420.
8. Menzies, F.M. *et al.* 2002. Mitochondrial dysfunction in a cell culture model of familial amyotrophic lateral sclerosis. *Brain* **125:** 1522–1533.
9. Wen, W. *et al.* 2006. Activated microglial supernatant induced motor neuron cytotoxicity is associated with upregulation of the TNFR1 receptor. *Neurosci. Res.* **55:** 87–95.
10. Pungerčar, J. *et al.* 1999. An aromatic, but not a basic, residue is involved in the toxicity of group-II phospholipase A2 neurotoxins. *Biochem. J.* **341:** 139–145.
11. Radvanyi, F. *et al.* 1989. A sensitive and continuous fluorometric assay for phospholipase A2 using pyrene-labeled phospholipids in the presence of serum albumin. *Anal. Biochem.* **177:** 103–109.
12. Smiley, S.T. *et al.* 1991. Intracellular heterogeneity in mitochondrial membrane potentials revealed by a J-aggregate-forming lipophilic cation JC-1. *Proc. Natl. Acad. Sci. USA* **88:** 3671–3675.
13. Harris, J.B. *et al.* 2000. The neurotoxicity of the venom phospholipases A2, notexin and taipoxin. *Exp. Neurol.* **161:** 517–526.
14. Dixon, R.W. & J.B. Harris. 1999. Nerve terminal damage by b-bungarotoxin: its clinical significance. *Am. J. Pathol.* **154:** 447–455.
15. Tseng, W.P. & S.Y. Lin-Shiau. 2003. Activation of NMDA receptor partly involved in b-bungarotoxin-induced neurotoxicity in cultured primary neurons. *Neurochem. Int.* **42:** 333–344.
16. Chen, Y.-J. 2005. Phospholipase A2 activity of b-bungarotoxin is essential for induction of cytotoxicity on cerebellar granule neurons. *J. Neurobiol.* **64:** 213–223.
17. Thompson, H.J. *et al.* 1992. Apoptosis in the genesis and prevention of cancer. *Cancer Epidemiol. Biomarkers Prev.* **7:** 597–602.
18. Hockenbery, D. 1995. Defining apoptosis. *Am. J. Pathol.* **146:** 16–19.

19. Susin, S.A. *et al.* 1997. The central executioner of apoptosis: multiple connections between protease activation and mitochondria in Fas/APO-1/CD95- and ceramide-induced apoptosis. *J. Exp. Med.* **186:** 25–37.
20. Moran, J. *et al.* 1999. Caspase-3 expression by cerebellar granule neurons is regulated by calcium and cyclic AMP. *J. Neurochem.* **73:** 568–577.
21. Martin, L.J. *et al.* 1998. Neurodegeneration in excitotoxicity, global cerebral ischemia, and target deprivation: A perspective on the contributions of apoptosis and necrosis. *Brain Res. Bull.* **46:** 281–309.
22. Ziegler, U. & P. Groscurth. 2004. Morphological features of cell death. *News Physiol. Sci.* **19:** 124–128.
23. Yagami, T. *et al.* 2002. Group IB secretory phospholipase A2 induces neuronal cell death via apoptosis. *J. Neurochem.* **81:** 449–461.
24. Yagami, T. *et al.* 2002. Human group IIA secretory phospholipase A2 induces neuronal cell death via apoptosis. *Mol. Pharmacol.* **61:** 114–126.
25. DeCoster, M.A. 2003. Group III secreted phospholipase A2 causes apoptosis in rat primary cortical neuronal cultures. *Brain Res.* **988:** 20–28.
26. Yagami, T. *et al.* 2003. Porcine pancreatic group IB secretory phospholipase A2 potentiates Ca^{2+} influx through L-type voltage-sensitive Ca^{2+} channels. *Brain Res.* **960:** 71–80.
27. Hernandez, M. *et al.* 2002. Secretory phospholipase A2 elicits proinflammatory changes and upregulates the surface expression of Fas ligand in monocytic cells: potential relevance for atherogenesis. *Circ. Res.* **90:** 38–45.
28. Fuentes, L. *et al.* 2002. Cooperation between secretory phospholipase A2 and TNF-receptor superfamily signaling: implications for the inflammatory response in atherogenesis. *Circ. Res.* **91:** 681–688.
29. Boilard, E. *et al.* 2003. Interaction of low molecular weight group IIA phospholipase A2 with apoptotic human T cells: role of heparan sulfate proteoglycans. *FASEB J.* **17:** 1068–1080.
30. Lee, C. *et al.* 2006. Secretory phospholipase A2 induces apoptosis through TNF-alpha and cytochrome *c*-mediated caspase cascade in murine macrophage RAW 264.7 cells. *Eur. J. Pharmacol.* **536:** 47–53.

Index of Contributors